Microbial Biotechnology in Horticulture

Volume 2

Microbial Biotechnology in Horticulture

Volume 2

Editors

Ramesh C. Ray
Central Tuber Crops Research Institute (Regional Center), Bhubaneswar, India
E-mail: *rc_rayctcri@yahoo.co.in*

Owen P. Ward
Department of Biology, University of Waterloo, Ontario, Canada
E-mail: *opward@sciborg.uwaterloo.ca*

Science Publishers

Enfield (NH) Jersey Plymouth

CIP data will be provided on request.

SCIENCE PUBLISHERS
An imprint of Edenbridge Ltd., British Isles.
Post Office Box 699
Enfield, New Hampshire 03748
United States of America

Website: *http://www.scipub.net*

sales@scipub.net (marketing department)
editor@scipub.net (editorial department)
info@scipub.net (for all other enquiries)

ISBN 978-1-57808-517-0

Published by Science Publishers, Enfield, NH, USA
An Imprint of Edenbridge Ltd.
Printed in India

Preface

Plant genetic engineering has revolutionized our ability to produce genetically improved plant varieties. Large proportions of our major crops have undergone genetic improvement through the use of recombinant DNA technology, and microorganisms play an important role in this development. The first three chapters in this volume focus on genetic engineering of plants. In the first chapter, Arvanitoyannis and Varzakas discuss the general applications of DNA technology, ethics, legislation and other issues related to transgenic crops, foods and feeds. In the second chapter, Arvanitoyannis and Mavromatis explain the impact of plant genetic engineering in developing high yielding elite cultivars of horticultural plants. This chapter focuses on resistance to diseases, insects and viruses, tolerance to herbicides, increased effectiveness of biocontrol agents, enhanced nutritional value and postharvest quality, altered photosynthetic activity, and metabolic production of pharmaceuticals and vaccines. In the subsequent chapter (Chapter 3), Khan examines the role of microorganisms, particularly *Agrobacterium tumefaciens*, in recombinant DNA technology and synthetic biology.

Bio-based agriculture involves the use of farm-generated inputs or applies biological systems, such as plant growth promoting rhizobacteria and arbuscular mycorrhizal fungi, for sustainable crop production by better exploitation of soil resources. Plant growth promoting rhizobacteria are free-living soil bacteria that colonize seeds and roots, and provide benefit to the host plant. These bacteria are distinct from both symbiotic bacteria such as Rhizobia that form N_2-fixing nodules on host legumes and biocontrol bacteria. This group of bacteria, which includes *Pseudomonas*, *Bacillus* and *Azospirillum* species, benefit their host plant by producing compounds, such as siderophores or enzymes (for example, ACC deaminase), that directly contribute to plant growth and yield. In Chapter 4, Patten and Glick define the significance of plant growth promoting rhizobacteria in horticulture.

Mycorrhizal fungi are potential candidates for replacement of chemical fertilizers, in whole or in part, and for protecting plants against environmental and cultivation stresses. The most important mycorrhizal response is an increased efficiency of mineral uptake, particularly phosphorus. Two chapters in this book deal with mycorrhizal response to horticultural plants. Gryndler in Chapter 5 addresses the different types of mycorrhizal fungi present in nature and their application in horticulture and forestry. He also describes the edible mycorrhizal fungi and contribution of gene technology to mycorrhizal biotechnology. In Chapter 6, Sharma and Adholeya emphasize the application of mycorrhiza in micropropagated fruits and ornamental plants, and the various strategies for large-scale inoculum production.

Baculoviruses are rod-shaped double-stranded DNA viruses which infect insects and other arthropods. These viruses have been used as biopesticides, as expression vectors and as mammalian cell gene transfer vectors. Mishra in Chapter 7 enumerates the molecular biology of baculoviruses and their development as biological pest control agents in agriculture/horticulture.

Natural pesticides (compounds of plants, microbial and animal origin) have elicited widespread interest in the development of alternatives to synthetic fungicides for controlling postharvest diseases. Currently, there are no natural antimicrobial compounds that are used on a commercial scale in the horticultural crops postharvest industry sector. Nevertheless, these compounds seem most promising for future application in the control of postharvest diseases. Schena and co-authors (Chapter 8) review the status of natural antimicrobials in controlling postharvest diseases of fruits.

Mycotoxins are low-molecular-weight secondary metabolites, which are toxic to humans and animals. Castoria and Logrieco in Chapter 9 provide an overview of the present knowledge on the occurrence and origin of mycotoxin contamination of fruits and fruit-derived products, which have been identified as major health issues. They also explain the fruit-fungus interactions responsible for the diseases causing mycotoxin contamination and provide strategies for solving mycotoxin problems in fruits and fruit-derived products.

Other aspects like wine biotechnology, coffee and cocoa fermentation, bioconversion of lignocellulosic wastes into single cell protein and mushrooms, etc will be covered in the third volume. We appreciate the promptness of the individual authors in providing and processing their manuscripts.

RAMESH C. RAY
OWEN P. WARD

Contents

About the Editors

Dr. Ramesh C. Ray, Principle Scientist (Microbiology), Central Tuber Crops Research Institute, Bhubaneswar, India is well known for his research in the field of bioprocessing of agriculture substrates and residues and starch based fermentation of root and tuber crops. He obtained Ph.D in Pesticide Microbiology in the year 1984 and thereafter did post-doctoral study in Germany. He has 75 research publications in reputed journals and has published 15 review articles and book chapters. He has developed several foods and industrial processes and is co-inventor on 3 patents. He is presently a member of the International Editorial Board of *Annals of Tropical Research,* published from philippines.

Dr. Owen P. Ward has a Ph.D. in Industrial Microbiology from the National University of Ireland. As Fermentation Development Manager for Biocon Inc. in Ireland, he developed and commercialized the company's first microbial enzyme fermentation processes. He was founding Head and Professor of the School of Biotechnology, Dublin City University and, since, 1986 has been Professor of Microbial Biotechnology at the University of Waterloo, Canada. He is author or co-author of 5 books, many chapters and almost 200 peer-reviewed publications. He is founder of three spin-off companies, has developed many industrial processes and is co-inventor of 6 patents.

The Contributors

Adholeya, Alok

Centre for Mycorrhizal Research, Division of Biotechnology and Management of Bioresources, The Energy and Resources Institute (TERI), Lodhi Road, New Delhi- 110 003, India.
Fax: 91-11-24682144
E-mail: aloka@teri.res.in

Arvanitoyannis, Ioannis S.

Department of Agriculture, Animal Production and Aquatic Environment, School of Agricultural Sciences, University of Thessaly, Fytokou Street, 38446 Nea Ionia Magnesias, Volos, Hellas, Greece.
Fax: +30 24210 93157
E-mail: parmenion@uth.gr

Castoria, Raffaello

Department of Animal, Plant and Environmental Sciences, University of Molise, via F. De Sanctis, 86100, Campobasso, Italy.
E-mail: castoria@unimol.it

Glick, Bernard R.

Department of Biology, University of Waterloo, Waterloo, Ontario, Canada, N2L 3G1.
Fax: 519-746-0614
E-mail: glick@sciborg.uwaterloo.ca

Gryndler, Milan

Institute of Microbiology CAS, Videnska 1083, CZ 142 20, Prague 4, Czech Republic.
E- Mail: gryndler@biomed.cas.cz

Ippolito, Antonio

Department of Plant Protection and Applied Microbiology, University of Bari, Via Amandola 165/A, 70126, Bari, Italy.
Fax: 0039-08-5442911
E-mail: ippolito@agr.uniba.it

Khan, Haseena

Department of Biochemistry and Molecular Biology, University of Dhaka, Dhaka 1000, Bangladesh.
Fax: 880-2-7161515
E-mail: haseena@bangla.net

Logrieco, Antonio F.

Institute of Sciences of Food Production, CNR, Via G. Amendola 122, 70125 Bari, Italy.
E-mail: antonio.logrieco@ispa.cnr.it

Mavromatis, Athanassios G.

Department of Agriculture, Crop Production and Rural Environment, School of Agricultural Sciences, University of Thessaly, Fytokou Street, 38446 Nea Ionia Magnesias, Volos, Hellas, Greece.

Mishra, Seema

National Institute of Biologicals, A32, Sector-62, Institutional Area, Noida 201 307, Uttar Pradesh, India.
Fax: +91-0120-2403014
E-mail: seema_nib@yahoo.com

Nigro, Franco

Department of Plant Protection and Applied Microbiology, University of Bari, Via Amandola 165/A, 70126, Bari, Italy.

Patten, Cheryl L.

Department of Biology, University of New Brunswick, Fredericton, New Brunswick, Canada, E3B 6E1.
E-mail: pattenc@unb.ca

Schena, Leonardo

Department of Plant Protection and Applied Microbiology, University of Bari, Via Amandola 165/A, 70126, Bari, Italy.

Present address: Department of Management of Agricultural and Forest Systems, Mediterranean University of Reggio Calabria, Località Feo di Vito, 89122 Reggio Calabria, Italy.

Sharma, Mahaveer P.

Centre for Mycorrhizal Research, Division of Biotechnology and Management of Bioresources, The Energy and Resources Institute (TERI), Lodhi Road, New Delhi- 110 003, India.

Present address: Microbiology Section, National Research Centre for Soybean (NRCS-ICAR), Khandwa Road, Indore- 452 017 (MP), India.
Fax: 91 731 2470520
E mail: mahaveer620@yahoo.com

Varzakas, Theodoros H.

Department of Processing of Agricultural Products, School of Agricultural Technology, Technological Educational Institute of Kalamata, Antikalamos 24100, Kalamata, Hellas, Greece.
E-mail: tvarzakas@teikal.gr

1

Plant Genetic Engineering: General Applications, Legislations and Issues

Ioannis S. Arvanitoyannis* and Theodoros H. Varzakas

INTRODUCTION – OBJECTIVES

In 2003, about 167 million acres (67.7 million hectares [ha]) grown by seven million farmers in 18 countries were planted with transgenic crops, the principal ones being herbicide- and insecticide-resistant soybeans, corn, cotton, and canola. Other crops grown commercially or field-tested are sweet potato resistant feathery mottle virus, which can reduce yields by 20-80% thus decimating most of the African harvest, rice with increased iron and vitamins that may alleviate chronic malnutrition in Asian countries, and a variety of plants able to survive weather extremes. In 2003, countries that grew 99% of the global transgenic crops were the US (63%), Argentina (21%), Canada (6%), Brazil (4%), China (4%), and South Africa (1%). Although growth is expected to plateau in industrialized countries, it is gradually increasing in developing countries. The next decade is anticipated to witness an exponential progress in GM (genetically modified) product development as researchers gain increasing and unprecedented access to genomic resources that are applicable to organisms beyond the scope of individual projects (http:/www.ornl.gov/sci/techresources/Human_Genome/elsi/gmfood. shtml).

*Corresponding Author

There seem to be at least four major objectives being pursued at this time in crop plant genetic engineering research (http://www.yale.edu/ynhti/curriculum/units/2000/7/00.07.02.x.html):

- To improve biological protection of crops against insects, weeds and fungi by inserting genes for the natural production of an insecticide (Feder, 1996), or for resistance to fungi or herbicide (Hinchee, 1988).
- To elevate levels of important nutrients, e.g., methionine levels in soybeans (Beardsley, 1996) so as to make crops more nutritious.
- To obtain better control of ripening and postharvest storage life to ensure that the produce is in peak condition when taken to the market (Maryanski, 1995).
- To specifically modify genomes so as to produce a specific product, e.g., caffeine-less coffee bean and edible vaccines in potatoes (Pollack, 2000).

This is the first of the three chapters in this volume focusing on genetic engineering of plants. In this chapter, the general applications, ethics, legislations and issues have been discussed.

CURRENT SITUATION IN PLANT GENETIC ENGINEERING (GE)

As discussed in the previous paragraphs, six principal countries grew 99% of the global GM crops in 2003. The USA grew 42.8 million ha (63% of global total), followed by Argentina with 13.9 million ha (21%), Canada with 4.4 million ha (6%), Brazil with 3.0 million ha (4%), China with 2.8 million ha (4%), and South Africa with 0.4 million ha (1%). Globally, the principal GM crops were soybeans (41.4 million ha; 61% of global area), maize (15.5 million ha; 23%), cotton (7.2 million ha; 11%), and canola (3.6 million ha; 5%) [International Life Sciences Institute (ILSI), 2003, 2004]. During the eight years since introduction of commodity GM crops (1996 to 2003), a cumulative total of over 300 million ha (almost 750 million acres) of GM crops were planted globally by millions of large- and small-scale farmers. Rapid adoption and planting of GM crops by millions of farmers around the world; growing global, political, institutional, and country support for GM crops and data from independent sources confirm and support the benefits associated with GM crops (James, 2003).

PREPARATION AND SYNTHESIS OF GENETICALLY MODIFIED ORGANISMS (GMOs)

Generally, there are two main approaches for GMO synthesis: one utilizing bacterial (usually *Agrobacterium tumefaciens*) and viral vectors, and the other utilizing either chemical or physical methods to introduce genes of interest into target cells. The viral gene transfer is a viral-mediated process referred to as an infection. The second, non-viral gene transfer, involves treatment of cells by chemical or physical means and the process itself is named transfection. Gene delivery by infection is more complicated than transfection. Infection requires more steps and more time than does transfection and biosafety issues may also arise, depending on the virus used. A much safer and faster alternative to infection-transfection requires only a few reagents including plasmid DNA containing the gene of interest under the control of a strong cell-specific promoter. However, inefficient gene delivery and poor sustained gene expression are its major drawbacks. Transfection can be categorized into two major types: transient and stable. Transient transfection is temporary, i.e., expression of foreign gene lasts for several days and is lost, as DNA never integrates into the host cell DNA. In contrast, stable transfection occurs with a lower frequency (10- to 100-fold lower), but expression is maintained for the long term because the foreign DNA does integrate into the host genome (Mitrovic, 2003).

Gene transfer methods can be distinguished in viral and non-viral ones. The former exploit recombinant viruses leading to recombinant viral short-term or long-term vectors. Non-viral gene systems have considerable advantages over the recombinant viruses in view of their highly immunogenic, non-infectious and non-toxic character in conjunction with accommodation of large DNA plasmid and large-scale production. However, their two main disadvantages reside in their lower gene transfer efficiency (*vis-a-vis* viruses) and transient gene expression. Non-viral gene transfer systems fall in the following categories: (1) chemical methods, and (2) physical methods. Chemical methods include utilization of positively charged polymers [dextran based – known as Diethyl Amino Ethyl Amino (DEAE) dextran], calcium phosphate co-precipitation with DNA and cationic lipids or liposomes. Physical methods comprise electroporation, particle guns, etc. (Chang et al., 1991; Chen et al., 2003; Cranshaw in http://www.ext.colostate,edu/pubs/insect05556.html).

GENERAL APPLICATIONS OF GENETIC ENGINEERING IN AGRICULTURE

Plants have evolved a large variety of sophisticated and efficient defence mechanisms to prevent the colonization of their tissues by microbial pathogens and parasites. The induced defence responses can be assigned to three major categories (**Table 1.1**).

Table 1.1. Categorization and characteristics of plant defence responses (Stuible and Kombrink, 2004; Kombrink and Somssich, 1995; Sticher, 1997)

Category	*Response Time*	*Expression Pattern*	*Result*
1. Hypersensitive response (HR)	Immediate	Early defence response in the invaded plant cell	Rapid death of challenged host cells
2. *De novo* synthesis of proteins	Subsequent to recognition	Local activation of genes in close vicinity of infection sites	Formation of phytoalexins, structural cell wall proteins, and other protective proteins
3. Systemic acquired resistance (SAR)	Temporarily delayed	Systemic activation of genes encoding pathogenesis related proteins	Immunity to secondary infections

The incorporation of disease resistance genes into plants was initially initiated with classical breeding which yielded high crop productivity. However, currently, plant biology is oriented towards clone identification and characterization of genes involved in disease resistance. The following approaches are adopted towards developing disease resistant transgenic plants (Punja, 2004):

- Expression of gene products that are directly toxic to, or which reduce growth of the pathogen including pathogenesis related proteins, such as hydrolytic enzymes, antifungal proteins, antimicrobial peptides and phytoalexins.
- Expression of gene products that destroy or neutralize a component of the pathogen (lipase inhibition).
- Expression of gene products that enhance the structural defences (high levels of peroxidase and lignin).
- Expression of gene products that release signals regulating plant defences such as salicylic acid.
- Expression of resistance gene products involved in race-specific avirulence interactions and hypersensitive response.

The first commercial product of recombinant(r) DNA technology was the Flavr Savr™ tomato characterized by prolonged shelf life. This was achieved by introducing an antisense polygalacturonase gene (Redenbaugh et al., 1995). One of the greatest contributions of GE to agriculture has been the introduction of viral genes to plants, including tomato, as a means of conferring resistance to viral diseases (Barg et al., 2001). Apart from the introduction of bacterial *Bt* (*Bacillus thuringiensis*) toxin genes in tomato, characteristics related to the quality of fruit (total soluble solids, fruit colour) were also genetically engineered.

Numerous genes of agronomic interest were introduced into *Brassica* spp., generally via *A. tumefaciens*-mediated transformation of oilseed rape (Murphy, 2002). In contrast to research with oilseed and vegetable brassicas, transformation research in forage brassicas is less extensive. Genetic manipulation in brassicas has already produced plants with increased insect resistance through use of *Bt* genes. In addition, expression of a chitinase gene in oilseed rape gave field tolerance to the fungal pathogens, *Leptosphaeria maculans, Sclerotinia sclerotiorum* and *Cylindrosporium concentricum* (Christey and Braun, 2001).

Due to its capacity for regeneration, carrot was one of the first species to be used in direct gene transfer studies and for transformation by *Agrobacterium*. Its excellent susceptibility to infection by *Agrobacterium* has made it the procedure of choice for stable transformation in most studies. Direct gene transfer is used for the analysis of transient gene expression or to transiently produce a gene product in regeneration studies. Since it has been proved quite easy to create protoplasts from carrots and it undergoes cell wall regeneration quickly and efficiently, carrot has been used in several transfection studies with electroporation and polyethylene glycol (Gallie, 2001). Soybean meals are currently used for animal feeding; however, feeding experiments need to be an integrated part of a GM food/ feed safety assessment. However, only 10 scientific (peer reviewed) feeding studies on animals have been published so far, most of which have been published by Pryme (2003). Only one feeding study involving fish has been published till date (Sanden et al., 2004). In this study, Atlantic salmons were fed with feed containing 17.2% GM soybean for six weeks, and the purpose was to investigate the fate of GM soy DNA fragments. No GM soy DNA fragments were detected in liver, muscle or brain tissue (Myhra and Dalmo, 2005). The recent achievements in the field of genetic engineering for brassicas are summarized in **Table 1.2.** Information is provided to the reader for the expressed gene product, and effect on disease/composition for the above mentioned vegetables.

The "second generation" of seeds now being introduced offer processor, end-user and consumer specific benefits, such as increased levels of protein, modified and healthier fats, modified carbohydrates, improved flavour characteristics and increased levels of desired phytochemicals. Modification of qualitative and quantitative characteristics, such as the composition of proteins, starch, fats or vitamins by modifications of metabolic pathway, could increase the nutritional status of the foods and might help to improve human health by addressing malnutrition and under-nutrition. For example, iron deficiency causes

Table 1.2. Expressed gene product, and its effect on disease/composition of *Brassica napus* L. and *B. juncea* L.

Expressed Gene Product	*Effect on Disease/Composition*	*Reference*
Agrobacterium	Resistance to *Xanthomonas campestris* pv. *campestris*	Braun et al., 2000
Barley oxidase oxalate	Not tested	Thompson et al., 1995
Bean chitinase	Reduced rate and total seedling mortality due to *Rhizoctonia solani*	Broglie et al., 1991
Ch Fat B2, a thioesterase DNA	Greater number of medium chain fatty acids	Dehesh et al., 1996
crylA gene from *Bacillus thuringiensis*	Resistance to *Helicoverpa zea* (Boddie) and *Spodoptera exigua* (Hubner)	Stewart et al., 1996
Hevia chitin binding lectin (hevein) (*B. juncea* L.)	Smaller lesion and reduced rate of development due to *Alternaria brassicae*	Kanrar et al., 2002
Lauroyl ACP thioesterase bay tree	Enhanced lauric acid content	Del Vecchio, 1996
Macadamia antimicrobial peptide	Reduced lesion size due to *Leptosphaeria maculans*	Kazan et al., 2002
Pea chitinase *PR 10.1* gene	No effect on *Leptosphaeria maculans*	Wang et al., 1999
Pea defence response gene (defensin)	Reduced infection and development of *Leptosphaeria maculans*	Wang et al., 1999, 2000
Phytoensynthase (daffodil)	Increased β-carotene content	Ye et al., 2000
Shiva protein	Resistance to *Xanthomonas campestris* pv. *campestris*	Braun et al., 2000
Tomato *Cf9* gene	Delayed disease development due to *Leptosphaeria maculans* Resistance against *Cladosporium fulvum*	Hennin et al., 2001; Punja, 2004
Tomato chitinase	Lower number of diseased plants due to *Cylindrosporium concentricum* and *Sclerotinia sclerotiorum*	Grison et al., 1996
Trichoderma endochitinase	Reduced lesion due to *Alternaria brassicae*	Mora and Earle, 2001
γ-tocopherol methyl transferase (*Arabidopsis*)	Enhanced vitamin E content	Shintani and De la Penna, 1998
δ-6 and δ-12 desaturases	Increased γ-linoleic acid content	Liu et al., 2002
δ-6 desaturase gene (*Mortierella*)	Increased ω-3-fatty acid content	Ursin, 2000; James, 2003

aneamia in pregnant women and young children. Aneamia has been identified as a contributing factor in over 20% of maternal deaths (after giving birth) in Asia and Africa (Conway and Toenniessen, 1999). Transgenic rice with elevated levels of iron has been produced using genes involved in the production of an iron-binding protein and in the production of an enzyme that facilitates iron availability in the human diet (Goto et al., 1999; Lucca et al., 2002).

There are four main approaches in which plants can be manipulated by GE. These are:

- Gene insertion using a bacterial vector—*Agrobacterium tumefaciens*. To achieve a genetic modification using this technique, a restriction enzyme is used to cut non-virulent plasmid DNA derived from *A. tumefaciens* and thus create an insertion point, into which the gene can be ligated. The engineered plasmid is then put into a strain of *A. tumefaciens*, which contains a "helper" plasmid and plant cells are treated with the recombinant bacterium (Dandekar and Fisk, 2005). The helper plasmid assists the expression of the new gene in the plant as these grow in culture. The utility of *A. tumefaciens*-mediated gene transfer is illustrated by its use to develop potato plants that are resistant to Colorado beetle (Perlak et al., 1993). The role of microorganisms in plant GE has been discussed in detail by Haseena Khan in Chapter 3 in this Volume.
- In microballistic impregnation or gene gun the target gene is applied to minute particles of gold or tungsten, which are fired into plant tissues at high velocity. Cells that incorporate the target and marker genes can then be selected in culture and grown (Baum et al., 1997; Christou, 1997).
- Electroporation is the application of a pulsed electric field that allows the transient formation of small pores in plant cells through which genes can be taken up, and in the form of naked DNA incorporated into the plant genome (Bates, 1995; Terzaghi and Cashmore, 1997).
- Gene neutralization by means of antisense technology, homologous recombination (gene knockout) and gene replacement. Antisense technology involves introduction of the non-coding strand sequence of a particular gene to inhibit the expression of the desired gene. The application of antisense technology has resulted in a consumer food product, the Flavr Savr™ tomato, in which the shelf life is enhanced by suppression of the enzyme polygalacturonase (Sitrit and Bennett, 1998). A similar increase in the shelf life of tomatoes can also be achieved by using antisense technology against the enzyme 1-amino-1-cyclopropanecarboxylic acid synthase (Hamilton et al., 1995). In traditional breeding technology, the development of a new variety was possible only by using a particular species and crossing its known types, whereas nowadays antisense technology can be much faster and more effectively employed towards this purpose.

GM DNA SEQUENCES IN DAIRY PRODUCTS

The possible transfer and accumulation of novel DNA and/or proteins in food for human consumption derived from animals receiving GM feed is at present the object of scientific dispute (Beever and Phipps, 2001; Phipps et al., 2002, 2003). Modified DNA or novel proteins consumed by animals should be investigated to find out whether they have the potential to affect animal health, or to enter the food chain. The actual amounts of DNA entering the digestive system of a typical dairy cow should be counted. Dairy cows can consume 24 kg of dry matter per day. If a dairy cow's ration includes 60% GM maize (as silage or grain) per day, then that cow eats approximately 60 grams of total DNA a day, 54 μg of which are transgenic DNA (Beever and Phipps, 2001). Dairy cows consume and process abundant quantities of raw materials and feed, producing a large volume of high quality milk. Environmental conditions may influence the intake of the nutrients, including GM molecules, and their transfer through the digestive system and blood circulation to the mammary glands (Schubbert et al., 1997; McAllan, 1982; Beever and Phipps, 2001; Poms et al., 2003). A number of studies failed to identify GM DNA in milk, meat, or eggs derived from livestock receiving GM feed ingredients.

Agodia et al. (2006) developed a valid protocol by polymerase chain reaction (PCR) and multicomponent analysis for the detection of specific DNA sequences in milk, focused on GM maize (Maximizer maize) and GM soybean feed (RoundUp Ready Soybeans), assessed the stability of transgenic DNA after pasteurization treatment and determined the presence of GM DNA sequences in milk samples collected from the Italian market. Results from the screening of 60 samples of 12 different milk brands demonstrated the presence of GM maize sequences in 15 samples (25%), and of GM soybean sequences in 7 samples (11.7%). Their screening methodology showed a very high sensitivity and an automatic identification of the amplified products increased specificity and reliability.

Moreover, it was demonstrated that the pasteurization process was not able to degrade the DNA sequences in spiked milk samples. Besides, due to the great stability of genomic DNA, sterilization or pasteurization could not be effective for the complete degradation of the molecule, which therefore, could maintain full functional integration. A number of studies have now been conducted in which GM DNA has not been detected in milk, meat, or eggs derived from livestock receiving GM feed ingredients (Phipps et al., 2002, 2003). Moreover, two studies determined the presence of GM DNA in different parts of the ruminant digestive tract (Faust, 2000; Phipps et al., 2003).

The rationale for choosing to amplify very short fragments (106 bp for GM maize and 145 bp for GM soybean) was that ingested DNA sequences would be degraded, although not completely, in the gastrointestinal (GI) tract. Recent studies suggested that the GI tract of mammals is not a complete barrier to the uptake of small fragments into the bloodstream (Schubbert et al., 1997; Poms et al., 2003) and Chowdhury et al. (2004) suggested that although feed-derived maize DNA was mostly degraded in the GI tract, yet fragmented DNA was detectable in the GI contents; as such, it could also serve as a possible source of transfer to milk *via* faecal contamination. A recent experimental model demonstrated that GM DNA sequences from feeds are detectable in milk subsequent to aerosol contamination with feed particles (Poms et al., 2003). The probability of feed particle contamination under real conditions is affected by the milking technologies employed. In Italy, modern milking plants use self-contained systems in which small particles can find entry only through insufficient air filters. Moreover, it was demonstrated that the pasteurization process could not degrade the DNA sequences in spiked milk samples. As such, the detection of GM sequences in milk could be interpreted either as a biomarker of transgene feeding *via* faecal contamination with GM DNA, or it may indicate the presence of GM feed in the vicinity of milking and milk-storage facilities. In the European Union (EU) the consumer is offered the option of choosing between traditional foods and those which contain or consist of GM organisms by legally regulating the labelling of foods and feeds. In addition, the use of GM organisms in organically grown products is prohibited. As such, the presence of GM sequences in organic milk may indicate poor-quality assurance by the farmers and/or the producers. One more issue, at least for the Cry sequences, is that one theoretical source of alternative contamination of milk could be the environment, i.e., *B. thuringiensis*, the soil microorganism from which the sequence has its natural origin. Likewise, *Agrobacterium* sp. from which the CP4 EPSPS sequence of GM soybean has been isolated, could, at least theoretically, contaminate the environment around milking and milk-storage facilities. The low level of concordance between the presence of endogenous and transgenic maize and soybean genes highlighted by their results, even on a limited number of milk samples, may support an alternative source of contamination, possibly recognizable in soil bacteria of the natural environment. Further studies performed on a larger number of milk samples are needed to understand the likely source of contamination of milk collected from the Italian market.

TOXICITY STUDIES PERFORMED WITH GM FOOD CROPS

The selection of animal studies is based on considerations including a molecule's structure, function and *in vitro* toxicity results, as well as ethical criteria. Of these three distinct approaches, evidence from animal tests is usually most indicative of potential toxic effects of a test substance in humans. However, the extrapolation of results from animal tests on humans is uncertain, and unpredictable differences can include inter-species and inter-individual differences in metabolism, physiological processes, and lifestyle. These uncertainties are usually addressed through the use of uncertainty factors (Konig et al., 2004).

The difficulties encountered in assessing the safety of foods derived from GM crops in bioassays such as animal tests, are well recognized [OECD (Organisation for Economic Cooperation and Development)], 1993, 1998, 2000 and 2002; LSRO (Limited Standard Reactor Operator), 1998; FAO/WHO, 2000). It has been pointed out on numerous occasions that animal feeding studies with whole foods or feeds should be designed and conducted with great care to avoid problems encountered with nutritional imbalance from overfeeding a single whole food, which itself can lead to adverse effects. In undertaking such tests, a balance must be struck between feeding enough of the test material to have the possibility of detecting a true adverse effect and, on the other hand, not inducing nutritional imbalance. In any event, what one would like to extrapolate, by means of animal testing, for anticipated human intake is simply not achievable for practical reasons, and enhanced margins of safety, of one to three times higher, have to be accepted (Hattan, 1996; Munro et al., 1996a; WHO, 1987). This limits the sensitivity of animal bioassays to detect small differences in composition, which may be more readily detected by analytical characterization (Novak and Halsberger, 2000).

According to two papers by Munro et al. (1996a, b) based on 120 rat bioassays (each of 90 days duration) of chemicals of diverse structure including food additives, pesticides, and industrial chemicals, it was found that the lowest observed adverse effect levels (LOAEL) ranged from 0.2 to 5,000 mg kg^{-1} body weight, with a median of 100 mg kg^{-1} and a 5th percentile of 2 mg kg^{-1}. To achieve the 5th percentile of exposure from a toxic constituent present in, say, a food crop in a rodent bioassay (at a food incorporation rate of 30%), the toxin would have to be present at around 80 μg g^{-1}. To achieve the median exposure of 100 mg kg^{-1} it would have to be present at 5,000 μg g^{-1}. These concentrations fall well within the range of existing analytical techniques for detection of inherent toxicants in food. The concentrations should also be readily detected during compositional analysis of the known toxicants in the host organism used to generate the improved nutrition crop. The broiler chicken has emerged as a useful animal model for assessing nutritional value of foods and feeds derived from GM crops. It should be noted, however, that, contrary to laboratory rodents, the rapidly growing broiler has been obtained through breeding efforts with the objective of creating an efficient food-producing animal. This may, therefore, not render it optimal for toxicological testing of foods and feeds. In fact, disorders such as "sudden death syndrome" and "ascites" (an abnormal accumulation of serous fluid in the abdominal cavity) are considered related to metabolic disorders associated with its rapid growth (Olkowski and Classen, 1995). On the other hand, broiler chickens have been optimized for growth relative to highly characterized diets such that small changes in nutrients or antinutrients in the diet are readily manifested in reduced growth. In addition, one of the first indications of an ill animal is loss of appetite or reduced growth rate (ILSI, 2004). **Table 1.3** summarizes some representative toxicity studies carried out on rats and mice (unless indicated) with GM vegetables.

The safety assessment of Novel Food, including GM biotechnology-derived crops, starts with the comparison of the Novel Food with a traditional counterpart that is generally accepted as safe based on a history of human food use. Substantial equivalence is established if no meaningful difference from the conventional counterpart is found, leading to the conclusion that the Novel Food is as safe and nutritious as its traditional counterpart. Engel et al. (1998) investigated glycoalkaloids in potatoes as example for the principle of substantial equivalence. In general, the non-significance of '*p*' value is used for the proof of safety. From a statistical perspective, the problems connected with such an approach are demonstrated, i.e., quite different component-specific false negative error rates result (Hothorn and Oberdoerfer, 2006). As an alternative, the proof of safety is discussed with the inherently related definition of safety thresholds. Moreover, parametric and non-parametric confidence intervals for the difference and the ratio to control (conventional line) are described in detail in the literature published by previous authors. The treatment of multiple components for a global proof of safety has also been explained in the literature.

CONSUMERS' BELIEFS TOWARDS GMO – DIFFERENCES AMONG COUNTRIES

GE has been used for many years in the production of consumer goods, such as pharmaceuticals and detergents. While consumers do not pay particular attention to these applications, the implementation of GE in food production is met with intense consumer mistrust. Nowadays, biotechnology is developing in an environment where public concern about food safety and environmental protection is steadily increasing. The

Table 1.3. Representative toxicity studies carried out on rats and mice (unless indicated) with GM vegetables

Vegetable	*Trait*	*Duration*	*Parameters*	*Reference*
Potato	Lectin (*Galanthus nivalis*)	10 d	Histopathology of intestines	Ewen and Pusztai, 1999
Potato	Cry1 endotoxin (*B. thuringiensis* var. *kurstaki* HD1)	14 d	Histopathology of intestines	Fares and El Sayed, 1998
Potato	Glycinin (soybean)	28 d	• Feed consumption • Body weight • Blood chemistry • Blood count • Organ weights • Liver and kidney histopathology	Hashimoto et al., 1999a, b
Tomato	Cry1 Ab endotoxin (*B. thuringiensis* var *kurstaki*)	91 d	• Feed consumption • Body weight • Blood chemistry • Organ weights • Histopathology	Noteborn et al., 1995
Tomato	Antisense polygalacturonase (tomato)	28 d	• Feed consumption • Body weight • Blood chemistry • Organ weights • Histopathology	Hattan, 1996

d = days

scientific claims related to biotechnological advantages in terms of social benefits are not accepted without criticism. The ethical issues arising from the foreseeable changes due to the applications of biotechnology in daily life can result in reluctance to adopt the new technology although questions may arise about the hazards caused by these changes (Kerr, 1999), these concerns may have less impact than the expected benefits.

The attempts to introduce GM food into the market must be supported by the scrupulous analysis of consumer evaluation procedures for GM products. The realization of the variations in consumer beliefs with respect to biotechnology is indispensable for decision-makers to foresee potential problems of biotechnology acceptance and to take consumer concerns into consideration in the development of novel products. Consumers will be the ultimate judges of success of the upcoming new foods produced from agricultural biotechnology (Saba et al., 1998), while they increasingly question whether any further technological treatment is beneficial or necessary, especially when it is considered to have a serious impact on food safety and environmental protection (Hoban, 1996a; Saba et al., 1998, 2000; Saba and Vassalo, 2002).

Attitudes towards GE crops were found to vary considerably among countries. American and Canadian consumers seem to be much more favourably disposed toward such products than their counterparts in Europe, and the Japanese are somewhere in the middle (Gaskell et al., 1998). It is rather surprising that the already existing negative attitude of European consumers towards GM foods was further reinforced from 1996 to 1999 (Gaskell et al., 2000), as opposed to Canadian consumers who are extremely interested in being well informed on GM foods (Magnusson and Hursti, 2002). However, even within the EU, Scandinavian consumers and, in particular, Swedish, appear to be the most sceptical and negatively inclined towards GM foods. Between the two sexes, women were shown to be even less willing to try these new GM products foods (Magnusson and Hursti, 2002).

One striking finding of the present survey is that the real awareness of the term 'GM' food among Greek consumers is very limited, despite their high stated awareness. Only one-third of consumers really understand the relevance of the term, with another 27% having a rough idea about it (Arvanitoyannis and Krystallis, 2005). In relation to the awareness level, the major discriminating factors among all consumer types regarding GM products, i.e., the 'unaware', the 'semi-aware' and the 'really aware', are their educational and income status. Provided that education is the most consistent and reliable measure of consumers' social class participation (Hupkens et al., 1997), this result indicates that consumers who are really familiar with the concept of GM food belong to upper average or high social classes. This profile corresponds to that of the Greek quality food buyer, such as the organic buyer, as described by earlier Greek surveys (Fotopoulos and Krystallis, 2001, 2003). Consumers in Greece appear to be in a state of confusion regarding GM food, which is confirmed through the emergence of some contradictory findings. For instance, the enhanced sensory and organoleptic characteristics appear to be the most important reason behind the perception that GM food is necessary, however, only a small minority of consumers believe that GM food is tastier compared to conventional food. This observation leads to the conclusion that even those who understand the term 'GM' as the modification of the genetic material of a plant or animal, are not really aware of the impacts of this transformation on the final product, and which of these products can be found on a supermarket shelf. GM meat products are regarded as less healthy compared to their GM plant counterparts, due to erroneous assumptions. Consumers directly relate the use of GE in animal-originated food production to potential food scares. However, the implementation of GE in the production of food is a relatively new phenomenon. Consumers do not have the occasion to regularly purchase GM foods (at least to their knowledge), simply because very few products are marketed with the relevant indication on the label. It is thus possible that Greek consumers rely on their preconceived (possibly negative) beliefs about food science in general, and food safety and healthiness in particular, than on perceptions regarding individual GM products (Zimmerman et al., 1994; Bredahl, 1999; Kerr, 1999; Baker and Burnhum, 2002; Lahteenmaki et al., 2002). This result can provide an adequate explanation about the certainty that GM foods of animal origin are less healthy. Consumers expressed their lack of trust of these products after the recent food scares related to meat (pre-conceived attitude). As meat products are perceived as potentially hazardous, GM meat products have the highest possibility of also being dangerous.

A substantial percentage of sample members think that GM foods stand nowadays for an everyday commodity, as they believe they have already consumed them despite their will. In other words, consumers are greatly concerned about loss of control over their food choice. Consequently, genetic modification appears to be a very influential factor in their food purchasing habits. However, the same substantial percentage of consumers exhibits an impressively realistic point of view by admitting that the use of GM food is now a necessity for a variety of reasons related to both improved properties of the novel GM foodstuffs (self-relevant benefits, Grunert et al., 2001) and long-term inefficiencies of the conventional food production system.

It is also very important to clarify the following controversy occurring in the present survey: sample members' purchasing intentions, attitudes towards, and beliefs about GM food do not vary among the three subgroups identified according to their awareness level, as the relevant variables are not statistically significant. Why do consumers of different social status provide the same answers in relation to GM foods? The most plausible explanation is that, regardless of their level of knowledge, consumers tend to perceive the production method of such foods as far more important than other product characteristics, which are considered crucial for the conventional foods. In this instance, one could possibly adopt the arguement by Grunert et al. (2001): 'the non-GM character of a foodstuff is itself the most valuable characteristic of the product'. The core finding of this study is that GM foods are perceived (by a percentage of really aware consumers) as dangerous, unhealthy, inappropriate for children unnatural and unethical. This fact is also highlighted by earlier surveys, with those of the Eurobarometer survey (European Commission, 1997; Verdurme et al., 2001; Springer et al., 2002) listing the most striking examples. The feeling of the necessity of GM foods is not enough on its own to outweigh the overall negative opinion about them, a fact clearly expressed by the particularly low percentages of sample members preferring GM to conventional food and being willing to purchase or to pay a premium for it. Regarding this last conclusion, it is worth mentioning that the perceptions of GM food risks seem to vary depending on the type of food product, with GM foods of animal origin perceived as the most hazardous, as demonstrated in previous literature (Zimmerman et al., 1994; Hamstra and Smink, 1996; Kuznesof and Ritson, 1996; Kerr, 1999; Mangusson and Hursti, 2002). At the same time, perceptions of the benefits of GM food consumption seem to vary considerably with the type of benefit. The benefit of improved taste, for instance, is not as important as compared to the advantage of increased shelf life. Overall, consumer beliefs about GM foods are built on the conflict between safety and benefits of their consumption, as is indicated by the results of factor analysis. Moreover, the perception of consumption benefit is not enough to outperform the perception of low safety, as previously published (Bredahl, 1999; Grunert et al., 2001; Lusk et al., 2002; Arvanitoyannis, 2003).

Additionally, a substantial percentage of consumers consider the provision of information on GM food labels to be a major issue, while they express their dissatisfaction with the amount of information presented by both the government and consumer organizations. The majority of respondents is in favour of the use of labelling as the most adequate means of information, and emphasize that such a labelling policy does not exist as yet, despite the fact that GM foods are already present on the Greek supermarket shelves. However, the percentages of those who (strongly) agree with the above findings are not particularly high, indicating that all really aware consumers do not unanimously hold these perceptions. The unethical dimension of GM implementation in food production, in particular, is much less believed than other risk dimensions, a finding also common in other EU countries (Saba and Vassalo, 2002). Additionally, the percentages of those who cannot express a firm opinion regarding the long-term unhealthy effects of GM food consumption, or the inappropriateness of GM food for children diet (32.6 and 37.5%, respectively) are relatively high. The percentages of those who cannot express a firm opinion regarding the need for information on GM food and the adequacy of the label as a means of information provision, are also relatively high. Furthermore, one should bear in mind that the opinions about the GM consumption benefits are unexpectedly positive for a number of product features (size, colour, shelf life, and shape).

The above observations support the clear-cut development of segments, based on consumers' beliefs about GM foods rather than on their socio-demographic profile, as crucial for drawing a firm conclusion. Several socio-demographic variables play an important role in the determination of consumers' awareness level regarding GM food, but do not appear as discriminating factors between specific consumer segments with the same awareness level, or with different purchasing behaviour and attitudes towards GM food. The weakness of socio-demographic variables as predictors of GM consumer behaviour is also indicated in the published literature (Hamstra and Smink, 1996; Kerr, 1999; Baker and Burnhum, 2002). Consumers who wish to avoid the consumption of GM food should be identified on the basis of what they believe rather than who they are, as indicated in the results of cluster analysis. Consumers who resist the purchase of GM products are possibly motivated by the same reasons that make them hesitant to purchase other novel products, perhaps because of their tendency to avoid risk and their low ability to adapt to changes. A marketing strategy targeting consumers who first adopt novel foods (the so-called 'early adopters', Baker and Burnhum, 2002), and who have the greatest potential to perceive consumption benefits of novel foods as more important than their inherent risks, has increased the possibilities of success in marketing GM food products. In any case, every marketing strategy should concentrate on GM consumption benefits that are directly relevant to the consumers and improve their quality of life, such as longer shelf life, better taste and enhanced nutritional content.

Huffman et al. (2006) have shown that preconceived notions and new information (both from interested parties and third-party sources) affect bidding behaviour of people who participated in auction market experiments for GM food. This contradicts the earlier findings of Viscusi (1997) and Tversky and Kahneman (1992) who argued that people frequently ignore base rates. One potential explanation for this difference of outcome is that instead of measuring prior beliefs as objective knowledge (e.g., monetary lotteries); lab participants were asked to give information about their prior knowledge about genetic modification. They were asked the following question: "How informed are you about genetic modification?" and they were given five options: extremely well informed, well informed, somewhat informed, not very informed, and not informed at all. This information was subjective, and was used as their subjective prior belief.

Furthermore, the use of prior beliefs was examined as well as new information, to give guidance on decision on willingness to pay for common food items available in grocery stores and supermarkets and not in a lottery. These results have implications for information policies by showing how both sceptics and proponents of new technologies (i.e., interested parties) might try to manipulate information to achieve private objectives. This is most likely to occur when firsthand scientific knowledge about the impacts of new technologies is insufficient or when third-party information is limited or unavailable (Milgrom and Roberts, 1986; Huffman and Tegene, 2002). Opponents to a new technology may try to target people who are relatively uninformed about the technology. Proponents of the technology may try to target people who have informative prior beliefs for maximum effectiveness. This reasoning might explain why the Council for Biotechnology Education (a pro-GM organization) funds TV commercials during family and sports programming, while the anti-GM groups invest in carrying out dramatic and spectacular shows of opposition for new technologies (e.g., with colourful and vocal demonstrations) that may be carried by a variety of media outlets. People who are uninformed seem less likely to use common media sources regularly. In this study a relatively blunt measure of prior beliefs about genetic modification has been used; future research might examine the objectiveness of these beliefs and attempt to illicit strengths of pro-biotechnology and anti-biotechnology prior beliefs and the way that they affect willingness to pay.

Eicher et al. (2006) have argued that agricultural biotechnology has the potential to help African small-holders, to confer benefits to consumers, ensure a clean environment, and better health of farmers and farm workers. However, many African decision makers are requesting more information on potential environmental and food safety issues related to GM products. There is a need to develop national capacity and regional dialogues to monitor potential health, environmental, distributive, and food safety risks and cross-border movement of GM cultivars, especially when neighbouring countries do not have functional regulatory systems in place.

Worldwide studies showed that consumers' concerns about GM foods are on the rise and acceptance of GM foods varies among countries (Bredahl, 1999; Gaskell et al., 1999; Curtis et al., 2004). Many consumers in European countries and Japan have difficulty in accepting GM foods (Hoban, 1997; Macer and Ng, 2000; Verdurme et al., 2001; Magnusson and Hursti, 2002; McCliskey and Wahl, 2003). However, the results of other studies show that the consumers are much less worried about GM foods in the US and many developing countries (Gaskell et al., 1999; Aerni, 2001; Hallman et al., 2003). Consumers' acceptance of GM food in the US ranged from 50% (Hallman et al., 2003) to 59% (IFIC, 2004), but has declined slightly over time (Hallman et al., 2004).

The findings from several recent consumer surveys in China are mixed. On one extreme, a study in Guangzhou, Shanghai, and Beijing by Greenpeace (2004) claimed that Chinese consumers did not generally accept GM foods. On the other extreme, Li et al. (2003) and Zhang (2002) showed that Chinese consumers were willing to pay a premium for GM foods. Zhang (2002) showed that the majority of consumers in Tianjin city were willing to pay up to 20% extra. A survey in Beijing concluded that consumers, on average, were willing to pay a 38% premium for GM rice versus non-GM rice (Li et al., 2003). There are also a number of recent surveys in different locations of China showing a large variation of consumers acceptance of GM foods ranging from about half in Tianjin (Wang, 2003) and Nanjing (Zhong et al., 2003) to about 80% in Beijing (Zhou and Tian, 2003).

The uncertainty about Chinese consumers' attitudes toward GM foods contributes to uncertainty for policy makers on how China should proceed with its future biotechnology policies in general and GM foods in particular. For example, although China has invested substantially in GM rice research and GM rice has been ready for commercialization since 2000 (Huang et al., 2004a), Chinese leaders have not yet decided whether production of GM rice should be allowed in China or not. At the same time, they have allowed imports of GM foods such as GM soybeans and maize.

China is an interesting case for several reasons. It is the world's most populous nation and has been one of the world leaders in promoting agricultural biotechnology research through public investment (Huang et al., 2002). China's final decision on whether it should commercialize GM rice will greatly influence what the rest of Asia does about GM food crops. The goal of this study is to conduct a comprehensive survey of consumers' attitudes toward GM foods in Urban China.

The evidence from the existing literature is mixed and sometimes contradictory. Huang et al. (2006) conducted a large in-depth face-to-face in-house survey to examine the consumers' awareness, acceptance of and willingness to buy GM foods in China. To achieve this objective, a well-designed consumer survey was conducted in 11 cities of five provinces in Eastern China in 2002 and 2003. The results indicated that despite the fact that inadequate information on GM foods was available publicly in China, more than two-thirds of consumers in urban areas had heard of GM foods. However, their knowledge on biotechnology was limited. Chinese consumers' acceptance of and willingness to buy GM foods was much higher than in other countries. Chinese consumers also demonstrated great variance in their acceptance of different GM foods. Information and prices of GM foods were two important factors affecting consumers' attitudes toward GM foods. Based on the findings of this study and given that the sample group is in the more developed eastern Urban China, it is concluded that the commercialization of GM foods is not likely to receive great resistance from consumers in China.

Europe is progressively losing not only its market share of GM foods, but also due to declining investments from large organizations, any near-term ability to be competitive in the field of biotechnology. The emerging gap between the EU and its main competitors will affect not only the European science and technology base, but also industry, consumers and farmers (Tencalla, 2006). There are many different measurable parameters to demonstrate the already considerable impact on European research. The number of notifications for field trials across the EU15 (JRC, 2004) peaked at 264 in 1997, and then fell to only 56 in 2002, and 23 in 2004. In contrast, there are around 900–1,000 notifications of GM trials in the US each year (Mitchell, 2003). According to a recent Commission survey of private biotech companies and public research institutes (European Commission, 2003), 39% of European research projects in this field have been aborted in recent years. In the private sector alone, the figure was as high as 61%.

Faced with overly strict political and regulatory frameworks, university and industry R&D is moving to other world areas such as the USA, Japan, and China where long-term strategies for exploiting the potential of plant genomics and strengthening the positions in related markets are in place. Each year, thousands of Europeans go to study in the US and over 70% remain there to pursue their careers. European biotechnology SMEs are more likely to enter into research collaborations with US companies rather than their EU counterparts and are also turning more and more to customers outside the EU (Bioscience Law Review, 2004). Furthermore, companies of all sizes are relocating their research activities and investments to these countries, as well as India and Argentina.

Recent examples include the Swiss agrochemical company Syngenta, which announced in July 2004 its intention to relocate its UK agricultural biotechnology research operations to North Carolina (US). Also, BASF, the world's largest chemical company, has threatened to move its GM crop research to the USA unless Europe becomes more receptive to new technologies.

Finally, the recalcitrant attitude towards biotechnology is also expected to affect the 15 million European farms in the 25 Member States and the European food industry. In a situation where agricultural output is less technologically competitive and Common Agricultural Policy subsidies are decreased and even eliminated, the growth of alternative niches such as the organic food market will not be able to compensate for a shrinking share of conventional and GM markets in Europe. Strong price competition coupled with loss of import protection and tariff reductions, could lead to a shift from local European products to imports, effectively limiting the range of EU consumer lifestyle and health choices (Tencalla, 2006).

In the near future, European consumers including farmers themselves risk being left with the choice of either buying local products at much higher prices or going for cheaper imports, all because of an unsubstantiated fear of GM products.

Rather than a leading position in the biotechnology and genomics field, Europe therefore, faces the deterioration of its R&D base, the loss of markets for European agricultural products and an increased dependence on food and feed imports (Mitchell, 2003). European industry's ability to contribute to agricultural innovations and a biotechnology-based economy may see itself severely restricted.

Europe is at crossroads and there are enormous opportunities to be grasped. Biotechnology and genomics, in particular relating to plants, have been identified as Welds for future growth, crucial for supporting the agricultural and food processing industry, Europe's main economic sector. However, in contrast to other world areas which are moving ahead and investing heavily, Europe has found itself 'bogged-down' in a heated, long-winded, and often acrimonious debate between opponents and supporters of these technologies. In the meantime, it is losing its competitive edge to other regions of the developed and developing world.

Acceptance or rejection of cutting-edge innovations depends on many interlinked political, economic, and social factors that create a favourable or unfavourable climate at a given time. Clearly more research is required on societal responses to the application of new technological innovations. Overall, a better understanding of people's attitudes and values is needed. The debate concerning GM crops has been dominated by the risks, while the positive economic, environmental, and health aspects have been ignored. The direct consequences are a lack of unanimous and consistent political support throughout the various European Member States, directly impacting new research initiatives, and market opportunities.

There is no doubt that as with all technologies and human endeavours, genetic modification carries the potential for risk, therefore safety assessments are required to be performed for every new trait, within the context of a sound international broadly harmonized legislative framework.

Innovation has continued to flourish under the US, Canada, Japan, and China, but not under Europe. If Europe is to remain a leader in this crucial area of innovation, the concerns of both critics and advocates need to be critically addressed and then Europe should move on. Used in conjunction with other methods, plant biotechnology can be a positive addition to the current agricultural portfolio. Europe should, therefore, proceed responsibly in developing biotechnology, while taking all reasonable steps to minimize adverse effects. It is important to remember that, in some cases, not proceeding may be just as costly as moving ahead. This can be measured by negative impacts for each of the parameters essential for global sustainability, i.e., health, environment, and socio-economic development.

Previous research has demonstrated that knowledge and awareness of biotechnology among Irish consumers is low. In addition, the Irish public distinguishes between different biotechnology applications, with medical uses generally perceived more favourably than food biotech innovation. Their primary concerns appear to stem from the unknown long-term effects of GM foods on the environment and human health (O'Connor, 2004). While previous research has been a source of much valuable information on Irish consumer reaction to biotechnology, the focus has been on GE as a somewhat abstract concept, and as such has not examined consumer reactions to specific products of the technology, a situation reminiscent of that in the UK in the early nineties (Frewer et al., 1996). A further criticism of previous work, which was survey based, is that it encouraged participants to reply in the role of a citizen rather than a consumer; in this role, the subjects had the freedom of using information or beliefs in reacting to the survey, which they would not actually use when making purchase decisions (Noussair et al., 2001a). In contrast, international studies have attempted to place respondents in the role of consumers, by presenting a survey audience with specific examples of GM foods and examining their intention to purchase such products (Frewer et al., 1996; Grunert et al., 2001; Moon and Balasubramanian, 2001; Noussair et al., 2001b; Koivisto Hursti et al., 2002).

The work described by O'Connor et al. (2006) adopted the approach of trying to place respondents in the role of consumers by presenting them with a hypothetical GM product, for example, yoghurt possessing anticancer properties. This paper examined Irish consumer acceptance of second-generation GM products, defined here as those which are expected to exhibit a specific consumer-oriented benefit. Conjoint analysis was used to determine Irish consumer preferences (n = 297) for attributes of a hypothetical GM yoghurt. Cluster analysis on the basis of the GM attribute revealed four segments of consumers. An "anti-GM" segment (24.4% of sample) was an outright rejecter of all GM foods since it perceived GM foods as being unnatural or in some way artificial, while a second cluster (33.4%) specifically rejected second-generation GM products. A further 20.5% of the sample segment was receptive to the notion of second-generation GM products. However, this group had a number of complex reservations, which needed to be resolved before they would truly accept such products. GM foods offering specific consumer benefits were found to be acceptable to 21.2% of the sample group, implying that these foods could represent a segment within the overall food market in the future. The "second-generation accepters" only saw the benefit offered by the product, and were not concerned about its naturalness.

The results of the present study indicate that while the majority of Irish yoghurt consumers continue to harbour an overall negative perception about GM foods, a sizeable sub-section of this population (about 40%) may be receptive to a second-generation product possessing an overt consumer health benefit. About 21% of the sample group was found to be receptive to second generation GM products, similar to results found in an earlier study in Germany which demonstrated that 27% of participants (n = 200) would buy GM foods which provide certain health functionality (Spetsidis and Schamel, 2001). A further 20% of the Irish sample segment was receptive to the notion of second-generation GM foods. The "second-generation rejecters" are an interesting group, as they reject second-generation GM products, yet do not hold very strong anti-GM feelings as indicated by their utility scores; the relative importance score reveals that fat content is the most important factor influencing choice of a yoghurt profile for this group.

The most regular consumers of yoghurt (50.8% consume it on a daily basis) were those who were receptive to the notion of a GM yoghurt with a health benefit ("conditional accepters"). On the other hand, those who demonstrated clear support for a second-generation GM yoghurt ("second-generation accepters") were less frequent consumers of yoghurt. The majority of this cluster (76.2%) consumed yoghurt three times a week or less (O'Connor et al., 2006).

Ethical concerns also provided some insight into the opinions held by the different segments. For example, animal welfare (''has been produced in a way that animals have not experienced pain'' and ''has been produced in a way that animals' rights have been respected'') was a more important food choice factor for ''conditional accepters'' and the ''anti-GM'' cluster, compared to the ''second-generation rejecters''. The ''anti-GM'' segment could, perhaps, associate GM foods with issues such as animal cloning and reject such foods on the basis of animal welfare concerns. In fact, the Eurobarometer surveys (http://www.europa.eu.int/comm/public_opinion) indicated that Irish respondents favoured applications which benefit human health and society at large, but had particular issued with GM animals even if such techniques may also benefit society. These considerations may also explain why the ''conditional accepters'' do not completely embrace second-generation GM products. Environmental protection emerged as a more influential food choice factor for the ''anti-GM'' group compared to ''second-generation rejecters'' and ''conditional accepters''. For example, 65.8% of the ''anti-GM'' group considered it very important that the food they eat on a typical day has been prepared in an environmental-friendly way, while in contrast, only 36.5% of ''second-generation accepters'' thought it important. It thus implies that the ''anti-GM'' segment may reject such foods on the basis of environmental concerns. Political values (''has been prepared in a way that does not conflict with my political values'', ''comes from a country I approve of politically'') emerged as more important for ''conditional accepters'' than ''second-generation rejecters''. It seems likely that a GM product would have to come from a country that this group approved of politically and which did not conflict with their political values in order for it to be fully accepted by this segment (O'Connor et al., 2006).

Consumers expressed most confidence in a seal of approval awarded by a European standards agency as opposed to a national body, consistent with the results of the latest Eurobarometer survey (Gaskell et al., 2003). Brand identity was found to be the least important factor in yoghurt choice in the present study.

Although obtained from a small sample segment, the results suggested that in general GM foods are not widely accepted by the Irish population. Irish yoghurt consumers rejected a GM product even when it offered an anticancer health benefit. The consumer focus groups indicated that yoghurt was perceived as a ''natural product'' with ''natural ingredients'', hence certain consumers may not like or agree with the use of GM ingredients in such a product. Moreover, consumers might be sceptical of the health claim associated with the product.

The debate over GMOs has reached heights of controversy in New Zealand as elsewhere in the world. A Royal Commission on Genetic Modification received more than 10,000 submissions, heard more than 300 witnesses, and published a 4-volume 1,000-page report at a cost of approximately NZ$ 6.5 million. The major conclusion was that "New Zealand should keep its options open. It would be unwise to turn their back on the potential advantages on offer, but they should proceed carefully, minimizing and managing risks" (Eichelbaum et al., 2001).

Much of the debate has centred on potential harm to New Zealand's country image in foreign markets for food products - particularly European markets (Eichelbaum et al., 2001). A common view in New Zealand appears to be that consumers in foreign markets are either 'for' or 'against' GMOs, mostly 'against'. New Zealand lifted its moratorium on GMOs in October 2003, and each application is to be dealt with on a case-to-case basis. In this evolving climate, it becomes imperative for a food exporting country such as New Zealand to know the likely impact of these moves on perceptions in foreign markets of food products from their country, in the event they introduce GMOs of various kinds.

Knight et al. (2005) conducted in-depth interviews with key distributors in the European food sector to ascertain factors that they consider important in determining the reputation of exporting countries, and to ascertain whether GM has an impact on such reputations. Highly negative consumer sentiment towards GM in Europe seems likely to continue to influence food industry buyers against importing GM food. However, no evidence was found regarding the concept that the presence of GM crops in a country causes negative perception of non-GM food imported from that country. Provided adequate steps are taken to avoid accidental contamination of conventional crops, producer countries do not appear at great risk of damaging their overall country image for food products if GM technology is introduced.

CONSUMER SAFETY – POST-MARKETING MONITORING

Consumer Safety

There has been a very heated debate over consumer safety issues *vis-a-vis* the consumption of GM foods. This controversy was primarily fuelled by the differences in terms of political orientation (see, legislation differences between EU and US) in conjunction with several incidents like the potato incident where a lectin called concanavalin A (known toxin) was identified in the GM potato. Another issue stands for the presence of a strong allergen in soybean. In this case, Brazil *nut* genes were transferred to soybeans (Leighton, 1999; Arvanitoyannis and Krystallis, 2005). Every side (EU and USA) adopted the respective pros and cons (social, environmental or economic nature) in order to justify their claims and adherence to that particular side **(Table 1.4).** If substantial equivalence can be established except for a single or few specific traits of the GM plant, further assessment focuses on the newly introduced trait itself. Demonstration of the lack of amino acid sequence homology to known protein toxins/allergens, and a rapid proteolytic degradation under simulated mammalian digestion conditions, was deemed to be sufficient to assume the safety of the new protein (FAO/WHO, 1996). However, according to Kuiper et al. (2001), there may be circumstances that require more extensive testing of the new protein, such as (i) the specificity and

Table 1.4. Potential economic and social pros and cons of genetically engineered (GE)/ modified (GM) crops and foods (Weick and Walchli, 2002; Mepham, 2000; The Royal Society, 2002, 2003; Engel et al., 2002)

Factor	*Pros*	*Cons*
Economic	• Cost savings and higher crop yields • Reduced need for spraying with pesticides/herbicides • Drought resistance • Precision agriculture • Resistant plants to soil, high metal content and salinity	• Initial high cost of GE grain (e.g.. Flavr Savr™ tomato) • HRCs* may be transformed into weeds acting as a source of pests and diseases undermining the principles of crop rotation • Transfer of genes from GM to non-GM crops may have undesirable effects upon the latter (e.g., development of *Bt* resistance through use in GM organisms might undermine organic farming)
Social	• Reduced usage of chemical pesticides used in grain production will reduce the environmental pollution • Enhanced food production may help avert world food shortage • Food products endowed with enhanced cooking properties and medicinal value will be available	• Long term environmental risks are unknown • GE crops are perceived as unnatural and unethical • Private organizations may place profit before safety and world needs • Risk of transferring antibiotic resistance from crop genes to humans will eventually compromise the effective treatment of patients • Consumers can scarcely understand the labelling wherever the latter is available • Production cost savings are not actually passed on to the consumers • Effective destruction of weeds may reduce the availability of habitats for various insects and invertebrates

*HRC: Herbicide Resistant Crop

biological function/mode of action of the protein is partly known or unknown; (ii) the protein is implicated in mammalian toxicity; (iii) human and animal exposure to the protein is not documented; or (iv) modification of the primary structure of naturally occurring forms. Bacterial Bt proteins are examples of proteins that have been introduced into crop varieties by genetic modification. Bt proteins (Cry proteins) from *B. thuringiensis* strains have been introduced into GM crop plants for their insecticidal properties in the larvae of target herbivoral insect species (Peferoen, 1997).

Post-marketing Monitoring-Allergenicity

Post-market monitoring systems have been established by several food companies for certain food products to act as early warning systems and to facilitate product recall in the event where health concerns might be associated with a specific food. The organization in charge of post-market monitoring is primarily the responsibility of the manufacturer of the food. Methods vary from establishing channels of communication in the firm to receive direct consumer feedback on the product, to the repurchase of products to determine the quality of the product on the supermarket shelf. Post-market monitoring programmes may serve to confirm the absence of specific adverse health effects of certain products after they have been marketed. The feasibility and validity of post-market monitoring depends on the health end point of interest and on the way the product is marketed (Konig et al., 2004).

The potential allergenicity of foods derived from biotechnology has been recognized as relevant for particular consideration because some of the modifications in these foods may include the expression of proteins not otherwise present in that food (Taylor and Hefle, 2002). The presence of proteins *per se* does not merit attention as a concern regarding the safety of foods derived from biotechnology. However, the current definition of food allergens (for example, proteins eliciting an IgE-mediated hypersensitivity response in sensitive individuals following consumption) dictates that an assessment for allergenic potential is in order when the biotechnology-derived food or food ingredient contains a novel protein that would otherwise not be present in food (ILSI, 1997, 2003, 2004; Goodman et al., 2005). A pre-market decision tree strategy (a number of questions organized in a tree diagram in order to facilitate decision taking for market issues) for assessing the allergenicity of GM foods has been recommended by FAO and WHO (FAO/WHO, 1996, 2000, 2001) and has proceeded to Step 8 of the Codex Alimentarius Commission's assessment of the safety of foods derived through biotechnology (Codex, 2002). According to ILSI (2004), the methodological considerations to be taken into account for an effective post-market monitoring are as follows:

(1) Measuring population exposure to foods and food ingredients.
(2) Tracking the disappearance of biotechnology-derived foods into the food supply.
(3) Integrating disappearance data with food consumption databases.
(4) Demonstration of causality.

The usage of GM foods for human consumption has raised a number of fundamental questions including the ability of GM foods to elicit potentially harmful immunological responses, as well as allergic hypersensitivity. To assess the safety of foods derived from GM plants including allergenic potential, the FDA, FAO, WHO, and the EU have developed approaches for evaluation assessment. One assessment approach that has been a very active area of research and debate is the development and usage of animal models to assess the potential allergenicity of GM foods. A number of specific animal models employing rodents, pigs, and dogs have been developed for allergenicity assessment. However, validation of these models is needed and consideration of the criteria for an appropriate animal model for the assessment of allergenicity in GM plants is required. A BALB/c mouse model has been employed by Prescott and Hogan (2006) to assess the potential allergenicity of GM plants. They demonstrated that this model was able to detect differences in antigenicity and identify aspects of protein post-translational modifications that can alter antigenicity. Furthermore, this model helped to examine the usage of GM plants as a therapeutic approach for the treatment of allergic diseases.

Techniques used as part of the weight of evidence approach (the assumption that if the protein's source is known to be allergenic, then the transferred protein is assumed to be an allergen unless scientifically proved otherwise) to assess the potential allergenicity of the newly expressed protein include consideration of the source of the gene, sequence similarity to known allergens, serum screens for IgE epitopes, ex vivo cell assays, and physico/chemical properties such as pepsin resistance and allergenicity in animal models (Metcalfe et al., 1996; Alinorm 03/34: Joint FAO/WHO Food Standard Programme, 2003; Goodman et al., 2005).

When the protein is not from an allergenic source, other assessment criteria including sequence homology analysis, IgE binding capacity and pepsin resistance are used. Animal models are also recommended for the purpose of screening proteins for allergenicity. Animal models are considered to be particularly useful when the GM plant expresses a protein that has not previously been consumed (Taylor and Hefle, 2002).

Collectively, many studies have demonstrated that IgE sensitivity could be transferred to the transgenic plants when an allergen is expressed in a non-native host through genetic modification. Several animal models have been developed, including models using the Brown Norway (BN) rat, BALB/c mouse, dogs, and pigs (Helm and Burks, 2002; Helm et al., 2003). It is unlikely that a single animal model will be able to definitively predict the potential allergenicity of the novel food antigens because food allergy in humans is a complex disorder that is dependent on multiple factors to promote allergic sensitization (Atherton et al., 2002). It is presumed that the use of a number of appropriate animal models, in conjunction with other assessment criteria, will help in identifying potential hazards.

BALB/c mice have previously been used in animal models of food hypersensitivity where they have been shown to develop many of the clinical symptoms of food allergy such as diarrhea, anaphylaxis, antibody responses (IgE and IgG1), eosinophil and mast cell accumulation (Li et al., 2000; Hogan et al., 2001; Brandt, 2003; Untersmayr et al., 2003). BALB/c mouse models developed for the assessment of GM plants for allergenicity have primarily measured antibody responses to the protein and not other clinical symptoms associated with food allergy (Dearman et al., 2001; Dearman and Kimber, 2001; Atherton et al., 2002).

Recently, Prescott and Hogan (2006) utilized the BALB/c mouse model to assess the potential allergenicity of two GM plants. Firstly, they examined a transgenic pea (*Pisum sativum* L.) expressing a common bean (*Phaseolus vulgaris* cv. Tendergreen) [transgenic-Pea-derived α-amylase inhibitor-1 (αAI)]. The transgenic peas were generated to protect the seeds from damage by inhibiting the α-amylase enzyme in old world bruchids (pea, cowpea, and azuki bean weevils) (Schroeder et al., 1995; Morton et al., 2000). The second plant was a narrow leaf lupin (*Lupinus angustifolius* L.) expressing the sunflower (*Helianthus annuus* L.) seed-derived albumin [SSA] (SSA-Lupin), which was generated to increase sulphur-rich amino acid content in livestock feed. It was shown that consumption of a GM plant expressing an allergen does not necessarily predispose to allergic hypersensitivity responses. In fact, it was demonstrated that consumption of GM plants could promote a protective regulatory T-cell response and protect against the subsequent development of allergic disease specific to that protein. However, it was also shown that diversity in translational and post-translational modification pathways between species could influence the molecular architecture of the expressed protein and subsequent cellular function and antigenicity. These findings highlight the need for further investigation into the usage of animal models for the assessment of antigenicity of GM plants and suggest that post-translational mechanisms may contribute to altered antigenicity.

Moneret-Vautrin (2006) reported experiments with hypoallergenic GM plants. The second generation of GM plants could improve the nutritional aspects of natural foods. Transgenic proteins could reach from 4 up to 8% of the total protein content in these foods. Any potential difference in allergenicity between second generation GM plants and the natural varieties must be examined with respect to the risk for food allergy caused by food products made from these plants and the risk for respiratory allergies in the people living near the crops caused by airborne pollen originating from the plants. WHO/FAO directives, as well as the Codex Alimentarius proposals and the European Food Safety Authority (EFSA) guidelines, recommend that transgenic proteins be screened for homology and cross-reactivity with known allergens, and examined carefully for modifications of host-plant proteomes. In vivo animal studies are also to be carried out to assess any potential immunogenicity. Lacking adequate safety data, the absence of potential allergenicity of transgenic plants cannot be ruled out. This is why data that do not meet the recommended safety criteria required for commercialization of GM plants do not allow to completely rule out the risk that may be associated with products that are going to be commercialized. Therefore, it is essential that commercialized GM plants be monitored. The establishment of public reference serum banks based on up-to-date WHO/FAO recommendations concerning the selection of sera according to precise criteria has been proposed, as well as the establishment of a system of allergovigilance linking national and European health and food safety agencies and a network of university hospital-based clinical and laboratory reference centres, together with a network of clinical allergists, responsible for the creation of the serum banks. Allergists working through these networks would be able to identify new sensitizations to transgenic foods in the population in the same manner as they currently identify new types of food allergies, which, in this case, would be GM foods. Such a project is presently being established in France.

Post-market monitoring and surveillance need to be established to assess the long-term health impacts of GM foods (Kuiper et al., 2001; Brent et al., 2003; EFSA, 2004). A post-market monitoring programme (PMM) complements the pre-marketing toxicological testing programme in order to confirm the pre-market risk assessment. The PMM for GM foods should generate information related to GM food consumption and any adverse effects on human health. According to Wal et al. (2003), a PMM addresses the questions regarding the predicted/recommended product use, the predicted side-effects and the induction of unexpected side-effects by the product. PMM should be carried out especially in those cases where there is no conventional counterpart.

Long-term effects should be accompanied by targeted epidemiological techniques beyond any post-marketing data collection. A very important consideration to be taken into account is that any GMO-derived products should not be launched in the market if there are questions still not answered in the pre-market assessment. Informative PMM systems should also be set up for products not monitored or surveyed until recently and are difficult to trace, but can be found in different food-production chains.

LEGISLATION FOR GM – LABELLING

Legislation

The introduction of rDNA technology of new genes into major crops consumed by animals has raised serious questions about the safety of novel feeds. The European Council Directive 2001/18/EC requires an assessment of risks for human, animals and the environment before viable seeds can be imported or the plant itself can be cultivated in Europe. Furthermore, the Novel Food and the Novel Food Ingredient Regulation (EC, 2003 http:/europa.eu.int/comm/food/fs/ssc/out327_en.pdf) covers the use of non-viable products of any GM plant intended for food purposes. In practice, compositional analysis of key nutrients and key toxicants used for comparing a GM plant with its conventional counterpart is the major source of data used for establishing substantial equivalence (MacMahon in http:/www.yale.edu/ynhti/curriculum/units/2000/7/00.07.02.x).

The current US process for assessing feed and food safety of crops produced by using technology, comprises the meticulous coordination of the USDA (United States Department of Agriculture), FDA (Federal Department of Agriculture), and EPA (Environmental Protection Agency) (for crops endowed with pesticidal properties). Reviews carried out by FDA for food safety have adopted the decision tree approach (several questions to be answered with a yes/no answer in order to make up a decision) for assessing host plants, gene donors, proteins introduced by donors, new or modified fats or oils, and carbohydrates (Faust, 2002). It is encouraging that various independent studies converge to the fact that there is no detrimental effect, as such, for livestock fed with biotechnology-derived crops, and no differences between livestock fed with GM plants and conventional crops (Aumaitre et al., 2002).

The UN Cartagena Protocol on Biosafety adopted in Montreal (January 2000) has already started to affect the international trade in GMOs and their products. Article 4 makes it clear that this protocol shall apply to the trans-boundary movement, transit, handling and use of all Living Modified Organisms (LMOs) that may have an adverse effect on the conservation and sustainable use of biodiversity, taking into account risks to human health as well (Gupta, 2000; Xue and Tisdell, 2000, 2002). LMOs are all living organisms that possess a novel combination of genetic material obtained through the use of modern biotechnology. According to the Protocol, trade in GMO requires approval by a specified competent authority in the importing country. The international movement of GMOs will be reviewed more rigidly not only from the point of view of agricultural production, but mainly from the perspective of environmental protection, human health, and other aspects. Increased cost will be involved in the export and import of GMOs. The risk assessment steps mainly include identification and documentation of any novel genotypic and phenotypic characteristics associated with LMOs that may adversely affect biodiversity and human health; direct use as food or feed and introduction into the environment (Xue and Tisdell, 2002).

Recently, Zedalis (2002) published a review referring to GMO food measures as "Restrictions" under the General Agreement on Tariffs and Trade (GATT) Article XI. GATT article XI (1) condemns the use of prohibited items imported and, thus, may attack GMO measures or flatly prevent importation. In case GMO is to be released to the environment, the following information will be required (EEC, 2001; Directive E.U. 2001/18/EC):

(1) Purpose of the release.
(2) Foreseen dates and duration of the release.
(3) Method by which the GM plants will be released.
(4) Method for preparing and managing the release site, prior to, during and post-release, including cultivation practices and harvesting methods.
(5) Approximate number of plants or plants per square metre.

Another EU Regulation (COM/2002/0085 - COD 2002/0046) aims at establishing a common system of notification and information for exports to third ccountries of GMOs in order to contribute towards ensuring an adequate level of protection in the field of the safe transfer, handling and use of GMOs that may have adverse effects on the conservation and the sustainable use of biological diversity, also taking into account risks to human health.

The current legislation (E.U. Directives and Regulations, and US Acts) for GMOs is described in **Tables 1.5** and **1.6**, respectively.

Table 1.5. E.U. Directives and Regulations (main points and comments) for GMOs

Directive – Title	*Main Points*	*Comments*
E.U. 90/219/EEC (entry into force 23/10/1991) Contained use of GM micro-organisms	• Measures for limited use of GM micro-organisms • Not applicable to certain techniques of genetic modification • Measures for avoidance of adverse effects on human health and environment • Emergency plan in case of an accident and regular inspections	Amendments ⇒Directive E.U. 98/81/EC • Additional elements for the articles. (entry into force 5/12/1998)
E.U. 90/220/EEC (entry into force 23/10/1991) Deliberate release into the environment of GMOs	• Protective measures for human health and environment • Not applicable to certain techniques of genetic modification • Measures for avoidance of adverse effects • Activities of Member States for deliberate release into the environment of GMOs for research, development and market placing purposes	Amendments ⇒Directive E.U. 97/35/EC • Additional elements for the disposal on the market of products which contain GMOs. ⇒Regulation (EC) No. 258/97 and Regulation (EC) No. 1139/98 • Labelling of food containing proteins or DNA derived from genetic modification. (entry into force 1/1/2002 and enforcement for 10 years)
E.U. 2001/18/EC (entry into force 17/4/2001) Deliberate release into the environment of GMOs	• Measures of authorization of the release and disposal on the market of GMOs • Obligatory controls after the disposal of GMOs on the market • Consultations with the public and labelling of GMOs	Repeal ⇒Directive E.U. 90/220/EEC since 17/10/2002 Amendment ⇒Regulation (EC) No. 1830/2003 (entry into force 7/11/2003)

(Table 1.5. Contd.)

(*Table 1.5. Contd.*)

E.U. 2004/204/EC (entry into force 23/3/2004) Arrangements for the operation of the registers for recording information on genetic modifications in GMOs.	• Lists of various information of genetic modification in GMOs • Lists should contain detailed report of documents • Lists are available to the public	
E.U. 2004/643/EC Placing in the market of a maize product (*Zea mays* L. line NK603) GM for glyphosate tolerance.	• Product should be as safe as conventional products (equivalence principle) • Handling, packaging and protection as per conventional products • Obligatory recordation of the code MON-00603-6 (unique) • Measures for labelling and traceability in all stages of the market promotion	
E.U. 2004/657/EC Placing in the market of a sweet corn from GM maize line Bt11 as a novel food or novel food ingredient.	• Product should be as safe as conventional products • Obligatory labelling as "GM sweet corn" • Obligatory recordation of the code SYN-*BTø*11-1 (unique) • No more controls after placing on the market	Replacement ⇒Directive E.U. 90/220/EC
Regulation (EC) No. 258/97 (entry into force 14/5/1997) Novel food and novel food ingredients.	• Placing on the market within the Community of foods and food ingredients, which have not been used for human consumption to a significant degree within the Community before • Not applicable to food additives, flavourings and extraction solvents • Specific requirements for labelling • Specific procedure for foodstuffs containing GMOs	

(*Table 1.5. Contd.*)

(*Table 1.5. Contd.*)

Regulation (EC) No. 1139/98 (entry into force 1/9/1998) The compulsory indication of the labelling of certain foodstuffs produced from GMOs.	• Application to food and food ingredients which are produced from GM soybean or GM corn • No application to food additives and condiments • No application to products which are legally produced, labelled and imported, commercialized in the Community	Replacement ⇒Regulation (EC) No. 1813/97 Amendments ⇒Regulation (EC) No. 49/2000 (entry into force 31/1/2000) ⇒Regulation (EC) No. 50/2000 (entry into force 31/1/2000) Additional elements for certain articles of the Regulation.
Regulation (EC) No. 1829/2003 (entry into force 7/11/2003) GM food and feed.	• Measures for human and animal health protection, Community procedures of approval, inspection and labelling of GM food and feed • Approvals are applicable for 10 years with the potential of renewal	Replacements ⇒Regulation (EC) No. 1139/98 ⇒Regulation (EC) No. 49/2000 ⇒Regulation (EC) No. 50/2000
Regulation (EC) No. 1830/2003 (entry into force 7/11/2003) Traceability and labelling of GMOs and traceability of food and feed products produced from GMOs.	• Traceability of products consisting of, or containing GMOs and foodstuffs, feed produced from GMOs • Application for all stages of disposal on the market • Specific demands on labelling • Inspection, control measures and sanctions in case of infringement	
Regulation (EC) No. 65/2004 (entry into force on the date of its publication in the *Official Journal of the European Union*) Establishment of a system for the development and assignment of unique identifiers for GMOs.	• Unique identifier for each GMO which is placed on the market • Not applicable to pharmaceuticals intended for human and veterinary use	

(*Table 1.5. Contd.*)

(Table 1.5. Contd.)

Regulation (EC) No. 641/2004 (entry into force 18/4/2004) The authorization of new GM food and feed, the notification of existing products and adventitious or technically unavoidable presence of GM material which has benefited from a favourable risk evaluation.	• Transformation of applications and statements in the applications • Requirements of input on the market of certain products • Transitional measures for adventitious or technically unavoidable presence of GM material which has benefited from a favourable risk evaluation
Proposal for a Regulation COM/2002/0085 – COD 2002/0046 (entry into force 27/10/2002) The transboundary movement of GMOs.	• Establishment of a notifying system and exchanging information on the exports of GMO to third countries • No application for pharmaceuticals for human use • Surveillance, submission of reports and imposition of sanctions for any infringement

Table 1.6. U.S. Acts (main points) for Genetically Engineered Food

Title	*Main Points*
Genetically Engineered Food Safety Act, 2003	• Definitions (genetically engineered organism, genetically engineered material, etc.) • Federal determination of safety of genetically engineered food, regulation as food additive • Rule making, effective date, previously unregulated marketed additives
Genetically Engineered Crop and Animal Farmer Protection Act (US), 2003	• Definitions (genetically engineered plant, genetically engineered animal, genetically engineered material, etc.) • Contract limitations regarding sale of genetically engineered seeds, plants and animals • Prohibition on labelling certain seeds as non-genetically engineered • Prohibition on certain non-fertile plant seeds
Genetically Engineered Food Right to Know Act (US), 2003	• Definitions (genetically engineered organism, genetically engineered material, etc.) • Requirements for labelling regarding genetically engineered material • Misbranding of food with respect to genetically engineered material
Genetically Engineered Pharmaceutical and Industrial Crop Safety Act (US), 2003	• A pharmaceutical crop or industrial crop is a plant that has been genetically engineered to produce a medical or industrial product, including a human or veterinary drug, biologic, industrial, or research chemical, or enzyme • Definitions (genetically engineered plant, genetically engineered animal, genetically engineered material, etc.) • Report to Congress on alternative methods to produce pharmaceutical and industrial crops

Labelling

The emergence of a new concept for food labelling has recently occurred: the consumers' right to information, allowing "informed choice" in full knowledge of the facts. This right has taken many forms: real or "perceived" safety information on ingredients and additives, philosophical or ethical concerns (mode of production, absence/presence of given ingredients, including GM foods); nutrition information, and declaration of potential allergens (Hollingsworth et al., 2003; Cheftel, 2005). Further information is in the offing, such as nutrition and health claims (with relevance to obesity and the risk of various diseases). The recent occurrence of several food crises has emphasized food safety and protection of consumers' health as main objectives for the food legislation. One of the most general rules of the European (and other) legislation can be stated as "no misleading the consumer" (the protection of consumers' interests is one of the principles of food law, as reiterated in Regulation EC 178/2002). This applies to information concerning the characteristics of foods (nature, identity, properties, composition, quantity, storage life, origin and method of production or manufacture). The label should not attribute to the food effects or properties which it does not possess, nor suggest that the food possesses special characteristics which are infact possessed by all other similar foodstuffs (Cheftel, 2005).

Labelling emerged as a major issue in the EU - US GMO confrontation. According to US authorities (FDA) and several reports published by FAO/WHO, ILSI and EU, GM foods could be consumed provided the principle of substantial equivalence is proved. However, in EU a recent series of directives and regulations aims at reassuring the consumer that the GMO consumption will continue to remain his choice, thanks to the strict traceability and labelling system [Regulation (EC) No.641/2004, Directive E.U. 2004/657/EC, Directive E.U. 2004/204/EC].

The proposal for a Regulation COM/2002/0085 - COD 2002/0046 aimed at controlling the transboundary movement of GMOs, in particular, to third world countries. Regulation (EC) No. 1829/2003 laid down the provisions for labelling food and feed. In fact, this regulation applies to:

(1) GMOs for food use.
(2) Food containing or consisting of GMOs.
(3) Food produced from or containing ingredients produced from GMOs.

Regulation (EC) No.1830/2003 is targeted at providing traceability and labelling of GMOs and traceability of food and feed products produced from GMOs. This Regulation applies at all stages of the placing in the market to:

(1) Products consisting of or containing GMOs placed in the market in accordance with community legislation.
(2) Food produced from GMOs, placed in the market in accordance with community legislation, and feed produced from GMOs, placed in the market in accordance with community legislation.

To be more specific, in terms of traceability, at the first stage of placing in the market of a product consisting of/or containing GMOs, including bulk quantities, operators shall ensure that the following information is transmitted in writing to the operator receiving the product:

(1) That it contains or consists of GMOs.
(2) The unique identifier assigned to those GMOs in accordance with the regulation.

As for labelling of products consisting of/or containing GMOs, operators shall ensure that:

(1) For pre-packaged products consisting of or containing GMOs, the words "This product contains GMOs" or "This product contains GM... (Name of organism)" appear on the label.

(2) For non-pre-packaged products offered to the final consumer, the words "This product contains GMOs" or "This product contains GM... (Name of organism)" shall appear on, or in connection with the display of the product.

Finally, a recent EU Directive No. 65/2004 has been enforced which aims at establishing a system for the development and assignment of unique identifiers for GMOs. According to this directive, consent or authorization for the placing of a GMO in the market is granted only if the following conditions are complied with:

(1) The consent or authorization shall specify the unique identifier for that GMO.

(2) The Commission, on behalf of the Community, or, where appropriate, the competent authority that has taken the final decision on the original application, shall ensure that the unique identifier for that GMO is communicated as soon as possible, in writing, to the biosafety clearing house.

(3) The unique identifier for each GMO concerned shall be recorded in the relevant registers of the commission.

Directive E.U. 2004/204/EC specifies the required arrangements for the operation of the registers for recording information on genetic modifications in GMOs. Furthermore, the registers shall be available to the public for inspection. The information recorded shall be divided into two sets: (1) data accessible to the public, and (2) data comprising additional confidential data, accessible only to the Member States, the Commission and the EFSA. Another more recent Regulation (EC) No. 641/2004 covers the topic of the authorization of new GM food and feed, the notification of existing products and adventitious or technically unavoidable presence of GM material which has benefited from a favourable risk evaluation. The application shall include the following:

- The monitoring plan.
- A proposal for labelling complying with the requirements of the regulation.
- A proposal for a unique identifier for the GMO in accordance with Commission Regulation.
- A proposal for labelling in all official community languages.
- A description of methods of detection, sampling and specific identification.

- A proposal for post-market monitoring regarding the use of the food for human consumption or the feed for animal consumption is not necessary.

And last but not least, is the Directive E.U. 2004/657/EC for placing in the market of sweet corn from GM maize line *Bt*11 as a novel food or novel food ingredient. Information to be entered in the Community Register of GM food and feed should comprise: (1) authorization holder (name, address, company), (2) designation and specification of the product (sweet maize, fresh or canned, whether it is a progeny from traditional crosses of traditionally bred maize with genetically modified maize line *Bt*11) and (3) labelling ("GM sweet corn"). The milestones of this seemingly approaching international consensus on the safety awareness of biotechnology derived foods are summarized in **Table 1.7.**

Anticipated objectives, strategies and suggested methodologies for achieving genetic and environmental traceability in the near future are summarized in **Table 1.8.**

The FDA has ruled that US food labels are not required to carry information about GM content unless the genetic modification significantly alters the properties of the food, e.g., introduces a potential allergen (U.S. Federal Register, 1992). However, in response to interest from various groups including food manufacturers and private certifying agencies, FDA has announced guidelines for food manufacturers who wish to voluntarily label their products as containing, or not containing, GM ingredients (U.S. Federal Register, 2001). While FDA does not require special labelling of GM foods, US consumers can turn to products certified organic by USDA to ensure products do not contain GM ingredients. Furthermore, several state legislatures have introduced bills during the past several years that would require state-level labelling of GM foods (Kohler and Naftzger, 2001) while Oregon voters introduced and defeated a state-level GM labelling initiative via public referendum (State of Oregon, 2002). It is well-known that mandatory labelling of foods with GM content is applied throughout the EU, Australia, New Zealand, Japan, South Korea and Brazil. The EU's insistence upon labelling GM foods has spurred the US's to file a suit against the EU to lift its moratorium on GM foods (Knight, 2003).

Roe and Teisl (2006) analyzed responses to a survey designed to elicit consumer reaction to various approaches to labelling GM foods. Consumers were shown sample labels that differed with respect to claims concerning the presence and potential effects of GM ingredients and the agency that certified these claims. A sample of 1998 US consumers rated 3,681 labels with regard to: (1) credibility and adequacy of the information content, (2) perceived health and environmental impacts of the product, and (3) purchase intent.

Table 1.7. Recent milestones in the international consensus on the safety assessment of biotechnology-derived foods (EU 2004/643EC; EU Regulation 1139/98/EC)

Year	Organization	Item Derived	Reference
2000	FAO/WHO	Expert consultation on safety assessment in general, including the principle of substantial equivalence	FAO/WHO 2000
2001	ILSI	Europe concise monograph series genetic modification technology and food consumer health and safety	ILSI, 2003
2001	EU	EU-sponsored Research on Safety of Genetically Modified Organisms. "GMO research in perspective." Report of a workshop held by External Advisory Groups of the "Quality of Life and Management of Living Resources Program".	EU 2001
2001	NZRC	New Zealand Royal Commission on Genetic Modification	NZRC 2001
2000–2003	FAO/WHO	• Guidelines for Codex Alimentarius committee, developed by Task Force for Foods Derived from Biotechnology. • Codex Ad Hoc Intergovernmental Task Force on Foods Derived from Biotechnology. • Food and Agriculture Organisation of the United Nations, Rome, Italy	FAO/WHO, 2001, 2002, 2003
2003	ILSI	Crop composition database (http://www.cropcomposition.org)	ILSI, 2003
2003	EU	GM food and feed (monitoring traceability and authenticity)	Regulation (EC) No. 1829/2003
2004	EU	Arrangements for the operation of the registers for recording information on genetic modifications in GMOs	Directive E.U. 2004/204/EC
2004	EU	The authorization of new GM food and feed, the notification of existing products and adventitious or technically unavoidable presence of GM material which has benefited from a favourable risk evaluation	EU Regulation (EC) No. 641/2004

Given that domestic and international forces could alter the US policy towards GM labelling, or alter the types of GM label messages that appear on US foods, it is important to evaluate how the various GM labels are viewed by consumers with regard to the credibility and adequacy of the information presented, and how the foods bearing the labels are viewed by consumers with regard to perceived health and environmental risks. Health and environmental risks involved in the use of GM technologies are often indicated as key concerns by consumers. (Teisl et al., 2003).

Table 1.8. Anticipated objectives, strategies and suggested methodologies for achieving genetic and environmental traceability

Yield	*Objectives*	*Strategies*	*Steps/Phases*	*Suggested Methodology*	*Possible Targets*
Genetic Traceability	Implement analytical methods and strategies to assist in certification of origin, authenticity, composition, processes. Provision of technical solutions in support of Common Policies and regulatory Governmental bodies	Develop diagnostics based on DNA, proteins, metabolites	Genomics Proteomics Metabolomics Molecular markers Genetic diversity and genotyping Influence of processing technology on isotopical values Statistical evaluation Establishing comprehensive databases	Access to products and raw materials "omics" technologies Panels of probes and molecular descriptors Genetic Identification (GI) qualitative and quantitative PCR	Differentiation between GM and non-GM plant products Species determination in mushroom products Species determination in berry and nut products High value EU import food products from outside the Community Absence of allergenic species Ethnical-religious food

(*Table 1.8. Contd.*)

(Table 1.8. Contd.)

Environmental Traceability	Assurance that chemicals, intentionally or accidentally present, do not pose unacceptable risks to human health and to the environment Nutritional imbalance due to primary production	Improved understanding of the type and nature of chemicals occurring in plants and their impact on public health. Dietary consideration of macro- micro-nutrients Risk analysis of chemicals in plants Develop stable isotope analysis with focus on light stable isotopes ($^{13}C/^{12}C$; $^{15}N/^{14}N$; $^{18}O/^{16}O$)	Systematic data collection Evaluation of nutritional and anti-nutritional components Evaluation of health promoting elements Evaluation of toxic compounds Determination and assessment of source of toxic elements contamination (e.g., soil, irrigation water, atmospheric pollution and packaging materials) in the food/plant chain	Spectroscopic methods (ICP, AAS, ICP-MS) Radioactivity determination, and isotope analysis Separation techniques (GC-LC, GC-MS, GC-MS, LC-MS) Mathematical models for simulation of pollutants behaviour Molecular monitoring on/off- line Biosensors Life Cycle Analysis of agroalimentary processes Monitoring inorganic nutrients content, and correlation with genotype and culture techniques Evaluation of Pb, Cr, Ni, Mn, V, Cd, As and Hg contents in food and beverages by Atomic Spectroscopy Techniques	Natural toxins Mycotoxins, aflatoxin, marine biotoxins Metal contaminants Nitrites, Nitrates Environmental pollutants Chemicals from production practices (plant protection) Hormone residues and "phyto-hormones" Added chemicals Food packaging contaminants Berries and nuts Mushrooms Vegetables and Fruits Cereals Herbs

The labels evaluated by Roe and Teisl (2006) differ in several ways, including the language that is used (text of the message) and the agency that certifies the label (messenger). The eight messages vary in terms of whether the consumer is informed of the absence *vs.* presence of GM content; whether a reason for the use of genetic modification is given; and whether a message conveys uncertainty surrounding the long term health or environmental effects of genetic modification. Seven messengers are considered: an unidentified messenger, three federal agencies, a health organization, a consumer organization, and an independent GM certifying organization. Of the two simple claims, the GM claim ("this product contains genetically modified ingredients") is more credible than the No-GM claim, though this difference is only significantly different for labels both with no certification and for USDA certified labels.

Another trend that is noticeable is that labels where genetic modification is mentioned as the means for implementing a more fundamental claim (e.g., 50% less fat or 50% fewer pesticides used), the credibility is higher. Hence, the explanation of the means of implementing promised improvements might help to establish credibility in the perception of the consumer. This is congruent with the findings forwarded by Roe et al. (2000): when labels communicated food safety improvements, consumers wanted to know how the stated improvements were achieved.

The question is whether a label that highlights the presence of GM ingredients or notes the absence of GM ingredients best serves society. While the relative costs of each type of label will depend upon the prevalence of GM and GM-free products and the technical requirements of product segregation and testing, the benefits may hinge on which message can be believed by consumers (Roe and Teisl, 2006).

Under current regulations, only USDA (via the organic certification programme) or private certifying firms can validate No-GM claims, which suggests that a change in labelling policy that might allow FDA to certify No-GM claims, could improve the label's credibility with consumers since FDA is more trusted by consumers and has a strengthening role in regulating the safety of food and drugs, and this aids in establishing credibility. Since FDA regulates the food industry, consumers may believe that their certification of the label also implies safety of the GM technique.

O'Fallon et al. (2005) collected data from 16,078 participants across 15 EU member countries from the Eurobarometer 53 and were examined. Using Univariate ANOVA, the results indicated that: (1) roughly 73% of the sample of the individuals residing in the 15 European countries are less likely to purchase a food product with a label indicating the existence of a GM ingredient; (2) women were less likely to purchase the GM product than men; and (3) those individuals who are more likely to purchase a GM food believed it is unnecessary to include complete information pertaining to the use of GM organisms in the production of food products. Specifically, individuals were significantly more likely to purchase a GM food product if the label provided only limited or no information about the use of GM organisms in the production of food products.

Gruere (2006) reported the results of a qualitative survey of GM and non-GM food labels in supermarkets in Canada and France, five months after the introduction of the new policies. It was found that there were almost nil GM labelled products in France and non-GM labelled products in Canada. Each policy tends to crowd out the targeted label attribute. However, Canadian consumers have the option of choosing between GM and non-GM organic products, whereas there are only non-GM products in French supermarkets. Recent political developments in Quebec (where the public opinion seems more opposed to GM food) suggest that the labelling landscape might change in Canada, either with an increase in the number of non-GM products at the retail level, or a transition towards a mandatory labelling policy like in France.

The labelling regulations in Canada and France seem to reflect at least partially the respective degree of consumer concern for GM food; simultaneously the labelling regulations have also reinforced the bias towards or against GM food. The labelling process may be based on consumer opinion *ex-ante*, but it also affects consumer choice and consumer opinion *ex-post*. Canada has voluntary labelling, which makes non-GM a niche market, and thus crowded out GM food in the short run. France and the EU have implemented a system that effectively crowded out GM food to an extent even greater than what the population wanted, and made GM products a target of negative campaigns. Mandatory labelling has acted as a hazard warning, and may have prevented many people from purchasing safe food products approved for consumption by the EU. There would be more visible labels if there were voluntary labelling in the EU and mandatory labelling in Canada. Perhaps the most important difference between the two labelling regulations is their effect on international production choices. Voluntary labelling is a market-driven regulation, and it can open opportunities for non-GM farmers in some niche markets, without distorting the production decision of others. The EU mandatory labelling policy obliges every exporter of food products to follow a strict traceability programme, and heavily influences the decisions of developing countries of whether to introduce GM food crops or not. For developing countries, the decision to adopt GM food crops is based not only on the potential risks and benefits of the crop, but also on the fact that any introduction of GM crops may ruin their chance to sell this crop to the European market. A simple field trial of a GM crop in a developing country may trigger EU food companies to ban imports of that crop because of the perceived or real incapacity of these countries to enforce a strict separation of GM and non-GM crops, as was the case for GM papayas in Thailand. In developed countries, these regulations also bar the introduction of future GM food crops (Gruere, 2006).

In recent years, requests for protection from imports has been on the increase from consumers over issues ranging from animal welfare concerns, employment of child labour, the use of growth hormones, differing environmental standards and GM foods. The current international trade regime is ill-suited to deal with consumer-based protectionism. Hobbs and Kerr (2006) developed a model that explicitly incorporates consumer concerns into an international trade model and compares the results with the standard treatment. Further, using the model incorporating consumer concerns, a labelling policy for imports is compared to an import embargo. The labelling policy, however, is found to be superior to an embargo.

This lies at the heart of the voluntary versus mandatory labelling arguement - do firms have the incentive to label GM products honestly if they fear a consumer backlash against products that do not contain the desired credence attribute? It is partly for this reason that mandatory labelling of, for example, GM products have been supported by a number of countries. In some European countries, the private sector has pre-empted regulatory moves to make labelling mandatory - a number of UK supermarkets publicly stated that they would not allow GM material in their own-label products or would not sell any products containing GMOs.

Moreover, the requirement that food be labelled as GM may impose relatively higher costs on producers of non-GM food. This is because it does not matter if GM canola might be "contaminated" with non-GM canola (Shewmaker et al., 1999). However, it matters greatly that non-GM canola not come into contact with GM canola (a nutritionally improved GM variety of canola) as it passes along the supply chain from farmer to consumer. Firms in the non-GM supply chain will incur higher transaction costs and may be less cost competitive as a result - of course, those higher costs could be offset by consumers being willing to pay a premium for non-GM products (Hobbs and Kerr, 2006).

ENVIRONMENTAL IMPACTS OF ENGINEERED CROPS

There are five arguments raised by many scientists:

(1) The engineered crops themselves could become weeds, i.e., plants with undesirable effects.

(2) The crops might serve as conduits through which new genes move to wild plants, which could then become weeds.

(3) Crops engineered to produce viruses could facilitate the creation of new, more virulent or more widely spread viruses.

(4) Plants engineered to express potentially toxic substances could present risks to other organisms like birds or deer.

(5) Crops may initiate a perturbation that may have effects that ripple through an ecosystem in ways that are difficult to predict. Finally, the crops might threaten centres of crop diversity.

Although problems of the nature listed above would be expected to occur within the next few years' time, the good news is that there have been no serious environmental impacts associated with the use of engineered crops in the US. This needs to be questioned at a European level. There has been a notable near miss with the monarch butterfly, a situation that has much to teach about the weaknesses of the US regulatory system.

In the spring of 2000, the US media were confronted with a preliminary report in *Nature* indicating that pollen from *Bt* corn could kill the larvae of monarch butterflies in laboratory studies. The *Nature* study was published after the EPA had approved several *Bt* corn varieties, and over 20 million acres of *Bt* corn were planted in the US. The big question was why the EPA had not addressed the threat to monarchs before approval of *Bt* corn. From a scientific standpoint, it is not surprising that a toxin aimed at the European corn borer (moth larvae) would also affect the larvae of the monarch butterfly. The tests required by the EPA prior to approval of *Bt* crops included a few trials in which *Bt* toxin was fed to honeybees and lacewings, among other organisms, but did not include tests on any non-pest moths and butterflies (EPA, 1997).

The storm of publicity eventually forced the government to do a thorough risk assessment of the threat. To its credit, the USDA organized a workshop for scientists with expertise in the many areas needed to evaluate the monarch issue and asked them to generate a multi-disciplinary research programme that would address the risks. It established a multi-stakeholder advisory committee to formulate a set of coordinated research projects to determine whether *Bt* corn is lethal to monarchs under field conditions. The department also provided funds—as did industry—to support the research. The results of the studies were published in five papers in September 2001 in the online version of the *Proceedings of the National Academy of Sciences (PNAS)* (Hellmich et al. 2001; Oberhauser et al. 2001).

The major conclusion of the research was that only one of several *Bt* corn varieties (Event 176) approved and planted for use in the US produced high enough levels of *Bt* toxin in pollen to be lethal to butterfly larvae. Fortunately, that variety of GM corn did not sell well and was not widely planted. Pollen from the two types of *Bt* corn which account for most of the *Bt* corn acreage (Mon 810 and *Bt*11) produce relatively low amounts of toxin and pose negligible risk to monarchs. If Event 176 had turned out to be popular, monarchs could have been in serious jeopardy. It was just a lucky break, and not governmental controls, that protected the monarch butterfly. However, these studies did not completely resolve the issue. Some scientists have pointed out that monarchs consume tissues from anthers—the pollen-producing parts of the corn flower—as well as pollen from *Bt* corn, as reported by Pleasants et al. (2001) and Sears et al. (2001). Since anthers have been shown to contain considerably more toxin than pollen, these scientists believe that the *PNAS* studies based on pollen alone may seriously underestimate the toxin dose consumed by monarch larvae in corn fields. These concerns are supported by other studies showing that a mixture of *Bt*11 pollen and anther fragments has a deleterious effect on monarch larvae (Stanley-Horn et al., 2001).

The *PNAS* studies also did not examine long-term effects of *Bt* corn, such as delayed development, impaired reproduction, and altered migration. The monarch story shows that these studies should be carried out *before* products are released and not after. Yet, there has been no interest in adopting the monarch research model to subsequent EPA risk assessments. The recent application for the approval of a new *Bt* corn variety directed against corn rootworms, for example, was not accompanied by research done in accordance with an agenda set by a multi-stakeholder group. EPA's risk assessment, which was heavily criticized, was carried out under strong pressure to quickly approve products. Until risk assessment procedures improve, the public will not have confidence that another monarch-like threat will be detected before it is too late.

Current investigations focus on: the potentially detrimental effect on beneficial insects or a faster induction of resistant insects; the potential generation of new plant pathogens; the potential detrimental consequences for plant biodiversity and wildlife, and a decreased use of the important practice of crop rotation in certain local situations; and the movement of herbicide resistance genes to other plants.

HUMAN HEALTH ISSUES

No major human health problems have emerged in connection with GM food crops, which have been consumed by a significant number of US consumers. As with environmental effects, it is likely that only dramatic effects easily connected to engineered foods would have been detected. Since GM foods are not labelled, people suffering from the ill effects would have difficulty in relating them to consumption of engineered products.

It is important to remember that only in the last three or four years herbicide- and insect-resistant soybeans and corn have been planted on millions of US acres and subsequently used in food processing. Over the past decade, food safety experts have identified several potential problems that might arise as a result of engineered food crops, including the possibilities of introducing new toxins or allergens into previously safe foods, increasing toxins to dangerous levels in foods that typically produce harmless amounts, or diminishing a food's nutritional value.

Among these potential impacts, scientists and regulators have been most worried about new allergens, and indeed, two events within the last decade legitimate that concern. First, predictions have been confirmed since 1996 that GM could transfer an allergen from a known allergenic food to another food. A few years earlier, scientists at Pioneer Hi-Bred Seed Company had successfully transferred a gene from Brazil nut into soybean to improve the grain crop's nutritional quality. Subsequent experiments showed that people allergic to Brazil nuts were similarly allergic to the transgenic soybean (Nordlee et al., 1996).

Second, in the late 1990s, reports that a *Bt* corn variety (StarLink) containing a potential allergen had illegally entered the food supply, had set off a tidal wave of controversy that ultimately reduced corn exports, disturbed the food industry, and created widespread doubts about the strength of the US regulatory framework. The EPA had not approved StarLink corn for human consumption because of scientific concerns that the *Bt* toxin might cause allergic reactions in some consumers.

GE FOOD IN US TRADE

Advocates of GM foods often assert that the processes of laboratory genetic engineering are really no different from those of plant and animal husbandry. This argument is not as convincing as they expect. Those who express concern about the safety of GM food claim that genetic engineering allows humans to do what nature will not – they worry that scientists cut and paste genes and can now transfer genes between species. This gene transfer raises new safety questions, making the production and marketing of GM foods a matter for consideration by public health authorities (http://www.mja.com.au/public/issues/172_04_210200/huppleed/leeder.html). Whenever official approval for the introduction of GM foods has been given in Europe or the US regulatory committees have invoked the concept of "substantial equivalence" (http://www.greenpeace.org.bt/). The EU, the World Health Organisation (WHO) and the Food and Agriculture Organisation (FAO) of the United Nations agreed on a methodology referred to as "substantial equivalence" as the most practical approach to assess the safety of GM foods and food ingredients (http://www.eufic.org/gb/food/pag/food37/food374.htm). The substantial equivalence concept was developed following an initial FAO/WHO consultation exercise (FAO/WHO, 1991) and further refined by OECD (1993) and FAO/WHO (1996). The concept has recently been re-evaluated at an OECD Workshop (OECD, 1998). Substantial equivalence is based on the premise that if a novel or modified food or food ingredient can be shown to be essentially equivalent in composition to an existing food or food ingredient, then it can be assumed that the new food is as safe as its conventional equivalent (http://www.mrc.ac.uk/pdf-strategy-gm_foods.pdf). "Substantial equivalence" focuses on the product rather than on the production process. It is a rigorous procedure including a detailed list of parameters and characteristics that need to be considered, including molecular characterization of the genetic modification, agronomic characterization, nutritional and toxicological assessments. The "substantial equivalence" approach acknowledges that the goal of the assessment cannot be to establish absolute safety (http://www.eufic.org/gb/food/pag/food37/food374.htm). Safety is established by the demonstration of the fact that there is no significant difference in a range of characteristics, including both phenotype and composition, between the new/modified food and its conventional equivalent. Any effect of genetic modification is considered in the context of the normal variation in phenotype and composition that exists in the conventional counterpart (http://www.mrc.ac.uk/pdf-strategy-gm_foods.pdf). However, if the GM product has new traits or characteristics that make it no longer substantially equivalent (such as a higher level of a vitamin), then an additional assessment is required (http://www.eufic.org/gb/food/pag/food37/food374.htm).

GM foods are placed in three classes, based on the results of substantial equivalence testing:

(1) the food is considered to be substantially equivalent in all respects and no more information is requested

(2) the food is considered to be different only in the GM characteristics – for example, the gene products that make the crop resistant to insects or tolerant to a herbicide

(3) the food is not considered to be substantially equivalent, so more toxicological and nutritional data are required. Scientists critical of this approach have previously argued that gross chemical

comparisons between GM foods and conventional counterparts are not sufficient to detect unexpected changes (http://www.the-scientist.com/news/20020206/04). The use of substantial equivalence is a "pseudo-scientific concept because it is a commercial and political judgement masquerading as if it were scientific. It is, moreover, inherently antiscientific because it was created primarily to provide an excuse for not requiring biochemical or toxicological tests. It therefore, serves to "discourage and inhibit informative scientific inquiry" (Millstone et al., 1999). One of the more difficult aspects of the substantial equivalence approach relates to unintended effects. This issue is addressed by the evaluation of comparative data on composition and phenotype. Current regulatory bodies take a decision on the extent of scientific data required to satisfy concern that a significant secondary effect might create a safety hazard. For any established crop plant, there may be known toxicological or nutritional concerns; for example, it may produce natural toxicants. In conventional plant breeding, this is an established safety issue and analysis of the levels of such substances in a GM plant provides reassurance that gene introduction has not created an unexpected change that might cause harm (http://www.mrc.ac.uk/pdf-strategy-gm_foods.pdf). Von Schomberg (1999) has argued that substantial equivalence will never serve the purpose of reassuring consumers and that toxicological and biochemical tests would always be required. Substantial equivalence has been accused of being a pseudo-scientific concept embodying commercial and political judgement and that the definition of the concept is too vague to serve as a benchmark for public health policy as a counter to this, it has been explained that substantial equivalence is not intended to be a substitute for safety assessment, but is simply a guiding principle which is a useful tool for regulatory scientists engaged in food safety assessments (Burke, 1999; Kearns and Mayers, 1999). The cornerstones in GM acceptance are given in **Table 1.9**. **Table 1.10** compares various approaches to risk assessment and safety testing. The first column lists requirements in ascending order of stringency, the second column indicates the types of testing required in each set, and the third column indicates the relative costs of the tests. Under the terms of the Cartagena Protocol on Biosafety – the framework for international regulation of GMOs – governments may ask the organization applying for approval to meet these costs directly. In relation to GM foods and crops, the types of tests required have changed over time. The practice followed in the US and the EU is represented in the first

Table 1.9. Cornerstones in GM acceptance

Year	*Institution*	*Verdict*	*Reference*
1990	UN Food and Agriculture Organization (FAO) and World Health Organization (WHO)	• The challenge of how to deal with the issue from consuming GM foods was first confronted in 1990 at an international at meeting.	http://www.greenpeace.org.br
Mid-1990s	World Trade Organization (WTO)	• Genetics and molecular biology were dominated by a set of assumptions generally presumed to be unproblematic. E.g., there was a one-to-one correspondence between genes and proteins, in other words, each gene directs the expression of a single protein.	http://www.twnside.org.sg/title2/service169.htm
Late 1990s	World Trade Organization (WTO)	• Compelling evidence emerged indicating that across all species, the number of proteins considerably exceeded the number of genes. This anomaly suggests that the expression of proteins may be controlled by combinations of genes, rather than just by individual genes.	http://www.twnside.org.sg/title2/service169.htm
1996	UN Food and Agriculture Organization (FAO) and World Health Organization (WHO)	• The concept of substantial equivalence was first introduced and subsequently endorsed.	http://www.greenpeace.org.br
1997	European Union (EU)	• Novel Food Regulation included a simplified procedure stipulating that when a novel food was considered to be substantially equivalent to an existing food, a company was required only to provide a justification for the claim of substantial equivalence, rather than a formal risk assessment. That procedure was used to authorize several GM foods for sale in the EU.	Levidow and Murphy, 2002

(Table 1.9. Contd.)

(*Table 1.9. Contd.*)

May 1999	UK Government's Chief Medical Officer (CMO) and Chief Scientific Adviser (CSA)	• There is no current evidence to suggest that genetic modification technologies used for producing food, are inherently harmful. • The precautionary nature and rigour of the current procedures used to assess the safety of individual GM foods are reassuring.	http;//www.food.gov.uk/gmdebate/aboutgm/gm_safety?view=GM+Microsite
May 2000	FAO, WHO Expert Consultation	• The application of the concept of substantial equivalence contributes to a robust safety assessment framework. The approach used to assess the safety of the GM foods that have been approved for commercial use is satisfactory.	http;//www.food.gov.uk/gmdebate/aboutgm/gm_safety?view=GM+Microsite
June 2000	Medical Research Council Expert Group	• Many of the potential effects of GM foods on human health also apply to food or food ingredients produced by conventional plant and animal breeding. • Current regulatory procedures, using the principle of substantial equivalence, addressed the theoretically possible health risks of known toxins and allergens in GM foods.	http;//www.food.gov.uk/gmdebate/aboutgm/gm_safety?view=GM+Microsite
2001	European Commission (EC)	• A revised Novel Food Regulation explained that this proposal does not include a notification (simplified) procedure or GM foods which are substantially equivalent to existing foods. The use of this regulatory short-cut for so-called 'substantially equivalent' GM foods has been very controversial in the Community in recent years and there is consensus at the international level that whilst substantial equivalence is a key step in the safety assessment process of genetically modified foods, it is not a safety assessment in itself.	CEC, 2001

(*Table 1.9. Contd.*)

(*Table 1.9. Contd.*)

2002	Royal Society	• There is no evidence at present that GM foods cause allergic reactions. • It seems reasonable to assume that this poses no significant risk to human health and eating GM DNA will have no ill effects.	http;//www.food.gov.uk/gmdebate/aboutgm/gm_safety?view=GM+Microsite
2003	European Union (EU)	• The EC indicated that chemical analyses would no longer be considered a sufficient basis for making a decision about the safety of GM foods, but that such analyses should provide a starting point for a more sophisticated approach.	EU, 2003, Recital 6

Table 1.10. Approaches to risk assessment and safety testing

Sl. No.	*Approaches*	*Test Options*	*Cost*	*Reference*
1	Recent practice for GM foods and crops in the United States and European Union (EU)	Coarse chemical analyses	Low	http://www.twnside.org.sg/title2/service169.htm
2	Current practice for GM foods and crops in the EU	Slightly finer chemical analysis and some short-term farm-animal feeding studies	Low	http://www.twnside.org.sg/title2/service169.htm
3	Officially envisaged future (in the EU) for GM foods and crops	Far finer chemical analyses: including proteomics and metabolomics as well as laboratory animal feeding studies, farm-scale cultivation trials for crops	Medium	http://www.twnside.org.sg/title2/service169.htm
4	Current practice for new food additives and pesticides	Chemical analyses plus toxicological tests with bacteria and studies on (400) live animals, and some immunological testing, but no human trials	High	http://www.twnside.org.sg/title2/service169.htm
5	Current practice for pharmaceutical products	Chemical analyses plus toxicological tests with bacteria and live animal studies, some immunological testing, and some clinical trials	Very high	http://www.twnside.org.sg/title2/service169.htm

two rows of the **Table 1.10** (first row - recent practice; second row - curent). Current testing requirements for GM foods and crops, especially in the EU, are more elaborate than those introduced in the late 1990s. The third row in the table refers to a set of testing requirements for GM foods, which EFSA's expert scientific advisory committee - the Scientific Panel on GM Organisms - is proposing for in the not-too-distant future (EFSA, 2004). The fourth row represents recent and current requirements for new food additives and new active ingredients for pesticide products. These chemicals are tested far more stringently than GM foods. The final row shows current requirements for pharmaceutical products - the most stringent of all those listed (http://www.twnside.org.sg/ title2/service169.htm). One of the first novel foods to be formally assessed in the UK was the mycoprotein `Quorn '. At roughly the same time, the safety of irradiated foods was assessed through a vast array of animal feeding studies. In both cases practical difficulties were encountered. In contrast to many non-nutritive substances, foods are intended to be consumed by man at levels which approach the maximum dosage that could be used in animal studies. Toxicological studies are designed to characterize the toxicological profile of individual chemical substances, and not complex substances such as foods. Long-term feeding of high levels of individual 'foods' to animals can result in nutritional imbalances, which make interpretation of such studies extremely difficult. **Table 1.10** compares the differences between the safety assessment of chemicals and foods (http://www.acnfp.gov.uk/ acnfppapers/inforelatass/toxrev?view=printerfriendly).

GE, the ability to insert a novel gene in an organism, is a developing science that offers possible benefits and hazards. GE foods present new issues of food safety. Given the consensus among the scientific community that GE can potentially introduce hazards, such as allergens or toxins; GE foods need to be evaluated on a case-to-case basis and cannot be presumed to be generally recognized as safe (http://www.mindfully.org/GE/GE4/ GE-Legislation-Rep-Ku cinich11jun02.htm). The FDA does not currently require Genetically Engineered Organisms (GEOs) to be identified because it has determined that GE foods are substantially equivalent to conventional crops and as such they are safe. There remain however, concerns that GE components could contaminate traditional food crops. Although regulation in this area continues to evolve, under current FDA rules, GE foods would require labelling only if the modifications altered food composition significantly, changed the nutritional value, or introduced an allergen (http://www.uschamber.com/issues/index/ agriculture/genetically engineeredfood.htm). FDA proposed regulations that would require a mandatory notification before a GE plant intended for food or feed is marketed, partially because FDA believes that the next generation of GE crops is more likely to raise new food-safety concerns. Although that proposal improves upon the current process by mandating a transparent agency review, it does not materially change the agency's scientific review and will not result in an official safety determination (http://www.cspinet.org/biotech/gefood_ approval.html).

GE Food Legislation

According to Genetically Engineered Food Safety Act (US), (2003), the Congress has made the following observations:

(1) GE is an artificial gene transfer process wholly different from traditional breeding.

(2) GE can be used to produce new versions of virtually all plant and animal foods. Thus, within a short time, the food supply could consist almost entirely of GE products.

(3) This conversion from a food supply based on traditionally bred organisms to one based on organisms produced through GE could be one of the most important changes in our food supply in this century.

(4) GE foods present new issues of safety that have not been adequately studied.

(5) Previously the Congress required that food additives be analyzed for their safety prior to their placement on the market.

(6) Adding new genes into a food should be considered equivalent to adding a food additive, thus requiring an analysis of safety factors.

(7) Federal agencies have failed to uphold congressional intent of the Food Additives Amendment Act of 1958 by allowing GE foods to be marketed, sold and otherwise used without requiring pre-market safety testing addressing their unique characteristics.

(8) The food additive process gives the FDA discretion in applying the safety factors that are generally recognized as appropriate, to evaluate the safety of food and food ingredients. The term "genetically engineered organism" means:

    (A) An organism that has been altered at the molecular or cellular level by means that are not possible under natural conditions or processes (including, but not limited to, recombinant DNA and RNA techniques, cell fusion, microencapsulation, macroencapsulation, gene deletion and doubling, introducing a foreign gene, and changing the positions of genes), other than a means consisting exclusively of breeding, conjugation, fermentation, hybridization, *in vitro* fertilization, or tissue culture.

(B) An organism made through sexual or asexual reproduction (or both) involving an organism described in clause (A) above, if possessing any of the altered molecular or cellular characteristics of the organism so described.

In the case of a genetic food additive, the factors considered by the Secretary regarding safety for use shall include (but not be limited to) the results of the following analyses:

(i) Allergenicity effects resulting from the added proteins, including proteins not found in the food supply.
(ii) Pleiotropic effects. The Secretary shall require tests to determine the potential for such effects.
(iii) Appearance of new toxins or increased levels of existing toxins.
(iv) Changes in the functional characteristics of food.
(v) Changes in the levels of important nutrients.

Generally, in the case of a genetic food additive, which in the US was in commercial use in food as of the day before the date on which the final rule under this subsection is promulgated, the amendments made by this Act apply to the additive upon the expiration of the two-year period beginning on the date on which the final rule is promulgated.

In Genetically Engineered Crop and Animal Farmer Protection Act (US), (2003), "genetically engineered animal" means an animal that contains a genetically engineered material or was produced with a genetically engineered material. An animal shall be considered to contain a genetically engineered material or to have been produced with a genetically engineered material if the animal has been injected or otherwise treated with a genetically engineered material or is the offspring of an animal that has been so injected or treated, and "genetically engineered plant" means a plant that contains a genetically engineered material or was produced from a genetically engineered seed. A plant shall be considered to contain a genetically engineered material if the plant has been injected or otherwise treated with a genetically engineered material. A biotech company that sells any genetically engineered animal, genetically engineered plant, or genetically engineered seed that the biotech company knows, or has reason to believe, will be used by the purchaser in the US to produce an agricultural commodity, shall provide written notice to the purchaser that fully and clearly discloses the possible legal and environmental risks that the use of the genetically engineered animal, genetically engineered plant, or genetically engineered seed may pose to the purchaser.

The provisions referred to in this subsection are any of the following:

(1) In the case of a sale of genetically engineered plants or genetically engineered seeds, a provision that prohibits the purchaser from either retaining a portion of the harvested crop for future crop planting by the purchaser.

(2) A provision that limits the ability of the purchaser to recover damages from the biotech company for a genetically engineered animal, genetically engineered plant, or genetically engineered seed that does not perform as advertised.

(3) A provision that shifts any liability from the biotech company to the purchaser.

(4) A provision that requires the purchaser to grant agents of the seller access to the purchaser's property.

(5) A provision that mandates arbitration of any disputes between the biotech company and the purchaser.

(6) A provision that mandates any court of jurisdiction for settlement of disputes.

(7) A provision that mandates that the purchaser pay liquidated damages of more than a technology fee or similar fee itself, plus interest.

(8) A provision that imposes any unfair condition upon the purchaser, as determined by the Secretary or a court.

A seed company or other person may not sell, or offer for sale, seeds for planting that are labelled as non-genetically engineered or otherwise represented as not containing genetically engineered material if the Secretary finds that any sample of the seeds contains genetically engineered material. Notwithstanding any other provision of law, effective 45 days after the date of the enactment of this Act, a person may not manufacture, distribute, sell, plant, or otherwise use any seed that is genetically engineered to produce a plant whose seeds are not fertile or are rendered infertile by the application of an external chemical inducer.

Furthermore, another Act (Genetically Engineered Food Right to Know Act [US], 2003) claims that "genetically engineered material" means material derived from any part of a genetically engineered organism, without regard to whether the altered molecular or cellular characteristics of the organism are detectable in the material. In the case of a recipient, who with respect to a food establishes a guaranty or undertaking in accordance with this subparagraph, the exclusion under such subparagraph from being subject to penalties applies to the recipient without regard to the use of the food by the recipient, including:

(i) Processing the food.
(ii) Using the food as an ingredient in a food product.
(iii) Repacking the food.
(iv) Growing, raising, or otherwise producing the food.

For purposes of this Act, a meat food is misbranded if it:

(a) Contains a genetically engineered material or was produced with a genetically engineered material.
(b) Does not bear a label that provides, in a clearly legible and conspicuous manner.

In the case of a recipient who establishes a guarantee or undertaking in accordance with this paragraph, the exclusion under such paragraph from being subject to penalties applies to the recipient without regard to the use of the meat food by the recipient (or the use by the recipient of the animal from which the meat food was derived, or of food intended to be fed to such animal), including:

(A) Processing the meat food.
(B) Using the meat food as an ingredient in another food product.
(C) Packing or repacking the meat food, or
(D) Raising the animal from which the meat food was derived.

A poultry product shall be considered to have been produced with a genetically engineered material if:

(I) The poultry from which the food is derived has been injected or otherwise treated with a genetically engineered material.
(II) The poultry from which the food is derived has been fed genetically engineered material, or
(III) The food contains an ingredient that is a food to which this paragraph applies.

For the purpose of Genetically Engineered Pharmaceutical and Industrial Crop Safety Act (US), (2003), "pharmaceutical crop" means a genetically engineered plant that is designed to produce medical products, including human and veterinary drugs and biologics. The term includes a crop intentionally treated with genetically engineered material that in turn, produces a medical substance. Pharmaceutical crops and industrial crops also pose substantial liability and other economic risks to farmers, grain handlers, food companies, and other persons in the food and feed supply chain. These risks include liability for contamination episodes, costly food recalls, losses in export markets, reduced prices for a contaminated food or feed crop, and loss of confidence in the safety of the American food supply among foreign importers and consumers of American agricultural commodities. No pharmaceutical crop or industrial crop may be grown, raised, or otherwise cultivated until the final regulations and tracking system required by this section are in effect. The USDA shall establish a tracking system to regulate the growing, handling, transportation, and disposal of all pharmaceutical and industrial crops and their by-products to prevent contamination. The maximum amount that may be accessed under this section for a violation may not exceed US$ 1,000,000. In determining the amount of the civil penalty, the Secretary shall take into account:

(i) The gravity of the violation.
(ii) The degree of culpability.
(iii) The size and type of the business.
(iv) Any history of prior offences under such section or other laws administered by the Secretary.

ATTITUDES TOWARDS GM FOOD

In Austria there are comparatively low levels of support for GE, with only 13% of Austrians surveyed in 1999 willing to buy GM fruits compared with 21% in all of Europe (Torgersen et al., 2001). However, their response patterns are somewhat unique, i.e., they also exhibit comparably low levels of perceived risks and factual knowledge. These factors usually correlate inversely with support in most European countries, but not in Austria. Torgersen and Seifert (1997) suggest that low genetic engineering acceptance in Austria may be related to a conservative attitude to new technologies that could be explained by historical experiences.

Germany has exhibited high levels of resistance to transgenic food in studies using examples such as GM yoghurt and beer (Bredahl, 1999). A number of theories have been proposed by Moses (1999) to explain Germany's high resistance, including: residue of a rejection of Nazi racial policy that rejects anything to do with genetics; a history of using a marketing strategy that has discussed food in terms of "100% security;" a tradition of strong anti-industry feeling towards chemical and pharmaceutical sectors, which may now extend to biotechnology; a conservative attitude towards novelty that asks "I have perfectly good food already; why try anything remotely doubtful?" .

In Spain—one of the nations said to have the highest levels of overall acceptance—there is a strong contrast between the population's general valuation of GE as a process and the population's attitudes toward applications related to food consumption. Spaniards may rank GM of plants high in terms of general benefit, but a clear majority says they would not consume fruit with flavour improved through GM (Lujan and Todt, 2000).

Differing views on the application of biotechnology to plants versus animals have been found among respondents in the UK (Frewer et al., 1997), and among students in the UK and Taiwan, where GE for growth enhancement was judged more negatively than GE for the purpose of disease or pest resistance (Chen and Raffan, 1999). In Europe generally, views towards the application of biotechnology to produce transgenic animals are far less supportive than towards plant modification; among bioindustry association sources surveyed in 1997, none saw the possibility of any involvement of transgenic animals in food production (Moses, 1999).

Frewer et al. (1997) found that the various applications of GE could be classified along two axes: one related to benefits (useful), and the other related to negative concerns, including risks, ethical concerns and consequences for the environment. Although they found that attitudes towards applications of GE were generally either positive or negative, some food-related applications were thought to have both benefits and negative concerns. Nowadays attitudes towards GM foods can best be described in terms of separate positive and negative components, using confirmatory factor analyses (CFA). In other words, positive and negative components of attitudes can be relatively independent from each other and are not necessarily related to each other in a hydraulic fashion. This reasoning is backed by growing evidence that positive and negative information in general is processed separately and through potentially different systems in the brain (Cacioppo et al., 1997).

People may find GM food useful for production enlargement in the third world, whereas at the same time they may find it useless in daily life. It would be interesting for future studies to investigate more closely where ambivalence towards GM food stems from: for instance, is it predominantly positive beliefs (cognitions) that conflict with negative feelings?

Past research on attitudes towards GM food has focused on measuring explicit attitudes. Spence and Townsend (2006) compared implicit attitudes towards GM foods with explicit attitudes towards GM foods. They used the Go No-Go association task (GNAT) to investigate context-free implicit evaluations of GM foods and compared these with evaluations made in the context of ordinary and organic foods. Semantic differential scales were used to evaluate explicit attitudes towards GM foods. As expected, explicit attitudes towards GM foods were found to be neutral. However, contrary to the hypotheses, participants were found to hold positive, rather than neutral, implicit attitudes towards GM foods when these were assessed in a context free manner. In addition, neutral implicit attitudes were found when attitudes were assessed in the context of ordinary or organic foods, again contrasting with the hypotheses. These results imply that implicit attitudes towards GM food are more positive than anticipated and may lead to approach behaviour towards such products. Thus, given the choice, consumers are likely to accept GM food although other incentives may be needed if alternative foods are available.

Implicit attitudes differ from that of explicit attitude measurements, i.e., responses measured are not consciously controlled; rather they are automatic or spontaneous. Various types of implicit attitude measures exist ranging from physiological measures, to examinations of non-verbal behaviour, and to the more frequently used reaction time tasks (Spence, 2005). In contrast, explicit attitude measures generally take the form of direct questions about how one feels about a particular topic. This means that explicit attitude measures are open to self-presentation effects and demand characteristics. Although implicit attitudes towards GM foods have not previously been studied, implicit attitudes towards non-GM foods have been measured. The effective priming task was recently found to be useful in identifying both strong and moderate attitudes towards different food stimuli (Lamote et al., 2004). In addition, the implicit association task (IAT) has been used to compare attitudes towards different foodstuffs. Maison et al. (2001) utilized the implicit association test (IAT) to measure attitudes towards fruit juices and sodas. The Eurobarometer series of studies is probably the largest investigation into explicit attitudes in Britain and across Europe. The most recent Eurobarometer report finds that the British population is ambivalent towards GM food (Gaskell, 2003; Poortinga and Pidgeon, 2003; Gaskell et al., 2003a). This finding is supported by the Public perceptions of Agricultural Biotechnologies in Europe (PABE) focus group study (Marris et al., 2001), which indicated that a key finding was that participants expressed arguments both for and against GM foods. Attitudes towards GM foods are found to vary greatly across the EU and Britain seems to be firmly in the middle of the spectrum of opinion polls in its ambivalence. Countries such as Spain, Portugal, Ireland and Finland are all quite positive towards GM food, whereas countries including France, Greece and Luxembourg are negative towards GM food (Gaskell et al., 2003). Differences in attitudes towards GM foods observed between countries are attributed to a variety of factors including culture, regulatory systems and local events including food scares.

The results of the GNATs indicate that implicit attitudes towards GM foods are positive when evaluated in a context free manner. Implicit attitudes, however, are positive when GM foods are measured in a context-free manner, indicating that approach behaviour is favourable towards GM foods (Spence and Townsend, 2006).

Hall and Moran (2006) investigated how members of anti-GM campaign groups and environment groups perceive the risks and benefits of GM technology in food and agriculture. The study targeted these groups as the most risk-averse sector of society when considering GM technology. Survey respondents were asked to rank the current and future risks and benefits of GM, and to rank GM risks against other health risks. Respondents appear to be unconvinced by the claims that future GM technologies will provide additional consumer (or environmental) benefit, since perceived future risks were ranked more highly than future benefits. Results support the claim that there is an inverse relationship between perceived risk and perceived benefit. Results also suggest that among the respondents there are differences of opinion regarding the degree of risk to health posed by the technology. The comparison among people living in both rural and urban areas, revealed that on an average, women in rural areas ranked risks more highly than men in urban areas.

Previous studies have revealed a diversity of consumer concerns relating to GM food, including unpredictable health risks (Lemkow, 1993; Subrahmanyan and Cheng, 2000; Isaacs, 2001; Olubobokun et al., 2001; Verdurme et al., 2001a), environmental safety risks (Isaacs, 2001; Olubobokun et al., 2001), and the structure of agri-business (Isaacs, 2001). There are fears that there may be long-term environmental effects (Lemkow, 1993), risks to future generations (Rosati and Saba, 2000; Poortinga and Pidgeon, 2003), and long-term food safety issues (Grove-White et al., 1997). These long-term effects are expected in many instances to be largely unpredictable.

There have been a number of studies that have emphasized the role of environmental action groups in the anti-GM debate. Purdue (2000) examined the role of anti-GM campaign groups in the UK in some depth. Although it is referred to the role of small, single issue NGOs, such as Genetics Forum, it also emphasizes the significance to the anti-GM debate of large membership NGOs such as Greenpeace. Environmental organizations are recognized as being predominant in the campaign against GE of food, with groups such as Friends of the Earth and Greenpeace taking public positions against GM technology in agriculture (Reisner, 2001).

Hall and Moran (2006) concluded that the public is likely to be more receptive to of GM food in the future when the products offer consumer benefits. However, respondents appear to be sceptical about the claims of future benefits of GM technology, suggesting that actually the perception of high risks will remain among the most risk-averse section of society. Nevertheless, the results of this study show that anti-GM campaigners and members of environment groups can be segmented into groups with different perceptions of the risks and benefits of GM food, and should not, therefore, be viewed as a homogeneous group.

There has been a limited focus on how individuals learn about the risks and benefits of GM food, along with the influence of information sources on the formation of both risk and benefits perceptions. Following a rational learning model, Costa-Font and Mossialos (2005) examined the determinants of risk and benefit perceptions. In doing so, risk and benefit perceptions are hypothesized as an expression of a latent and unobserved variable and thus it is tested whether perceptions of risk and benefits are simultaneously determined. A UK sample of the Eurobarometer survey 52.1 for 1999 as well as several model specifications, are employed that account for simultaneity and endogeneity, such as the two-stage least square (2SLS) and the three-stage least square (3SLS) regressions. It was shown that risk and benefit perceptions are not independent, both appear endogenously, and are simultaneously determined. Furthermore, the impact of information determinants for risk and benefit learning processes are specification dependent.

The importance of public and private information sources has been indicated. In particular, knowledge of science significantly affects perceptions of benefits, whereby the larger the individual knowledge of biotech-related facts generally, the larger the perceived benefits of GM food. This points out that there might be a 'fear of the unknown' underlying individual perceptions of GM food, especially resulting from a failure of different stakeholders to provide a clear view of what the public can gain or lose from the extension of GM food.

Understanding the socio-cultural construction of risk is important for improving risk communication and policy development about GM foods. To improve understanding of risk construction, systematic examination into the socio-cultural basis of different risk perceptions is needed and it is suggested that Douglas and Wildavsky's (1982) cultural theory is a useful starting point for structuring examinations of socio-cultural factors that orient and motivate individuals. Cultural theory has been highly influential in the debate on risk perception, providing a parsimonious account of the complexities underlying what people fear and why. To the extent that judgements of risk are influenced by such ''non-technical'' factors as cultural values and belief systems, attempts to communicate about risk will be improved by models that describe how people use socially embedded worldviews to navigate a complex, uncertain, and sometimes dangerous world (Slovic and Peters, 1998).

Finucane and Holup (2005) described how socio-psychological and cultural factors might affect public perceptions of the risk of GM foods. The psychological, sociological, and anthropological research on risk perception was presented as a framework for understanding cross-national differences in reactions to GM food. Differences in the cultural values and circumstances of people in the US, European countries, and the developing world are examined. The implications of cultural theory for risk communication and decision making about GM food are discussed and directions for future research highlighted.

Both qualitative and quantitative methods (from psychological, sociological, anthropological, and other disciplines) have pointed to several socio-cultural factors as important determinants of GM food risk perceptions. However, further research is still needed to improve understanding of the context-dependent and constructed nature of cultural values that affect the uptake of complex new technologies and their products.

According to USDA and EPA, herbicide tolerant (Roundup Ready or RR) and insect resistant (*Bt*) corn has "no significant impact" on human health and environmental integrity. In Europe, GM maize strains - the identical Bt and RR biotech crops used in the USA - are banned by a "safeguard clause" that allows any member state of the EU to impose limited term restrictions on an approved imported or exported product. To understand these different policies, an explanatory model that analyzes political culture as a recursive phenomenon that impacts, and is influenced by, regulations has been considered by Guehlstorf and Hallstrom (2005). The way governments regulate modern biotechnology is not necessarily a reflection of how their political culture perceives the new scientific technology, but how their existing regulatory structure can create a political culture of acceptance or rejection for contested technological advancements. This comparative study casts doubts on interpreting agricultural biotechnology decisions solely on equations of risk analysis, and offers a detailed cultural analysis of the regulatory differentiation of modern agricultural policy between USA and Europe.

In both the USA and Europe, there is a significant difference between the exacting scientific standards of evidence for risk assessment and the less demanding standards used for predicting regulatory estimates of harm in the same evaluation.

By comparing a single GM product between two culturally dissimilar regulatory bodies, it could be demonstrated that culture can be identified as something between a single causal essence and a list of abstract features. Moreover, one could state that the American regulatory regime is not open to environmental democracy because it is an aggressive economic system with elite social and political goals that must maintain the dominant positions of no consumer labeling and only industry safety testing of GM foods in order to control USA and Canadian markets.

Additionally, if the EU can recover from the crises of legitimacy during the 1990s and affect a culturally-oriented return of trust in both EU regulation and science, it is possible for the EU to release the ban on all GM foods and fulfil its goals of economic integration.

GENETIC MODIFICATION AND ETHICS

Ethics can be defined as the branch of philosophy concerned with how one may decide what is morally right or wrong. It is a specific discipline which attempts to analyze the concepts and principles used to justify our moral choices and actions in particular situations. Ethics is a specific discipline trying to probe the reasoning behind one's moral life. The first question one is asked is whether biotechnology really raises new ethical questions in food production. One has to take into account that ethical implications can be divided in the so-called pre-farm and post-farm gate. The former comprises environmental ethics (principle of non-interference with nature), biodiversity threat (very few plant varieties will continue to grow), sustainability (spread of weedy plants brought from overseas), animal rights (young animals suffering from viruses or malformation), and socio-economic impacts (threat to small farm survival and impact on developing countries economy) (Thompson, 1997). The post-farm gate is focused on issues like relevant principles pertaining to the responsibility of the producer (provide consumer with correct information), consumers' rights (to be informed to take the risk to consume a novel food), right to be informed (proper labelling), right to choose (related to religion, morality taste, and opinion) (Straughan, 1998). The second bioethics issue of great importance refers to the environment. In fact, application of the precautionary principle (Rio Declaration) stating "in order to protect the environment, the precautionary approach shall be widely applied by States according to their capabilities". Where there are threats of serious or irreversible damage, lack of full scientific certainty shall not be used as a reason for postponing cost-effective measures to prevent environmental degradation". Environment sustainability and biodiversity are the two major issues being held at stake with the introduction of GM products (Cockburn, 2002).

Anselm Jappe, in his attempt to look at the modern commercial societies with a critical eye, cites the following: "Someday we will wonder why the radical criticism of the society at the beginning of the XXI century was more concerned about the Balkans war than companies such as Novartis and Nestle or biotechnologies in general" (Jappe, 2006). Although his questioning is opposite to what Jeremy Rifkin believes, there might be a point of contact between them: this century is the century of biotechnology. Jeremy Rifkin's famous book entitled *The Century of Biotechnology* (Rifkin, 1998) encourages the acceptance of biotechnology and the adopting of ecologic objections as well. However, its reading focuses mainly on the defence of moral-evolutionary views so as to underestimate conflicting arguments for the benefit of ruling authoritative practices.

The approach of a phenomenon such as biotechnology, close to concepts such as bioethics, biopolitics, biomedicine and eugenics, could be carried out either using existing prejudgements of the lifeworld, as they are met in the public sphere or in one's personal experience, or using the goodwill of the inhibition of the pre-existing knowledge so that the object could be seen or looked at in one's conscience setting up the problem of its historicity as a new but different problem. Even if this view of matters is a prerequisite of the cognitive requirements for validity in decision making, given the participative observation in their reification, it is already well-known from the relations with the world that this intuition is not adequate. There is always a question/matter for the way the knowledge is correlated with action and technique for the lifeworld that is leading up. As far as the GMOs issue is concerned, all the above mentioned seem to be valid in the following way: no positivistic cognitive certainty or disproval from observation, no secure view of the world from a distance so that positivistic knowledge could be produced, no knowledge derived from consequences, which would certify the existence of all the objects that preoccupy us.

CONCLUSION AND FUTURE TRENDS

The release of GMOs into the environment and the marketing of GM foods have resulted in a public debate in many parts of the world. This debate is likely to continue, probably in the broader context of other uses of biotechnology and their consequences to human societies. Even though the issues under debate are usually very similar (costs and benefits, safety issues), the outcome of the debate differs from country to country. On issues such as labeling and traceability of GM foods as a way to address consumer concerns, there is some consensus to date expressed by the two regulations. Significant progress has been made today on the harmonization of views concerning risk assessment with the guidance on GMOs published by EFSA.

The approach of a phenomenon such as biotechnology and GMOs, close to concepts such as bioethics, biopolitics, biomedicine and eugenics, could be carried out either using existing prejudgements of the lifeworld, as they are met in the public sphere. Evolutionary ethics is also concerned with epigenetic rules, meaning external cultural factors which modify the gene expression through epigenetic rules. The immediate result is the formation of ethical rules produced from the impact of genes and culture. As far as GMO issue is concerned, the following was discussed: no positivistic cognitive certainty or disproval from observation, no secure view of the world from a distance so that positivistic knowledge could be produced, no knowledge derived from consequences, which would certify the existence of all the objects that preoccupy us.

The use of GMOs raises several questions on the morality of the technology, the balance of risk in society, public involvement in decision making and the appropriateness of using patents in an area linked to life processes (Atkinson, 1998). Although the usage of GM plants appears to be a promising route out of the imminent nutritional crisis in view of the occurring world population explosion, especially in the third world countries, there seems to be no solid experimental data on the long-term effects of GE. There are still several basic questions unanswered:

- Will people over time develop allergic reactions to the transgenic proteins produced from GE?
- Will the horizontal movement of genetic materials have a negative impact on ecosystems?
- Will some virulent new pathogen develop from the transfer and transformation of microbial DNA made available by GE?
- Will humans be able to make the correct ethical choices so that all of humanity may share in the potential benefits of GE?

It is crystal clear that such answers can be given in the near future, whereas the GM production will continue to increase exponentially fuelled by US, Canada, Argentina and more recently China as already stated in the relevant statistics. Although EU states seemed to be very sceptical initially and a series of EU directives reflected this tendency, there has been a change in the latest Directives Regulation (EC) No.641/2004 and EU 2004/657/EC regarding the permission of placing in the market of sweet corn from GM maize line *Bt*11 as a novel food or novel food ingredient. These new directives could be regarded as an effort from the EU part towards reaching a consensus with US and Canada *vis-a-vis* the production and consumption of GM foods.

REFERENCES

Aerni, P. (2001). Public attitudes towards agricultural biotechnology in developing countries: A comparison between Mexico and the Philippines. STI/CID policy discussion paper no. 10, Harvard University, Cambridge, MA, USA.

Agodia, A., Barchittaa, M., Grillob, A. and Sciaccac, S. (2006). Detection of genetically modified DNA sequences in milk from the Italian market. Int. J. Hyg. Environ. Health 209: 81-88.

Arvanitoyannis, I.S. (2003). Genetically engineered/modified organisms in foods. Applied Biotechnology, Food Science and Policy, 1: 3-12.

Arvanitoyannis I.S. and Krystallis, A. (2005). Consumers' beliefs, attitudes and intentions towards GM foods. Int. J. Food Sci. Technol. 40: 343-360.

Atherton, K.T., Dearman, R.J. and Kimber, I. (2002). Protein allergenicity in mice: a potential approach for hazard identification. Ann. NY Acad. Sci. 964: 163-171.

Atkinson, D. (1998). Genetically modified organisms. Available at http://www.sac.ac.uk/info/External?publications/GMO.ASP. Scottish Agricultural College.

Aumaitre, A., Aulrich, K., Chesson, A., Flachowsky, G. and Piva, G. (2002). New feeds from genetically modified plants: Substantial equivalence, nutritional equivalence, digestibility, and safety for animals in the food chain. Livestock Prod. Sci. 74: 223-238.

Baker, G.A. and Burnhum, T.A. (2002). The market for genetically modified foods: consumer characteristics and policy implications. International Food and Agribusiness Management Review 4: 351-360.

Barg, R., Shabtai, S. and Salts, Y. (2001). Transgenic Tomato. In: Biotechnology in Agriculture and Forestry 47. Transgenic Crops II (ed.) Y.P.S. Bajaj, Springer, Berlin, Germany, pp. 212-233.

Bates, G.W. (1995). Electroporation of plant protoplasts and tissues. Methods Cell Biol. 50: 363-73.

Baum, K., Groning, B. and Meier, I. (1997). Improved ballistic transient transformation conditions for tomato fruit allow identification of organ-specific contributions of I-box and G-box to the RBCS2 promoter activity. Plant J. 12: 463-469.

Beardsley, T. (1996). Advantage: Nature. Scientific American 274: 33.

Beever, D.E. and Phipps, R.H. (2001). The fate of plant DNA and novel proteins in feeds for farm livestock: a United Kingdom perspective. J. Anim. Sci. 79: E290-E295.

Brandt, P. (2003). Overview of the current status of genetically modified plants in Europe as compared to the USA. Plant Physiol. 160: 735-742.

Braun, R.H., Reader, J.K. and Christey, M.C. (2000). Evaluation of cauliflower transgenic for resistance against *Xanthomonas campestris* pv. *campestris*. Acta Hortic. 539: 137-143.

Bredahl, L. (1999). Consumers' cognitions with regard to genetically modified foods: results of a qualitative study in four countries. Appetite 33: 343-360.

Brent, P., Bittisnich, D., Brooke-Taylor, S., Galway, N., Graf, L., Healy, M. and Kelly, L. (2003). Regulation of genetically modified foods in Australia and New Zealand. Food Control 14: 409-416.

Broglie, K., Chet, I. and Holliday, M. (1991). Transgenic plants with enhanced resistance to the fungal pathogen *Rhizoctonia solani*. Science 254: 1194-1197.

Burke, D. (1999). No GM conspiracy. Nature 40: 640-641.

Cacioppo, J.T., Gardner, W.L. and Berntson, G.G. (1997). Beyond bipolar conceptualizations and measures: The case of attitudes and evaluative space. Personality and Social Psychology Review 1: 3-25.

Chang, C.D., Chassy, M.B., Saunders, A.J. and Sowers, E.A. (1991). Guide to Electroporation and Electrofusion. Academic Press, London, UK.

Chang, M.M.. Chiang, C.C., Martin, M.W. and Hadviger, L.A. (2001). Expression of a pea disease resistance response gene in the potato cultivar Shepody. Am. Potato J. 70: 635-647.

Cheftel, J.C. (2005). Food and nutrition labeling in the European Union. Food Chem. 93: 531-550.

Chen, F., Duran, A., Blount, J.A., Sumner, L.W. and Dixon, R.A. (2003). Profiling phenolic metabolites in transgenic alfalfa modified in lignin biosynthesis. Phytochemistry 64: 1013-1021.

Chen, S. and Raffan, J. (1999). Biotechnology: student's knowledge and attitudes in the UK and Taiwan. J. Biol. Educat. 34(1): 17-23.

Chowdhury, E.H., Mikami, O., Murata, H., Sultana, P., Shimada, N., Yoshioka, M., Guruge, K.S., Yamamoto, S., Miyazaki, S., Yamanaka, N. and Nakajima, Y. (2004). Fate of maize intrinsic and recombinant genes in calves fed genetically modified maize Bt11. J. Food Prot. 67: 365-370.

Christey, M.C. and Braun, R.H. (2001). Transgenic vegetable and forage *Brassica* species. In: Biotechnology in Agriculture and Forestry 47: Transgenic Crops II. (ed.) Y.P.S. Bajaj, Springer, Berlin, Germany, pp. 87-101.

Christou, P. (1997). Rice transformation: bombardment. Plant Mol. Biol. 35: 197-203.

Cockburn, A. (2002). Assuring the safety of GM food. J. Biotechnol. 98: 79-106.

Conway, G. and Toenniessen, G. (1999). Feeding the world in the twenty-first century. Nature 402: C55-C58.

Costa-Font, J. and Mossialos, Elias. (2005). Are perceptions of 'risks' and 'benefits' of genetically modified food (in)dependent? Food Qual. Pref. 18(2): 173-182.

Cranshaw, W.S. "Bacillus Thuringiensis" in http://www.ext.colostate.edu/pubs/insect 05556.html. Accessed 2004 December 21. Cocci, C., Mezetti, B. and Rosati, P. (1994). Regeneration and transformation of strawberry. In: VIIIth Int. Congr. of Plant Tissue and Cell Culture, Firenze, Italy, pp. 152.

Curtis, K.R., McCluskey, J.J. and Wahl, T.I. (2004). Consumer acceptance of genetically modified food products in the developing world. AgBioForum, 7(1 and 2): 70-75.

Dandekar, A.M. and Fisk, H.J. (2005). Plant transformation: *Agrobacterium*-mediated gene transfer. Methods Mol. Biol. 286: 35-46.

Dearman, R.J. and Kimber, I. (2001). Determination of protein allergenicity: studies in mice. Toxicol. Lett. 120: 181-186.

Dearman, R.J., Caddick, H., Stone, S., Basketter, D. A. and Kimber, I. (2001). Characterization of antibody responses induced in rodents by exposure to food proteins: Influence of route of exposure. Toxicology 167: 217-231.

Dehesh, K., Jones, A., Knutzon, D.S. and Voelker, T.A. (1996). Production of high levels of 8:0 and 10:0 fatty acids in transgenic canola by over expression of *Ch FatB2*, a thioesterase cDNA from *Cuphea hookeriana*. Plant J. 9: 167-172.

Del Vecchio, A.J. (1996). High laurate canola. How Calgene's program began, where it's headed. [INFORM] International News on Fats, Oils and Related Materials 7: 230.

Douglas, M. and Wildavsky, A. (1982). Risk and culture: an essay on the selection of technological and environmental dangers. Berkeley: University of California Press, USA.

EFSA (European Food Safety Authority) (2004). Guidance document on the scientific panel on genetically modified organisms for the risk assessment of genetically modified plants and derived food and feed. The EFSA Journal 99: 1-94.

Eichelbaum, T., Allan, J., Fleming, J. and Randerson, R. (2001). Report of the Royal Commission on Genetic Modification. Wellington, New Zealand.

Eicher, C.K., Maredia, K. and Idah, S.N. (2006). Crop Biotechnology and the African Farmer. Food Policy. In press.

Engel, K.H., Frenzel, T. and Miller, A. (2002). Current and future benefits from the use of GM technology in food production. Toxicol. Lett. 127: 329-336.

Engel, K.H., Gerstner, G. and Ross, A. (1998). Investigation of glycoalkaloids in potatoes as example for the principle of substantial equivalence. In: Novel Food Regulation in the EU - Integrity of the Process of Safety Evaluation. Federal Institute of Consumer Health Protection and Veterinary Medicine, Berlin, Germany, pp. 197-209.

EPA (Environmental Protection Agency) (1997). Pesticide fact sheet: *Bacillus thuringiensis* CryIA(b) delta endotoxin and the genetic material necessary for its production (plasmid vector pCIB4431) in corn. Washington, D.C.: Office of Pesticide Programs.

Ewen, S.W.B. and Pusztai, A. (1999). Effect of diet containing genetically modified potatoes expressing *Galanthus nivalis* lectin on rat small intestine. Lancet 354: 1353-1354.

Fares, N.H. and El Sayed, A.K. (1998). Fine structural changes in the ileum of mice fed on delta-endotoxin-treated potatoes and transgenic potatoes. Nat. Toxins 6: 219-233.

Faust, M.A. (2000). Livestock products - Corn composition and detection of transgenic DNA/proteins. Symposium held in conjunction with American Dairy Science Association and American Society of Animal Science Meeting, Baltimore, USA.

Faust, M.A. (2002). New feeds from genetically modified plants: the US approach to safety for animals and the food chain. Livestock Prod. Sci. 74: 239-254.

Feder, B.J. (1996). Geneticists Arm Corn Against Corn Borer, Pest May Still Win. The New York Times, Tuesday, July 23, 1986, p. C1.

Finucane, M.L. and Holup, J.L. 2005. Psychosocial and cultural factors affecting the perceived risk of genetically modified food: an overview of the literature. Social Sci. Medicine 60: 1603-1612.

Fotopoulos, C. and Krystallis, A. (2001). The Quality Labels a Real Marketing Advantage? A Conjoint Application on Greek PDO Protected Olive Oil. J. Int. Food Agribusiness Marketing 12(1): 1-22.

Fotopoulos, C. and Krystallis, A. (2003). Wine produced by organic grapes in Greece: Using means - End chains analysis to reveal organic buyers' purchasing motives in comparison to the non-buyers. Food Qual. Pref. 14(7): 549-566.

Frewer, L. J., Howard, C. and Shepherd, R. (1996). The influence of realistic product exposure on attitudes towards genetic engineering of food. Food Qual. Pref. 7(1): 61-67.

Frewer, L. J., Howard, C. and Shepherd, R. (1997). Public concerns in the United Kingdom about general and specific applications of genetic engineering: Risk benefit and ethics. Science, Technology and Human Values 22: 98-124.

Gallie, D.R. (2001). Transgenic carrot. In: Biotechnology in Agriculture and Forestry 47. Transgenic Crops II. (ed.) Y.P.S. Bajaj, Springer, Berlin, Germany, pp. 147-159.

Gaskell, G., Bauer, M.W. and Durant, J. (1998). Public perceptions of biotechnology in 1996: Eurobarometer 46.1 In: Biotechnology in the Public Sphere: A European Sourcebook (eds.) J. Durant, M.W. Bauer and G. Gaskell, Science Museum, London.

Gaskell, G., Bauer, M. W., Durant, J. and Allum, N. C. (1999). Worlds apart? The reception of genetically modified foods in Europe and the US. Science, 16: 384-387.

Gaskell, G., Bauer, M.W., Durant, J., Allansdotir, A., Bonfadeli, H., Boy, D., de Cheveigne, S. and Sakellaris, G. (2000). Biotechnology and the European public. Nat. Biotechnol. 18: 953-958.

Gaskell, G., Allum, N. and Stares, S. (2003). Europeans and Biotechnology in 2002, Eurobarometer 58.0 (2nd Edition: March 21st 2003). A report to the EC Directorate General for Research from the project "Life Sciences in European Society QLG7-CT-1999-00286. Available at http://www.europa.eu.int/comm/public_opinion/.

Gaskell, G., Allum, N., Bauer, M., Jackson, J., Howard, S. and Lindsey, N. (2003a). Ambivalent GM nation? Public attitudes towards biotechnology in the UK, 1991–2002 Life Sciences in European Society report. London: London School of Economics, UK.

Goodman, R.E., Hefle, S.L., Taylor, S.L. and van Ree, R. (2005). Assessing genetically modified crops to minimize the risk of increased food allergy: a review. Int. Arch. Allergy Immunol. 137: 153-166.

Goto, F., Yoshihara, T., Shigemoto, N., Toki, S. and Takaiwa, F. (1999). Iron fortification of rice seed by the soybean ferritin gene. Nat. Biotechnol. 17: 282-286.

Grison, R., Grezes-Besset, B., Scheider, M. and Toppan, A. (1996). Field tolerance to fungal pathogens of *Brassica napus* constitutively expressing a chimeric chitinase gene. Nat. Biotechnol. 14: 643-646.

Grove-White, R., Macnaghten, P., Mayer, S. and Wynne, B. (1997). Uncertain world: Genetically Modified Organisms, Food and Public Attitudes in Britain. IEPPP, Lancaster University, Lancaster, UK.

Gruere, G.P. (2006). A preliminary comparison of the retail level effects of genetically modified food labeling policies in Canada and France. Food Policy 31: 148-161.

Grunert, K.G., Lahteenmaki, L., Nielsen, N.A., Poulsen, J.B., Ueland, O. and Astrom, A. (2001). Consumer perceptions of food products involving genetic modification - results from a qualitative study in four Nordic countries. Food Qual. Pref. 12: 527-542.

Guehlstorf, N.P. and Hallstrom, L.K. (2005). The role of culture in risk regulations: a comparative case study of genetically modified corn in the United States of America and European Union. Environmental Science and Policy 8: 327-342.

Gupta, A. (2000). Governing trade in GMO: The Cartagena protocol on biosafety. Environment 40(4): 23-33.

Hall, C. and Moran, D. (2006). Investigating GM risk perceptions: A survey of anti-GM and environmental campaign group members. J. Rural Studies 22: 29-37.

Hallman, W.K., Hebden, W.C., Auino, H.L., Cuite, C.L. and Lang, J.T. (2003). Public perceptions of genetically modified foods: National study of American's knowledge and opinion. Food Policy Institute, Cook College, Rutgers, the State University of New Jersey, New Brunswick, NJ. Available at: http://www.foodpolicyinstitute.org.

Hallman, W.K., Hebden, W.C., Cuite, C.L., Auino, H.L. and Lang, J.T. (2004). Americans and GM food: Knowledge, opinion and interest in 2004. Food Policy Institute, Cook College, Rutgers, the State University of New Jersey, New Brunswick, NJ. Available at: http://www.foodpolicystitue.org.

Hamilton, A.J., Fray, R.G. and Grierson, D. (1995). Sense and antisense inactivation of fruit ripening genes in tomato. Curr. Top. Microbiol. Immunol. 197: 77-89.

Hamstra, A.M. and Smink, C. (1996). Consumers and biotechnology in the Netherlands. British Food J. 98(4 and 5): 34-38.

Hashimoto, W., Momma, K., Katsube, T., Ohkawa, Y., Ishige, T., Kito, M., Utsumi, S. and Murata, K. (1999a). Safety assessment of genetically engineered potatoes with designed soybean glycinin: Compositional analyses of the potato tubers and digestibility of the newly expressed protein in transgenic potatoes. J. Sci. Food Aric. 9: 1607-1612.

Hashimoto, W., Momma, K., Yoon, H-J., Ozawa, S., Ohkawa, Y., Ishige, T., Kito, M., Utsumi, S. and Murata, K. (1999b). Safety assessment of transgenic potatoes with soybean glycinin by feeding studies in rats. Biosci. Biotechnol. Biochem. 63: 1942-1946.

Hattan, D. (1996). Evaluation of toxicological studies on Flavr Savr tomato. In: Food Safety Evaluation, Organization for Economic Cooperation and Development, Paris, France, pp. 58-60.

Hellmich, R.L., Siegfried, B.D., Sears, M.K., Stanley-Horn, D.E., Daniels, M.J., Mattila, H.R., Spencer, T., Bidne, K.G. and Lewis, L.C. (2001). Monarch larvae sensitivity to *Bacillus thuringiensis*-purified proteins and pollen. Proc. Natl. Acad. Sci., USA 98: 11925-11930; published online 9/14/01.

Helm, R.M. and Burks, A.W. (2002). Animal models of food allergy. Curr. Opin. Allergy Clin. Immunol. 2: 541-546.

Helm, R.M., Ermel, R.W. and Frick, O.L. (2003). Nonmurine animal models of food allergy. Environ. Health Perspect. 111: 239-244.

Hennin, C., Hofte, M. and Diederichsen, E. (2001). Functional expression of *Cf9* and *Avr9* genes in *Brassica napus* induces enhanced resistance to *Leptosphaeria maculans*. Mol. Plant Microbe Interact. 14: 1075-1085.

Hinchee, M.A.W. (1988). Production of transgenic soybean plants using *Agrobacterium*-mediated DNA transfer. Biotechnology 6: 915-922.

Hoban, T.J. (1996). Anticipating public reaction to the use of genetic engineering in infant nutrition. Am. J. Clin. Nutr. 63: 657-662.

Hoban, T.J. (1997). Consumer acceptance of biotechnology: An international perspective. Nat. Biotechnol. 15: 232–234.

Hobbs, J.E. and Kerr, W.A. (2006). Consumer information, labelling and international trade in agri-food products. Food Policy 31: 78–89.

Hogan, S.P., Mishra, A., Brandt, E.B., Royalty, M.P., Pope, S.M. and Zimmermann, N. (2001). A pathological function for eotaxin and eosinophils in eosinophilic gastrointestinal inflammation. Nat. Immunol. 2: 353-360.

Hollingsworth, R.M., Bjeldanes, L.F., Bolger, M., Kimber, I., Meade, B.J., Taylor, S.L. and Wallace, K.B. (2003). The safety of genetically modified foods produced through biotechnology. Report of the Society of Toxicology. Toxicol. Sci. 71: 2-8.

Hothorn, L.A. and Oberdoerfer, R. (2006). Statistical analysis used in the nutritional assessment of novel food using the proof of safety. Regulatory Toxicology and Pharmacology 44: 125-135.

Huang, J., Rozelle, S., Pray, C. and Wang, Q. (2002). Plant biotechnology in China. Science 295: 674-677.

Huang, J., Hu, R., Rozelle, S. and Pray, C. (2004a). GM rice in farmer fields: Assessing productivity and health effects in China. Working paper, Center for Chinese Agricultural Policy, Chinese Academy of Sciences, Republic of China.

Huang, J., Qiu, H., Bai, J. and Pray, C. (2006). Awareness, acceptance of and willingness to buy genetically modified foods in Urban China. Appetite 46: 144-151.

Huffman, W.E. and Tegene, A. (2002). Public acceptance of and benefits from agricultural biotechnology: a key role for verifiable information. In: Market Development for Genetically Modified Food (eds.) V. Santaniello, R.E. Evenson and D. Zilberman, CAB International, Wallingford, UK, pp. 179-190.

Huffman, W.E., Rousu, M., Shogren, J.F. and Tegene, A. (2006). The effects of prior beliefs and learning on consumers' acceptance of genetically modified foods. J. Econ. Behav. Org. In press.

Hupkens, C.L.H., Knibbe, R.A. and Drop, M.J. (1997). Social class differences in women's fat and fibre consumption: a cross-national study. Appetite 28: 131-149.

Isaacs, J.C. (2001). Acceptance of Genetically Modified Foods and Environmental Attitude. Louisiana Dept of Wildlife and Fisheries, Louisiana, USA.

James, C. (2003). Preview: Global Status of Commercialized Transgenic Crops: 2003. ISAAA Briefs Nr 30. Ithaca, NY: International Service for the Acquisition of Agribiotech Applications. Available at: http://www.isaaa.org/. Accessed 2005 17 February.

Jappe, A. January (2006). Genes, values and agricultural rebellions, translated by Kapa, A. Greek Newspaper Vavilonia, p.12. (republished from Krisis Magazine, 24/2001).

Kanrar, S., Venkateswari, J.C. and Kirti, P.B. (2002). Transgenic expression of hevein, the rubber tree lectin, in Indian mustard confers protection against *Alternaria brassicae*. Plant Sci. 162 (3): 441-448.

Kazan, K., Rusu, A., Marcus, J.P., Coulter, K.C. and Manners, J.M. (2002). Enhanced quantitative resistance to *Leptosphaeria maculans* conferred by expression of a novel antimicrobial peptide in canola. Molecular Breeding 10: 63-70.

Kearns, P. and Mayer, P. (1999). Substantive equivalence is a use tool. Nature 40: 640.

Kerr, W.A. (1999). Genetically modified organisms, consumer scepticism and trade law: Implications for the organisation of international supply chains. Supply Chain Management 4: 67-74.

Knight, J.G. (2003). Trade war looms as US launches challenge over transgenic crops. Nature 423: 369.

Knight J.G., Mather, D.W. and Holdsworth, D.K. (2005). Impact of genetic modification on country image of imported food products in European markets: Perceptions of channel members. Food Policy 30: 385–398.

Kohler, C. and Naftzger, D. (2001). Labelling genetically modified foods. Report issued by the National Conference of State Legislatures, October, 2001.

Koivisto Hursti, U.-K., Magnusson, M.K. and Algers, A. (2002). Swedish consumers' opinions about gene technology. British Food J. 104(11): 860-872.

Kombrink, E. and Somssich, I.E. (1995). Defense responses of plants to pathogens. Adv. Bot. Res. 21: 1-34.

Konig, A., Cockburn, A., Crevel, R.W.R., Debruyne, E., Grafstroem, R., Hammerling, U., Kimber, I., Knudsen, I., Kuiper, H.A., Peijnenburg, A.A.C.M., Penniks, A.H., Poulsen, M., Schauzu, M and Wal, J.M. (2004). Assessment of the safety of foods derived from genetically modified (GM) crops. Food and Chemical Toxicology 42: 1047-1088.

Kuiper, H.A., Kleter, G.A., Noteborn, H.P.J.M. and Kok, E.J. (2001). Assessment of the food safety issues related to genetically modified foods. The Plant J. 27(6): 503-528.

Kuznesof, S. and Ritson, C. (1996). Consumer acceptability of genetically modified foods with special reference to farmed salmon. British Food J. 98 (4 and 5): 39-47.

Lahteenmaki, L., Grunert, K., Ueland, Ø., Astrom, A., Arvola, A. and Bech-Larsen, T. (2002). Acceptability of genetically modified cheese presented as real product alternative. Food Qual. Pref. 13: 523-533.

Lamote, S., Hermans, D., Baeyens, F. and Eelen, P. (2004). An exploration of affective priming as an indirect measure of food attitudes. Appetite 42: 279-286.

Leighton, J. (1999). Science, medicine and the future: Genetically modified food. British Med. J. 318: 581-584.

Lemkow, L. (1993). Public Attitudes to Genetic Engineering: Some European Perspectives. Office of the Official Publications of the European Communities, Brussels.

Levidow, L. and Murphy, J. (2002). The Decline of Substantial Equivalence: How Civil Society Demoted a Risky Concept. Paper for Conference on Science and Citizenship in a Global Context: Challenges from New Technologies, at the Institute of Development Studies, University of Sussex. 12-13 December. Available at: http://www.ids.ac.uk/ids/env/biotechpaperrev1Peter.pdf.

Li, Q., Curtis, K.R., McCluskey, J.J. and Wahl, T.I. (2003). Consumer attitudes toward genetically modified foods in Beijing, China. AgBioForum 5(4): 145-152.

Li, X.M., Serebrisky, D., Lee, S.Y., Huang, C.K., Bardina, L. and Schofield, B.H. (2000). A murine model of peanut anaphylaxis: T- and B-cell responses to a major peanut allergen mimic human responses. J. Allergy Clin. Immunol. 106: 150-158.

Liu, Q., Singh, S. and Green, A. (2002). High-oleic and high-stearic cottonseed oils: Nutritionally improved cooking oils developed using gene silencing. J. Am. Coll. Nutr. 21: 205S-211S.

Lucca, P., Hurrell, R and Potrykus, I. (2002). Fighting iron deficiency anemia with iron-rich rice. J. Am. Coll. Nutr. 21: 184S-190S.

Lujan, J.L. and Todt, O. (2000). Perceptions, attitudes and ethical valuations: the ambivalence of the public image of biotechnology in Spain. Public Understanding of Science 9: 383-392.

Lusk, J., Roosen, J. and Fox, J. (2002). Demand for beef from cattle administered growth hormones or fed genetically modified corn: A comparison of consumers in France, Germany, the United Kingdom, and the United States. Am. J. Agric. Econ. 85(1): 16-29.

Macer, D. and Ng, M.A.C. (2000). Changing attitudes to biotechnology in Japan. Nat. Biotechnol. 18: 945-947.

MacMahon, R.R. Genetic Engineering of Crop Plants. Available at: http://www.yale.edu/ynhti/curriculum/units/2000/7/00.07.02.x. Accessed 2005 April 21.

Magnusson, M.K. and Hursti, U.K. (2002). Consumer attitudes towards genetically modified foods. Appetite 39: 9-24.

Maison, D., Greenwald, A. and Bruin, R. (2001). The implicit association test as a measure of implicit consumer attitudes. Polish Psychological Bull. 32: 61-69.

Marris, C., Wynne, B., Simmons, P. and Weldon, S. (2001). Public perceptions of agricultural biotechnologies in Europe. PABE final report. Available at: http://www.lancs.ac.uk/depts/ieppp/pabe/docs/pabe_finalreport.pdf.

Maryanski, J.H. (1995). FDA'S policy for foods developed by biotechnology. In: Genetically Modified Foods: Safety Issues (eds.) T. Engels and K. Teranishi, Am. Chem. Soc. Symposium Series No. 605: pp. 12-22.

McAllan, A.B. (1982). The fate of nucleic acids in rum infants. Proc. Nutr. Soc. 41: 309-317.

McCliskey, J. and Wahl, T. (2003). Reacting to GM foods: Consumer response in Asia and Europe. IMPACT highlights. International Marketing Program for Agricultural Commodities and Trade, College of Agriculture and Home Economics, Washington State University, USA.

Mepham, B. (2000). A framework for the ethical analysis of novel foods: the ethical matrix. J. Agric. Environ. Ethics 12: 165-176.

Metcalfe, D.D., Astwood, J.D., Townsend, R., Sampson, H.A., Taylor, S.L. and Fuchs, R.L. (1996). Assessment of the allergenic potential of foods derived from genetically engineered crop plants. Crit. Rev. Food Sci. Nutr. (36(S)): S165-S186.

Milgrom, P. and Roberts, J. (1986). Relying on the information of interested parties. RAND J. Econ. 17: 18-32.

Millstone, E., Brunner, E. and Mayer, S. (1999). Beyond substantial equivalence. Nature 401: 525-526. Available at: http://www.nature.com/.

Mitchell, P. (2003). EU biotech firms drastically cut back research. Nat. Biotechnol. 21(5): 468-469.

Mitrovic, T. (2003). Gene Transfer Systems. Facta Universitatis 10(3): 101-105. Available at: http://www.facta.junis.ni.ac.yu/facta/mab/mab200303/mab200303-01.pdf.

Moneret-Vautrin, A.D. (2006). Transgenic plants (GM plants): What we do and don't know about the risks of allergenicity. Revue Française d'Allergologie et d'Immunologie Clinique 46: 85-91.

Moon, W. and Balasubramanian, S.V. (2001). Public perceptions and willingness to pay a premium for non-GM foods in the US and the UK. AgBioForum 4(3 and 4): 221–231. Available at: http://www.agbioforum.org/.

Mora, A.A. and Earle, E.D. (2001). Resistance to *Altenaria brassicola* in transgenic broccoli expressing a *Trichoderma harzianum* endochitinase gene. Molecular Breeding 8: 1-9.

Morton, R.L., Schroeder, H.E., Bateman, K.S., Chrispeels, M.J., Armstrong, E. and Higgins, T.J. (2000). Bean alpha-amylase inhibitor 1 in transgenic peas (*Pisum sativum*) provides complete protection from pea weevil (*Bruchus pisorum*) under field conditions. Proc. Natl. Acad. Sci., USA 97: 3820-3825.

Moses, V. (1999). Biotechnology products and European consumers. Biotechnol. Adv. 17(8): 647-678.

Munro, I.C., McGirr, L.G., Nestmann, E.R. and Kille, J.W. (1996a). Alternative approaches to the safety assessment of macronutrient substitutes. Regul. Toxicol. Pharmacol. 23: S6-S14.

Munro, I.C., Ford, R.A., Kennepohl, E. and Sprenger J.G. (1996b). Correlation of structural class with no-observed-effect levels: A proposal for establishing a threshold of concern. Food Chem. Toxicol. 34: 829-867.

Murphy, D.J. (2002). Biotechnology and the improvement of oil crops - genes, dreams and realities. Phytochem. Rev. 1: 67-77.

Myhra, I. and Dalmo, R. (2005). Introduction of genetic engineering in aquaculture: Ecological and ethical implications for science and governance. Aquaculture. 250: 542-554.

Nordlee, J.A., Taylor, S.L., Townsend, J.A., Thomas, L.A. and Bush, R.K. (1996). Identification of a Brazil-nut allergen in transgenic soybeans. New Engl. J. Med. 334: 688-692.

Noteborn, H.P.J.M., Bienenmann-Ploum, M.E., van den Berg, J.H.J., Alink, G.M., Zolla, L., Reynerts, A., Pensa, M. and Kuiper, H.A. (1995). Safety assessment of the *Bacillus thuringiensis* insecticidal crystal protein CRY1A(b) expressed in transgenic tomatoes. In: Genetically Modified Foods. Safety Issues (eds) K.H. Engel, G.R. Takeola and R. Teranishi, ACS Symposium Series 605, Washington DC, USA, pp. 134-147.

Noussair, C., Robin, S. and Ruffieux, B. (2001a). Genetically modified organisms in the food supply: Public Opinion vs. Consumer Behaviour. Paper No. 1339 January 2001. Krannert Graduate School of Management, Purdue University, West Lafayette, USA.

Noussair, C., Robin, S. and Ruffieux, B. (2001b). Do consumers not care about biotech foods or do they just not read the labels? Paper no. 1142 February 2001. Krannert Graduate School of Management, Purdue University West Lafayette, Indiana. Available at: http://www.mgmt.purdue.edu/faculty/.

Novak, W.K. and Halsberger, A.G. (2000). Substantial equivalence of antinutrients and inherent plant toxins in genetically modified novel foods. Food Chem. Toxicol. 38: 473-483.

O'Connor, E. (2004). Second generation GM foods: Perspectives on likely future acceptance by Irish consumers. Unpublished thesis, University College Dublin, Ireland.

O'Connor, E., Cowan, C., Williams, G., O'Connell, J. and Boland, M.P. (2006). Irish consumer acceptance of a hypothetical second-generation GM yoghurt product. Food Qual. Pref. 17: 400-411.

O'Fallon, M.J., Gursoy, D. and Swanger, N. (2005). To buy or not to buy: Impact of labeling on purchasing intentions of genetically modified foods. Hospitality Management. In press.

Oberhauser, K.S., Prysby, M.D., Mattila, H.R., Stanley-Horn, D.E., Sears, M.K., Dively, G., Olson, E., Pleasants, J.M., Lam, W.-K.F. and Hellmich, R.L. (2001). Temporal and spatial overlap between monarch larvae and corn pollen. Proc. Natl. Acad. Sci., USA 98: 11913-11918; published online 9/14/01.

Olkowski, A.A. and Classen, H.L. (1995). Sudden death syndrome in broiler chickens: a review. Poult. Avian Biol. Rev. 6: 95-105.

Olubobokun, S., Phillips, P. and Hobbs, J.E. (2001). Analysis and differentiation of consumers' perceptions of genetically modified foods. Paper presented at the Fifth International Conference of the ICABR on 'Biotechnology, Science and Modern Agriculture: A New Industry at the Dawn of the Century', Ravello, Italy, 15-18 June 2001.

Peferoen, M. (1997). Progress and prospects for field use of Bt genes in crops. Trends Biotechnol. 15: 173-177.

Perlak, F.J., Stone, T.B., Muskopf, Y.M., Petersen, L.J., Parker, G.B., McPherson, S.A., Wyman, J., Love, S., Reed, G. and Biever, D. (1993). Genetically improved potatoes: protection from damage by Colorado potato beetles. Plant Mol. Biol. 22: 313-321.

Phipps, R.H., Beever, D.E. and Humphries, D.J. (2002). Detection of transgenic DNA in milk from cows receiving herbicide tolerant (CP4 EPSPS) soyabean meal. Livestock Prod. Sci. 74: 269-273.

Phipps, R.H., Deaville, E.R. and Maddison, B.C. (2003). Detection of transgenic and endogenous plant DNA in rumen fluid, duodenal digesta, milk, blood, and feces of lactating dairy cows. J. Dairy Sci. 86: 4070-4078.

Pleasants, J.M., Hellmich, R.L., Dively, G.P., Sears, M.K., Stanley-Horn, D.E., Mattila, H.R., Foster, J.E., Clark, P. and Jones, G.D. (2001). Corn pollen deposition on milkweeds in and near cornfields. Proc. Nat. Acad. Sci., USA 98: 11919-11924, published online 9/14/01.

Pollack, A. (2000). New Ventures Aim to Put Farms In Vanguard of Drug Production. New York Times, May 14, 2000, p.1.

Poms, R.E., Hochsteiner, W., Luger, K., Glossl, J. and Foissy, H. (2003). Model studies on the detectability of genetically modified feeds in milk. J. Food Prot. 66: 304-310.

Poortinga, W. and Pidgeon, N. (2003). Public perceptions of risk, science and governance. Main findings of a British survey of five risk cases. Centre for Environmental Risk, University of East Anglia, Norwich.

Prescott, V.E. and Hogan, S.P. (2006). Genetically modified plants and food hypersensitivity diseases: Usage and implications of experimental models for risk assessment. Pharmacology and Therapeutics. In press.

Pryme, I. (2003). In-vivo studies on possible health consequences of GM food and feed. Nutr. Health 17: 1-8.

Punja, Z.K. (2004). Genetic engineering of plants to enhance resistance to fungal pathogens. In: Fungal Disease Resistance in Plant (ed.) Z.K. Punja, Haworth Press, Inc. New York, USA, pp. 207-258.

Purdue, D.A. (2000). Anti-Genetix: The Emergence of the Anti-GM Movement. Ashgate Publishing Ltd., Aldershot.

Redenbaugh, K., Hiuatt, W., Martineau, B. and Emlay, D. (1995). Determination of the safety of genetically engineered crops. In: Genetically Modified Foods. Safety Issues (eds.) K.H. Engel, G.R. Takeola and R. Teranishi, ACS Symp Ser 605. Washington DC, USA, pp. 72-87.

Reisner, A. (2001). Social movement organisations' reactions to genetic engineering in agriculture. American Behavioural Scientist 44 (8): 1389-1404.

Rifkin, J. (1998). The biotechnology century. Genetic trade and the daylight of a wonderful new world, translated by Alavanou, A., Livanis publishers.

Roe, B., Teisl, M.F., Levy, A.S., Boyle, K., Messonnier, M.L., Riggs, T.L., Herrmann, M.J. and Newman, F.M. (2002). Consumers' assessment of the food safety problem for meals prepared at home and reactions to food safety labeling. J. Food Products Marketing 6: 9-26.

Roe, B. and Teisl, M.F. (2006). Genetically modified food labeling: The impacts of message and messenger on consumer perceptions of labels and products. Food Policy. In press.

Rosati, S. and Saba, A. (2000). Factors influencing the acceptance of food biotechnology. Italian J. Food Sci. 4(12): 425-434.

Saba, A., Moles, A. and Frewer, L.J. (1998). Public concerns about general and specific applications of genetic engineering: a comparative study between the UK and Italy. Nutr. Food Sci. 1: 19-29.

Saba, A., Rosati, S. and Vassalo, M. (2000). Biotechnology in agriculture: Perceived risks, benefits and attitudes in Italy. British Food J. 102: 114-121.

Saba, A. and Vassalo, M. (2002). Consumer attitudes towards the use of gene technology in tomato production. Food Qual. Pref. 13: 13-21.

Sanden, M., Bruce, I.J., Rahman, M.A. and Hemre, G. (2004). The fate of transgenic sequences present in genetically modified plant products in fish feed; investigating the survival of

GM soybean DNA fragments during field trials in Atlantic salmon, *Salmo salar* L. Aquaculture 237: 391-405.

Schroeder, H.E., Gollasch, S., Moore, A.E., Tabe, L.M., Craig, S. and Hardie, D.C. (1995). Bean α-amylase inhibitor confers resistance to the pea weevil (*Bruchus pisorum*) in transgenic peas (*Pisum sativum* L.). Plant Physiol. 107: 1233-1239.

Schubbert, R., Renz, D., Schmitz, B. and Doerfler, W. (1997). Foreign (M13) DNA ingested by mice reaches peripheral leukocytes, spleen, and liver via the intestinal wall mucosa and can be covalently linked to mouse DNA. Proc. Natl. Acad. Sci., USA 94: 961-966.

Sears, M.K., Hellmich, R.L., Stanley-Horn, D.E., Oberhauser, K.S., Pleasants, J.M., Mattila, H.R., Siegfried, B.D. and Dively, G.P. (2001). Impact of Bt corn pollen on monarch butterfly populations: A risk assessment. Proc. Natl. Acad. Sci., USA 98: 11937-11942; published online 9/14/01.

Shewmaker, C.K., Sheely, J.A., Daley, M., Colburn, S. and Ke, D.Y. (1999). Seed-specific over expression of phytoene synthase: Increase in carotenoids and other metabolic effects. Plant J. 22: 401-412.

Shintani, D. and de La Penna, D. (1998). Elevating the vitamin E content of plants through metabolic engineering. Science 282: 2098-2100.

Sitrit, Y. and Bennett, A.B. (1998). Regulation of tomato fruit polygalacturonase mRNA accumulation by ethylene: a re-examination. Plant Physiol. 116: 1145-1150.

Slovic, P. and Peters, E. (1998). The importance of worldviews in risk perception. Risk Decision and Policy 3(2): 165-170.

Spence, A. (2005). Using implicit tasks in attitude research: A review and a guide. Social Psychological Review 2-17.

Spence, A. and Townsend, E. (2006). Implicit attitudes towards genetically modified (GM) foods: A comparison of context-free and context-dependent evaluations Appetite 46: 67-74.

Spetsidis, N.M. and Schamel, G. (2001). A survey over consumers cognitions with regard to product scenarios of GM foods in Germany. Contributed paper for the 71st EAAE Seminar – The Food Consumer in the Early 21st Century, 19th–20th April 2001, Zaragoza, Spain.

Springer, A., Mattas, K., Papastefanou, G. and Tsioumanis, A. (2002). Comparing consumer attitudes towards genetically modified food in Europe. Paper Presented in the Xth EAAE Congress 'Exploring Diversity in the European Agri-food System'. 28-31 August, Zaragoza.

Stanley-Horn, D.E., Dively, G.P., Hellmich, R.L., Mattila, H.R., Sears, M.K., Rose, R., Jesse, L.C.H., Losey, J.E., Obrycki, J.J. and Lewis, L. (2001). Assessing the impact of Cry1Ab-expressing corn pollen on monarch butterfly larvae in field studies. Proc. Natl. Acad. Sci., USA 98: 11931-11936; published online 9/14/01.

Stewart, C.N., Adang, M.J. and All, J.N. (1996). Genetic transformation, recovery, and characterization of fertile soybean transgenic for a synthetic *Bacillus thuringiensis cry1Ac* gene. Plant Physiol. 112 (1): 121-129.

Sticher, L. (1997). Systemic acquired resistance. Annu. Rev. Phytopathol. 35: 235-270.

Straughan, R. (1998). Moral concerns and the educational function of ethics. In: Genetic Modification in the Food Industry (eds.) S. Roller and S. Haslander, Blackie, London, UK, pp. 180-189.

Stuible, H.-P. and Kombrink, E. (2004). The hypersensitive response and its role in disease resistance. In: Fungal Disease Resistance in Plants (ed.) Z.K. Punja, Haworth Press, Inc. New York, USA, pp. 57-92.

Subrahmanyan, S. and Cheng, P.S. (2000). Perceptions and attitudes of Singaporeans toward genetically modified food. J. Consumer Affairs 34(2): 269-290.

Taylor, S.L. and Hefle, S.L. (2002). Genetically engineered foods: Implications for food allergy. Curr. Opin. Allergy Clin. Immunol. 2: 249-252.

Teisl, M.F., Gardner, L., Roe, B. and Vayda, M. (2003). Labeling genetically modified foods: How do U.S. consumers want to see it done? AgBioForum 6: 48-54.

Tencalla, F. (2006). Science, politics, and the GM debate in Europe. Regulatory Toxicology and Pharmacology 44: 43-48.

Terzaghi, W.B. and Cashmore, A.R. (1997). Plant cell transfection by electroporation. Methods Mol. Biol. 62: 453-462.

Thompson, C., Dunwell, J.C.E., Lay, V., Ray, J., Schmitt, M., Watson, H. and Nisbet, G. (1995). Degradation of oxalic acid by transgenic oilseed rape plants expressing oxalate oxidase. Euphytica 85: 169-172.

Thompson, P.B. (1997). Food Biotechnology in Ethical Perspective. Blackie Academic and Professional, London, U.K.

Torgersen, H. and Seifert, F. (1997). Aversion preceding rejection: results of the Eurobarometer survey 39.1 on biotechnology and genetic engineering in Austria. Public Understanding of Science 6: 131-142.

Torgersen, H., Egger, C., Grabner, P., Kronberger, N., Seifert, F., Weger, P. and Wagner, W. (2001). Austria: narrowing the gap with Europe. In: Biotechnology 1996–2000: the years of controversy (eds.) G. Gaskell and M.W.Bauer, Science Museum, London, UK, pp. 131-144.

Tversky, A. and Kahneman, D. (1992). Advances in prospect theory: cumulative representation of uncertainty. J. Risk and Uncertainty 5: 297-323.

Untersmayr, E., Scholl, I., Swoboda, I., Beil, W.J., Forster-Waldl, E. and Walter, F. (2003). Antacid medication inhibits digestion of dietary proteins and causes food allergy: A fish allergy model in BALB/c mice. J. Allergy Clin. Immunol. 112: 616-623.

Ursin, V. (2000). Genetic modification of oils for improved health benefits. Presentation at conference, Dietary Fatty Acids and Cardiovascular Health: Dietary Recommendations for Fatty Acids: Is There Ample Evidence? Reston Va.: American Heart Association, 5-6 June 2000.

Verdurme, A., Gellynck, X. and Viaene, J. (2001). Consumer's acceptance of GM food. Paper presented at the 71st EAAE Seminar on 'The Food Consumer in the Early 21st Century', Zaragoza, Spain, 19–20 April 2001.

Verdurme, A., Gellynck, X. and Viaene, J. (2001a). Consumers' acceptability of GM food. Contributed paper at 71st EAAE Seminar, 'The Food Consumer in the Early 21st Century', Zaragoza, Spain, 19-20 April 2001.

Viscusi, W.K. (1997). Alarmist decisions with divergent risk information. The Economic J. 107: 1657-1670.

Von Schomberg, R. (1999). GMO-releases: from risk-based to uncertainly based regulation. In: Past, Present and Future Considerations in Risk Assessment when using GMOs. (ed.) G.E. de Vries, Commission on Genetic Modification, Bilthoven, The Netherlands.

Wal, J.M., Hepburn, P.A., Lea, L.J. and Crevel, R.W.R. (2003). Post-market surveillance of GM foods: applicability and limitations of schemes used with pharmaceuticals and some non-GM novel foods. Regulatory Toxicol. Pharmacol. 38: 98-104.

Wang, H.Z., Zhao, P.J. and Zhou, X.Y. (2000). Regeneration of transgenic *Cucumis melo* and its resistance to virus diseases. Acta Phytophylactica Sinica 27: 126-130.

Wang, P., Nowak, G., Culley, D., Hadwiger, L.A. and Fristensky, B. (1999). Constitutive expression of a pea defense gene DRR206 confers resistance to blackleg disease in transgenic canola (*Brassica napus*). Mol. Plant Microbe Interact. 12: 410-418.

Wang, Z. (2003). Knowledge of food safety and consumption decision: An empirical study on consumer in Tianjing, China. China Rural Economy 4: 41-48.

Weick, C.W. and Walchli, S.B. (2002). Genetically engineered crops and foods; back to the basics of technology diffusion. Technol. Soc. 24: 265-283.

Xue, D. and Tisdell, C. (2000). Safety and socio-economic issues praised by modern biotechnology. Int. J. Soc. Econ. 27(7/8/9/10): 699-708.

Xue, D. and Tisdell, C. (2002). Global trade in GM food and the Cartagena protocol on biosafety: consequences for China. J. Agric. Environ. Ethics 15: 337-356.

Ye, X., Al-Babili, S., Klöti, A., Zhang, J., Lucca, P., Beyer, P. and Potrykus, I. (2000). Engineering the provitamin A (beta-carotene) biosynthetic pathway into (carotenoid-free) rice endosperm. Science 287(5451): 303-305.

Zedalis, R.J. (2002). GMO food measures as Restrictions under GATT article XI (1). Eur. Environ. Law Rev. Jan 2002: 16-28.

Zhang, X. (2002). Chinese consumers' concerns over food safety. Working paper, Agricultural Economics Institute, Hague, The Netherlands. Available at: http://mailman.greenpeace.org/mailman/listinfo/pressreleases.

Zhong, F., Marchant, M., Ding, Y. and Lu, K. (2003). GM foods: A Nanjing case study of Chinese consumers' awareness and potential attitudes. AgBio-Forum 5(4): 136-144.

Zhou, F. and Tian, W. (2003). Consumer perceptions and attitudes toward genetically modified foods and their determinants: A Beijing case study. China Agric. Econ. Rev. 1(3): 266-293.

Zimmerman, L., Kendall, P., Stone, M. and Hoban, T. (1994). Consumer knowledge and concern about biotechnology and food safety. Food Technol. 48: 71-77.

Alinorm 03/34: Joint FAO/WHO Food Standard Programme. 30 June-5 July, 2003. C.A.C.: Appendix III: Guideline for the conduct of food safety assessment of foods derived from recombinant-DNA plants; and Appendix IV: Annex on the assessment of possible allergenicity, Rome, Italy. Codex Alimentarius Commission. pp. 47-60.

Bioscience Law Review (2004). Europe's biotech vision, so far so good? http://pharmalicensing.com/features/disp/1075973829_40220ec5d94f5. Accessed 2004 August.

Codex (2002). Codex Alimentarius Commission. Report of the third session of the Codex ad hoc intergovernmental task force on foods derived from biotechnology. Yokohama, Japan 4-8 March 2002. Alinorm 03/34. Available at: ftp://ftp.fao.org/codex/alinorm03/A103_34e.pdf. Accessed 2003 Jun 24.

Commission Regulation (EC) No. 49/2000 of January 10, 2000 amending Council Regulation (EC) No. 1139/98, concerning the compulsory indication on the labeling of certain foodstuffs produced from genetically modified organisms of particulars other than those provided for in Directive 79/112/EEC.

Commission Regulation (EC) No. 50/2000 of January 10, 2000 on the labeling of foodstuffs and food ingredients containing additives and flavourings that have been genetically modified or have been produced from genetically modified organisms.

EU 2004/204/EC (entry into force 23/3/2004) Arrangements for the operation of the registers for recording information on genetic modifications in GMOs.

EU 2004/643/EC Placing on the market of a maize product (*Zea mays* L. line NK603) GM for glyphosate tolerance.

EU 2004/657/EC Placing on the market of a sweet corn from GM maize line Bt11 as a novel food or novel food ingredient.

EU 90/219/EEC (entry into force 23/10/1991) Contained use of G.M. Micro-organisms.

EU 90/220/EEC (entry into force 23/10/1991) Deliberate release into the environment of GMOs.

EC (2001). Directive 2001/18/CE of the European Parliament and of the Council of 12 March 2001 on the deliberate release into the environment of genetically modified organisms and repealing Council Directive 90/220/EEC.

EC (2003). Commission calls on EU Member States to intensify eiforts in life sciences and biotechnology. European Commission press release. http://www.europa.eu.int/rapid/press Releases Action do ?reference D IP/03/313andformat D HTMLandaged D 1andlanguage D EN andguiLanguage=en. Accessed 2004 August.

EC (2003). Guidance document for the risk assessment of genetically modified plants and derived food and feed, health and consumer protection. Directorate-General, European Commission. Available at: http://europa.eu.int/comm/food/fs/sc/ssc/out327_en.pdf. Accessed 2005 April 2.

EU (2001). EC-sponsored Research on Safety of Genetically Modified Organisms; 5th Framework Program - External Advisory Groups. "GMO research in perspective." Report of a workshop held by External Advisory Groups of the "Quality of Life and Management of Living Resources Program". Available at: http://europa.eu.int/comm/research/quality-of-life/gmo/index.html http://europa.eu.int/comm/research/fp5/eag-gmo.html. Accessed 2003 July 22.

EU 2001/18/EC (entry into force 17/4/2001) Deliberate release into the environment of GMOs.

EU 2004/204/EC (entry into force 23/3/2004). Arrangements for the operation of the registers for recording information on genetic modifications in GMOs.

EU 2004/643/EC Placing in the market of a maize product (*Zea mays* L. line NK603) GM for glyphosate tolerance.

EU 2004/657/EC Placing in the market of sweet corn from GM maize line Bt11 as a novel food or novel food ingredient.

EU 90/219/EEC (entry into force 23/10/1991) Contained use of GM microorganisms.

EU 90/220/EEC (entry into force 23/10/1991) Deliberate release into the environment of GMOs.

EU Regulation COM/2002/0085 - COD 2002/0046 (entry into force 27/10/2002) The transboundary movement of GMOs.

EU Regulation (EC) No. 1139/98 (entry into force 1/9/1998) The compulsory indication of the labeling of certain foodstuffs produced from GMOs.

EU Regulation (EC) No. 1829/2003 (entry into force 7/11/2003) GM food and feed.

EU Regulation (EC) No. 1830/2003 (entry into force 7/11/2003) Traceability and labeling of GMOs and traceability of food and feed products produced from GMOs.

EU Regulation (EC) No. 641/2004 (entry into force 18/4/2004) The authorization of new GM food and feed, the notification of existing products and adventitious or technically unavoidable presence of GM material which has benefited from a favourable risk evaluation.

EU Regulation (EC) No. 65/2004 (entry into force on the date of its publication in the Official Journal of the European Union) Establishing a system for the development and assignment of unique identifiers for GMOs.

FAO/WHO (1991) Report of the Nineteenth Session of the Joint AO/WHO Codex Alimentarius Commission Rome, 1 - 10 July 1991, Joint FAO/WHO Expert Committee on Food Additives (JECFA) & Joint FAO/WHO Meeting on Pesticide Residues (JMPR).

FAO/WHO (1996). Biotechnology and food safety. Report of a joint FAO/WHO consultation; 30 Sept. - 4 Oct. 1996. FAO Food and Nutrition Paper 61. Rome, Italy: Food and Agriculture Organization of the United Nations. Available at: ftp://ftp.fao.org/es/esn/food/biotechnology.pdf. Accessed 2003 July 22.

FAO/WHO (2000). Safety Aspects of Genetically Modified Foods of Plant Origin. Report of a Joint FAO/WHO Expert Consultation on Foods Derived from Biotechnology, 29 May–2 June, 2000. Rome, Italy: Food and Agriculture Organization of the United Nations. Available at: ftp://ftp.fao.org/es/esn/food/gmreport.pdf. Accessed 2005 June 5.

FAO/WHO (2001). Evaluation of Allergenicity of Genetically Modified Foods Report of a Joint FAO/WHO Expert Consultation on Allergenicity of Foods Derived from Biotechnology, Rome, Italy, 22–25 January 2001. (http://www.who.int/foodsafety/publications/biotech/en/ec_jan2001.pdf).

FAO/WHO (2001). Safety Assessment of Foods Derived from Genetically Modified Microorganisms. Report of a Joint FAO/WHO Expert Consultation on Foods Derived from Biotechnology; 24–28 Sept., 2001. Geneva, Switzerland: World Health Organization/Food and Agriculture Organization of the United Nations. Available at: http://www.who.int/fsf/Documents/GMMConsult_Final_.pdf.

FAO/WHO (2002). Report of the third session of the Codex Ad Hoc Intergovernmental Task Force on Foods Derived from Biotechnology (Alinorm 01/34). Rome, Italy: Codex Ad Hoc Intergovernmental Task Force on Foods Derived from Biotechnology, Food and Agriculture Organization of the United Nations. Available at: ftp://ftp.fao.org/codex/alinorm03/Al03_34e.pdf. Accessed 2005 March 22.

Genetically Engineered Crop and Animal Farmer Protection Act (US), 2003.

Genetically Engineered Food Right to Know Act (US), 2003.

Genetically Engineered Food Safety Act (US), 2003.

Genetically Engineered Pharmaceutical and Industrial Crop Safety Act (US), 2003.

Greenpeace (2004). Public's perception of genetically engineered food: Summary of report. Available at: http://www.greenpeace.org.hk.

http://www.acnfip.gov.uk/acnfppapers/inforelatass/toxrev?view=printerfriendly.

http://www.cspinet.org/biotech/gefood_approval.html

http://www.eufic.org/gb/food/pag/food37/food374.html.

http://www.europa.eu.int/comm/public_opinion

http://www.food.gov.uk/gmdebate/aboutgm/gm_safety?view=GM+Microsite

http://www.greenpeace.org.br/.

http://www.ornl.gov/sci/techresources/Human_Genome/elsi/gmfood.shtml

http://www.mindfully.org/GE/GE4/GE-Legislation-Rep-Kucinich11jun02.htm

http://www.mja.com.au/public/issucs/172_04_210200/huppleed/leeder.html.

http://www.mrc.ac.uk/pdf-strategy-gm_foods.pdf

http://www.twnisde.org.sg/title2/service169.htm

http://www.twnside.org.sg/title2/service169.htm

http://www.yale.edu/ynhti/curriculum/units/2000/7/00.07.02.x.html

IFIC [International Food Information Council] (2004). Support for food biotechnology stable despite news on unrelated food safety issues. Available at: http://www.ific.org/research/biotechres03.cfm.

ILSI (1997). Europe Novel Foods Task Force. The safety assessment of novel foods. Food Chem. Toxicol. 34: 931-940.

ILSI (2003). Best practices for the conduct of animal studies to evaluate crops genetically modified for input traits. Washington, DC: International Life Sciences Institute.

ILSI (2004). ILSI crop composition database. Washington, DC: International Life Sciences Institute. Available at: http://www.cropcomposition.org. Accessed 2004 25 June. Indiana. Available at: http://www.mgmt.purdue.edu/faculty/.

JRC (Joint Research Center), 2004. Deliberate field trials. http://www.biotech.jrc.it/deliberate/dbcountries.asp. Accessed 2004 August.

LSRO [Limited Standard Reactor Operater] (1998). Alternative and traditional models for safety evaluation of food ingredients. (eds.) D.J. Raiten and M. Bethesda, Life Sciences Research Office, American Society for Nutritional Sciences. Prepared for Center for Food Safety and Applied Nutrition, Food and Drug Administration, Dept. of Health and Human Services, Washington, DC.

NZRC (2001). Report of the Royal Commission on Genetic Modification. New Zealand Royal Commission on Genetic Modification. Available at: http://www.gmcommission.govt.nz/index.html. Accessed 2005 May.

OECD (1993). Safety evaluation of foods derived by modern biotechnology: Concepts and principles. Paris, France: Organization for Economic Co-operation and Development.

OECD (1998). Report of the OECD workshop on the toxicological and nutritional testing of novel foods. Organization for Economic Co-operation and Development, Paris, France.

OECD (2000). Report of the OECD task force for the safety of novel foods and feeds. Organization for Economic Co-operation and Development, Okinawa, Japan.

OECD (2002). Report of the OECD workshop on nutritional assessment of novel foods and feeds. Organization for Economic Co-operation and Development, Ottawa, Canada. Paris, France, 5-7 February 2001.

Proposal for a Regulation COM/2002/0085-COD2002/0046 (entry into force 27/10/2002). The transboundary movement of GMOs.

Regulation (EC) No. 178/2002 Of the European parliament and of the Council of 28 January 2002 laying down the general principles and requirements of food law, establishing the European Food Safety Authority and laying down procedures in matters of food safety.

Regulation (EC) No. 1139/98 (entry into force 1/9/1998) The compulsory indication of the labeling of certain foodstuffs produced from GMOs.

Regulation (EC) No. 1829/2003 (entry into force 7/11/2003) GM food and feed Regulation (EC) No.1830/2003 (entry into force 7/11/2003) Traceability and labeling of GMOs and traceability of food and feed products produced from GMOs.

Regulation (EC) No. 258/97 (entry into force 14/5/1997) Novel food and novel food ingredients.

Regulation (EC) No. 641/2004 (entry into force 18/4/2004) The authorization of new GM food and feed, the notification of existing products and adventitious or technically unavoidable presence of GM material which has benefited from a favourable risk evaluation.

Regulation (EC) No. 65/2004 (entry into force on the date of its publication in the *Official Journal of the European Union*) Establishment of a system for the development and assignment of unique identifiers for GMOs.

Royal Society (2002). Genetically modified plants for food use and human health—an update. London, UK: The Royal Society. Available at: http://www.royalsoc.ac.uk/policy/index.html. Accessed on 2003 July 22.

Royal Society (2003). GM crops, modern agriculture and the environment. Report of a Royal Society Discussion Meeting held on 11 February 2003. London, UK: The Royal Society. Available at: http://www.royalsoc.ac.uk/files/statfiles/document-222.pdf. Accessed 2003 July 22.

State of Oregon, Division of Elections (2002). Online Voter's Guide: General Election, November 5, 2002. Available at: http://www.sos.state.or.us/elections/nov52002/guide/measures/m27.htm.

United States Federal Register, January 18, 2001, 66, 4839.

United States Federal Register, May 29, 1992, 57, 22984.

WHO (1987). Principles of the safety assessment of food additives and contaminants in food. Environmental Health Criteria, Nr 70. Geneva Switzerland: International Program on Chemical Safety. WHO.

2

Plant Genetic Engineering: Horticultural Applications

Ioannis S. Arvanitoyannis* and Athanassios G. Mavromatis

INTRODUCTION

The nutrient value of vegetables and fruits is of crucial importance for human health. Human nutrition greatly depends on storage products of plants abundant in roots, tubers or seeds, and includes starch, carbohydrates, oils, lipids and proteins, as well as secondary metabolites, minerals and vitamins. The knowledge of metabolic pathways and isolation of genes greatly associated with plant physiology and biochemical interactions is necessary for plant improvement. Altering protein and vitamin levels, changing the composition of fatty acids, starch and sugars and/or incorporating most beneficial compounds in a balanced way will hopefully result in production of a lower number of fruits and vegetables enriched with advantageous properties, thus reducing the cost for users and improving the processes in agro-based industry.

Recombinant DNA technology is the widely used term for a series of techniques applied to modify the basic genetic "make up" of a living organism. This could happen by inserting or removing or silencing genes that carry the hereditary information. Plants, animals or microorganisms which originated from this human intervention, are Genetically Modified

*Corresponding Author

Organisms (GMOs) or transgenics. Therefore, GMO food products are all those containing or consisting of/or produced from GMO organisms. Production of GMOs includes the use of molecular methods and techniques that was referred to as "genetic engineering". These techniques include the following:

(a) detection and isolation of desirable genes from an organism
(b) insertion of these genes into the DNA of improved organism
(c) control of gene's action into the new background of the modified organism.

All these processes allow the transfer of "new genes" from an organism to another even if they come from different living structures (e.g., gene from a bacterium to a plant).

The enhancement of desired traits has traditionally been undertaken for years through conventional plant breeding. Nowadays, plant breeding methods are theorized to be time consuming and are often not very accurate. On the other hand, genetic engineering is a promising technique in terms of producing plants very rapidly and with high accuracy, and endowed with the exact desired traits or specialized properties (Gasson and Burke, 2001). Furthermore, the extra power of genetic engineering comes from its ability to introduce genetic material from unrelated species overcoming the barriers between organisms (Slater et al., 2003). This is the critical point for discussion and the main difference from the classical hybridization techniques applied in plant breeding.

Genetic engineering offers plant breeders access to an infinitely wide array of novel genes and traits, which can be inserted through a single event into high yielding elite cultivars of horticultural plants. The impact is focused on:

(i) resistance to diseases, insects and viruses
(ii) tolerance to herbicides and abiotic stress factors
(iii) increased effectiveness of biocontrol agents
(iv) enhanced nutritional value and postharvest quality
(v) altered photosynthetic activity, which in consequence has an increase in yield and other components like sugar and starch production
(vi) knowledge of metabolic pathways and production of pharmaceuticals and vaccines.

Historically, the transgenic plants were first created in the early 1980s, by two groups working independently in the University of St. Louis (Washington) and University of Gent (Belgium), which managed to insert bacterial genes into tobacco plants, thereby making them resistant to kanamycin (Hartl et al., 1983) and methotrexate (Herrera-Estrella et al.,

1983). At the same time, the scientists of Monsanto Inc. had also produced petunia plants that were resistant to kanamycin (Fraley et al., 1983) and the group of J. Kemp and T. Hall from the University of Wisconsin had inserted a gene from pea (*Pisum sativum*) in to sunflower (*Helianthus annus*) through genetic transformation. The first commercially grown genetically modified horticultural crop was the Flavr Savr™ tomato produced by Calgene Inc., USA in 1992 which was characterized by prolonged shelf life. This capacity was achieved by introducing an antisense polygalacturonase gene from a bacterium into tomato's genome (Smith et al., 1990; Kou et al., 2004).

However, the application of GMOs in farming could find a place in history as one of mankind's greatest scientific and industrial breakthroughs. From another point of view this could prove a high risk experiment, similar to that of extensive use of mobile telephones.

METHODS FOR GMO PRODUCTION

A number of methods are currently available for production of transgenic plants. The oldest method used for genetic transformation was

(i) *"recombinant DNA technique (rDNA)"* and relies on biological vectors like bacteria (*Agrobacterium tumefaciens*) or viruses **(Fig. 2.1)** (discussed in Chapter 3 in this Volume by Haseena Khan).

Other gene transfer techniques are:

(ii) the *"electroporation method"* which allows the entry of new genes by creating pores or small holes in the cell membrane of the modified organism with the application of a weak electric current **(Fig. 2.2)**

(iii) the *"osmochemical poration method"* which follows the same process, but can be conducted by bathing the host cells in special chemical solutions like polyethylene glycol (PEG)

(iv) the *"microinjection method"*, where the genetic material containing the new gene is simply injected into the recipient cell. The injection should be carried out with a fine-tipped glass needle under the microscopy **(Fig. 2.3)**

(v) the *"bioballistic (particle gun acceleration) method"* that uses metal slivers to deliver the "new gene" into the interior of the target cell. The small slivers are coated with the genetic material and exposed into the target cells by using a shot gun **(Fig. 2.4)**. A performed metal plate stops the shell cartridge but allows the slivers to pass through the living cells. Once in the cell, the genetic material is transported into the nucleus where it is incorporated among the host genes.

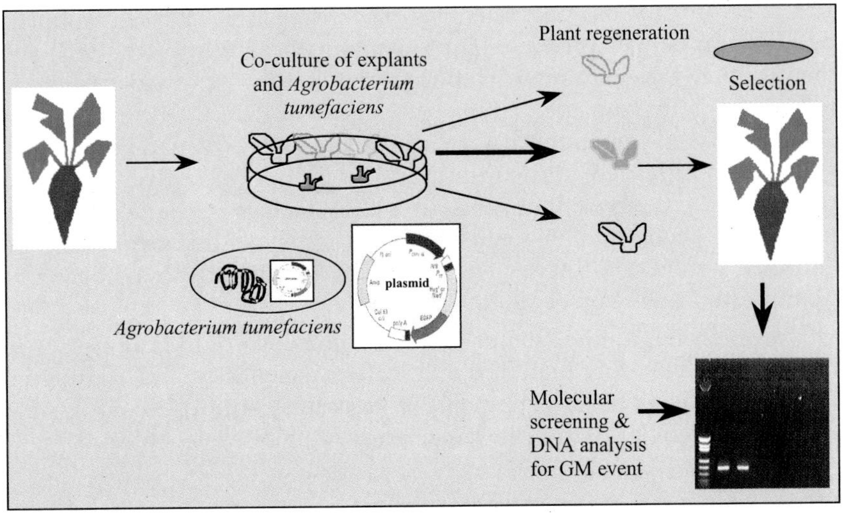

Fig. 2.1. Plant genetic transformation using *Agrobacterium tumefaciens* method

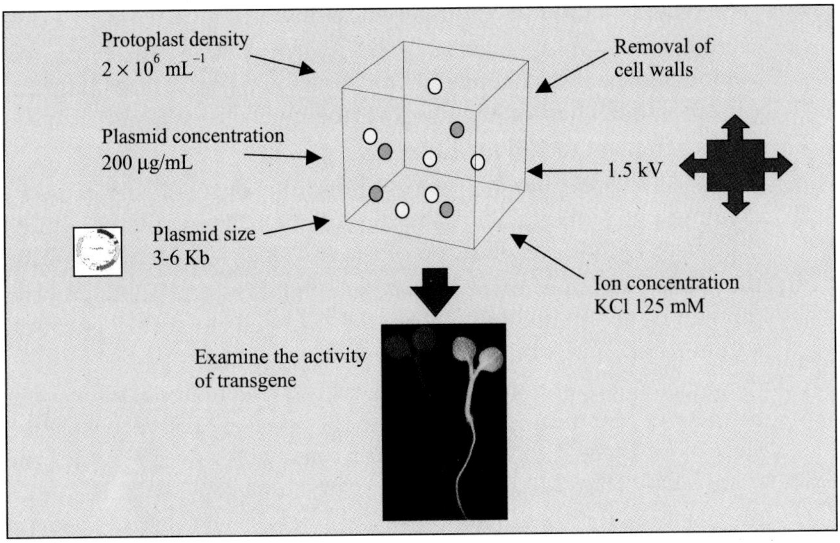

Fig. 2.2. Plant genetic transformation using "*electroporation method*"

BENEFITS OF GMOs

The world population has topped six billion people and is predicted to double within the next 30-50 years. Ensuring adequate food supply for this booming population is going to be a major challenge in the years to

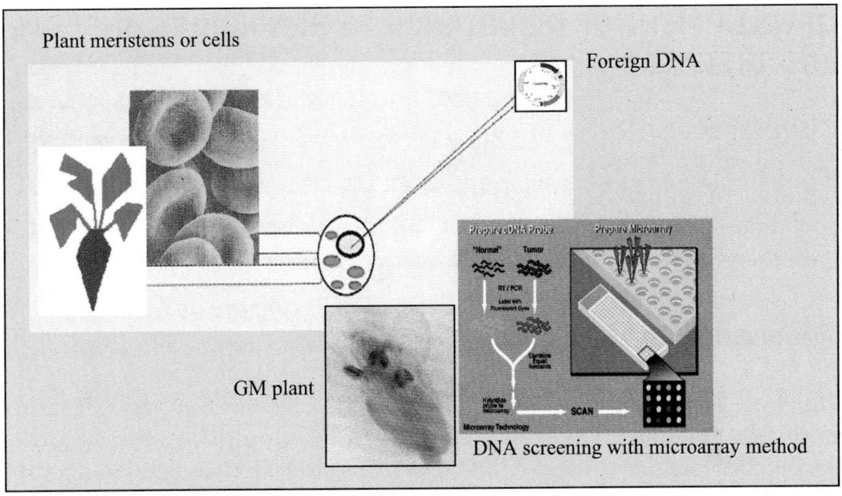

Fig. 2.3. Plant genetic transformation using "*microinjection method*" and screening with the microarray method

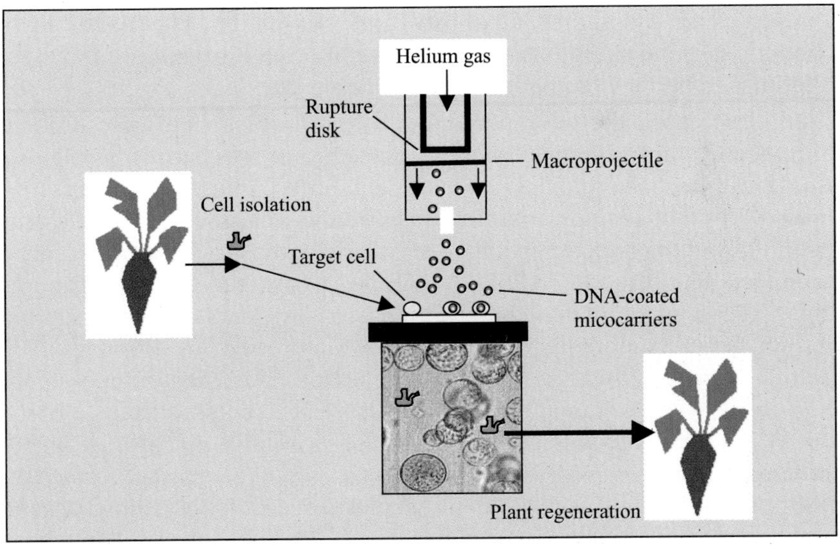

Fig. 2.4. Plant genetic transformation using the "*bioballistic (particle gun acceleration) method*"

come. Nowadays, nearly 1.2 billion people live in a state of absolute poverty, out of which 800 million people live under uncertain food security and 160 million children suffer from malnutrition (Daily et al., 1998). Genetically modified crops promise to meet this need as follows:

DEVELOPMENT OF RESISTANCE TO HERBICIDES, INSECTS AND DISEASES

Farmers greatly invest in agrochemicals, whereas consumers oppose consumption of food containing high amounts of these substances because of potential health hazards. GM crops partake benefits to consumers through affordable food and feed that require less pesticides, and, therefore, a more sustainable environment.

Resistance to Herbicides

During a 10-year period (1996-2005) since the first commercial GMO crop, herbicide tolerance has consistently been the dominant modified trait followed by insect resistance (FAO/WHO, 2001). Herbicide tolerant crops were put forward as a powerful new tool available for integrated weed management. The most commonly cited benefits to the producers include, a broader spectrum of weeds controlled with less herbicide carryover, the use of more environment friendly herbicides and crop management characterized by higher flexibility and simplicity. The well-known chimeric genes used for genetic transformation in horticultural crops for tolerance to herbicides are presented in **Table 2.1.**

In most cases, the tolerance refers to the herbicides glyphosate and glufosinate ammonium, which are broad-spectrum herbicides. Imidazolinone herbicides, which include imazapyr, imazapic, imazethapyr, imazamox, imazamethabenz and imazaquin, control weeds by inhibiting the enzyme acetohydroxyacid synthase (AHAS), also called acetolactate synthase (ALS). AHAS is a critical enzyme for the biosynthesis of branched-chain amino acids in plants. Several variant AHAS genes conferring imidazolinone tolerance were discovered in plants through mutagenesis and selection. Imidazolinone-tolerant horticultural crops (tomato and potato) have been commercialized based on AHAS gene variants (Tan et al., 2006). Similarly, *bar* and *pat* genes isolated from *Streptomyces hygroscopicus* and *S. viridochromogenes*, respectively, have been inserted into plants to encode phosphinothricin N-acetyltransferase to detoxify glufosinate. Glufosinate-tolerant crops have been commercialized using one of these two transgenes (De Block et al., 1987; Tan et al., 2006).

Glyphosate-resistant (GR) crops have been sold commercially in the USA since 1996. The use of glyphosate alone, or with conventional pre- and post-emergence herbicides with different modes of action, gives growers many options for affordable, safe, easy and effective wide-

Table 2.1. Transgenic horticultural crops for resistance to herbicides

Inserted Gene	*Gene Product*	*Gene Source*	*GM Species*	*Engineered Useful Trait*	*Reference*
PESPS	3-Phosphate synthase	*Petunia hybrida/Zea mays* *Salmonella typhimurium*	Potato, tomato and canola	Resistance to glyphosate	Zhou et al., 2006 Wang et al., 2003
CP4-EPSPS	5-Nolpyruvishikimate	*Agrobacterium* sp.	Tomato and cabbage	Resistance to glyphosate	Tan et al., 2006
gox -200	Glyphosate oxidoreductase	*Ochrobactrum anthropi*	Potato and tomato	Resistance to glyphosate	Tan et al., 2006
ALS/crs1-1	Acetolactate synthase	*Nicotiana tabacum/Oryza sativa, Zea mays*	Tomato	Resistance to chlorsulphuron	Tan et al., 2006
AHAS	Inhibitor of acetohydroxy acid synthase		Lettuce and canola	Tolerance to imidazolins	Tan et al., 2006
bxn	Bromoxynil specific nitrilase	*Klebsiella ozaenae*	Beet and carrot	Resistance to bromoxynil	Stalker et al., 1988 www.agbios.com
psbA	Chloroplastic energizer of QB enzyme (photosynthetic quinine-binding protein)	*Amaranthus hybridus*	Canola	Resistance to atrazines (diuron and tebuthiuron)	Mengistu et al., 2005
bar gene	Phosphinothricin acetyltransferase enzyme	*Streptomyces viridochromogenes*	Potato and chicorium	Resistance to glufosinate	Tan et al., 2006
pat gene	Phosphinothricin N-acetyltransferase	*Streptomyces hygroscopicus*	Potato	Resistance to bialaphos	De Block et al., 1987

spectrum weed control. Glyphosate affects aromatic amino acid biosynthesis by inhibiting 5-enolpyruvylshikimate-3-phosphate synthase (EPSPS).

A *gox* gene isolated from *Ochrobactrum anthropi* has been employed to encode glyphosate oxidoreductase to detoxify glyphosate in tomato and potato plants (Tan et al., 2006). Moreover, glyphosate-tolerant tomato and potato plants with EPSPS transgene alone, or both EPSPS and *gox* transgenes, have been commercialized (Tan et al., 2006).

The gene (*bar*) inserted into the radicchio lines (*Chichorium intybus*) encodes for phosphinothricin acetyltransferase (PAT) enzyme which inactivates phosphinothricin, the active component in the herbicide glufosinate (Tan et al., 2006).

Resistance to Insects

The representative example for resistance to insects is the insertion of *Bt* gene from *Bacillus thuringiensis* in a wide range of crop and horticultural plants **(Table 2.2)**. *B. thuringiensis* is a soil bacterium whose spores contain a crystalline (Cry) protein bearing deleterious effects on insects. The responsible gene for production of this protein (Cry I) was isolated and inserted into the plant's genome. When an insect is fed with this kind of plant tissues, this protein breaks down into the insect's gut to release a toxin, known as delta endotoxin. This toxin causes paralysis in the digestive system of the insects. Different versions of the *Cry* modified genes were identified and used against Lepidoptera, Coleoptera and Diptera **(Table 2.2)**.

The first transgenic plants with *Bt* genes were produced in 1987 (Barton et al., 1987). While most of the transgenic plants resistant to insects were produced by means of the *Bt δ-endotoxin* genes, many studies initiated for genes greatly affected protease inhibitors (CpTi), chitinases, secondary plant metabolites (α-amylase inhibitor genes) and lectins (Hinder and Boulter, 1999).

Transgenic plants expressing insecticidal proteins from the bacterium *Bt* were grown on over 13 million ha in the US and 22.4 million ha worldwide in 2004. Among vegetables, tomatoes are susceptible to damage from many insects, nematodes, and fungal, viral, and bacterial pathogens. Tobacco budworm (*Heliothis virescens*) and tomato fruitworm (*Helicoverpa zea*) are destructive pests of tomato. An insect-resistant tomato line was developed using rDNA techniques to express the insecticidal protein, Cry1Ac, encoded by the *cry1Ac* gene from the soil bacterium *Bt* subsp. *kurstaki* strain HD73 (Delannay et al., 1989). Three transgenic

Table 2.2. Transgenic horticultural crops for resistance to pests (insects)

Inserted Gene	*Gene Product*	*Gene Source*	*GM Species*	*Engineered Useful Trait*	*Reference*
Bt (*cry1*Aa)	Crystalline protein	*Bacillus thuringiensis* var. *kurstaki*	Tomato, canola,	Resistance to Lepidoptera	Zhao et al., 2005
Bt (*cry1*Ac)	Crystalline protein	*B. thuringiensis* var. *kurstaki*	Broccoli, chickpea, potato, walnut, vitis	Resistance to Lepidoptera	Zhao et al., 2005
Bt (*cry1*Ab)	Crystalline protein	*B. thuringiensis*	Potato	Resistance to Lepidoptera	Jansens et al., 1995
Bt (*cry1*B)	Crystalline protein	*B. thuringiensis*	Tomato, melon	Resistance to Lepidoptera	Fischhoff, 1989 Delannay et al., 1989
Bt (*cry1*C)	Crystalline protein	*B. thuringiensis*	Broccoli, squash, cabbage	Resistance to Lepidoptera	Zhao et al., 2005 Cho et al., 2001
Bt (*cry1*D)	Crystalline protein	*B. thuringiensis*	Strawberry	Resistance to Lepidoptera	Shelton et al., 2000
Bt (*cry1*F)	Crystalline protein	*B. thuringiensis*	Coffee tree, melon	Resistance to Lepidoptera, Diptera	Douches et al., 2004
Bt (*cryIII*A)	Crystalline protein	*B. thuringiensis* var. *kumamotoensis*	Eggplant, potato	Resistance to Coleoptera	Perlak et al., 1993
Bmk	Scorpion toxin	*Buthus martensii*	Rapeseed, potato	Resistance to Lepidoptera	Wang et al., 2005
C-II/PHV	Inhibitor of serine (protease)	*Glycine max*	Potato	Resistance to Lepidoptera, Coleoptera	Perlak et al., 1993
CpTi	Inhibitor of trypsine (protease)	*Phaseolus vulgaris*	Tomato, potato, sweetpotato, lettuce, Chinese cabbage, strawberry	Resistance to Lepidoptera, Coleoptera	Yang et al., 2002
aAI-Pv	Inhibitor of α-amylase	*P. vulgaris*	Bean, pea	Resistance to Coleoptera	Prescott et al., 2005
p-Lec	Lectin	*Pisum sativum*	Potato	Resistance to Homoptera, Coleoptera	Melander et al., 2003
BCH	Bean chitinase	*P. vulgaris*	Potato	Resistance to Homoptera, Lepidoptera	Down et al., 2001
GNA	Insecticidal lectin	*Galanthus nivalis*	Potato	Resistance to *Myzus* spp.	Down et al., 2003

potato cultivars ('Atlantic', 'Superior' and NewLeaf®), were genetically engineered to be resistant to attack by Colorado potato beetle (*Leptinotarsa decemlineata*). These lines were developed by introducing the *cry3A* gene, isolated from the common soil bacterium *Bt* subsp. *tenebrionis* (*Btt*) and inserted into the potato genome by *Agrobacterium*-mediated transformation (Douches et al., 2004). Furthermore, transgenic potato lines were genetically engineered to produce their own insecticide developed by introducing the *cry3A* gene, into the potato cultivar 'Russet Burbank' (Perlak et al., 1993). In several studies plant lectins proved promising as transgenic resistance factors against various insect pests. Certainly, pea seed lectin is a potential candidate for use against pollen beetle, a serious pest of *Brassica* species (Melander et al., 2003).

A novel insect-resistant gene combination, chitinase (*chi*) and BmkIT (*Bmk*), containing an insect-specific chitinase gene and a scorpion insect toxin gene was introduced into *Brassica napus* cultivar via *Agrobacterium*-mediated transformation for resistance to *Plutella maculipenis* (Wang et al., 2005).

Preventing or slowing down the evolution of resistance by insects is critical for the sustainable use of Bt crops. Plants containing two dissimilar *Bt* toxin genes in the same plant ("pyramided") displayed the potential to delay insect resistance. However, the advantage of pyramided Bt plants for resistance management may be compromised if they share similar toxins with single-gene plants that are simultaneously deployed. Zhao et al. (2005) combined broccoli plants transformed to express different Cry toxins (Cry1Ac, Cry1C, or both) and a synthetic population of the diamondback moth (*Plutella xylostella*) carrying resistant genes to *cry1*Ac and *cry1*C. When only two-gene plants were used in the selection, either none or few insects survived on one-or two-gene Bt plants, indicating that concurrent use of transgenic plants expressing a single and two *Bt* genes will be selected for resistance to two-gene plants more rapidly than the use of two-gene plants alone (Zhao et al., 2005).

Resistance to Diseases

Vegetable crops are grown worldwide as an indispensable source of nutrients and fibre in the human diet. Fungal plant pathogens can cause extensive devastation in these crops under appropriate environmental conditions. Vegetable producers confronted with the challenges of managing fungal pathogens have the opportunity of using fungi and yeasts as biological control agents (Punja, 2003). The potential of disease resistant plants is enormous, bearing in mind the human genome project about gene therapy. Thus, it has been under investigation during the last couple of years. Plants have evolved a large variety of sophisticated and effective defence mechanisms to prevent the colonization of their tissues with microbial pathogens (bacteria and mycorrhizae) and parasites (fungi and viruses). The incorporation of disease resistance genes into plants was initially carried out by means of classical breeding, thus yielding high crop productivity. However, currently, plant genetics and molecular biology are oriented towards clone identification and characterization of genes involved in disease resistance. Representative examples of GM horticultural cultivars are presented in **Table 2.3** and mechanisms of resistance were referred to are:

(a) Expression of gene products which are directly toxic to the pathogen (including hydrolytic enzymes, antifungal proteins, antimicrobial peptides and phytoalexins).

(b) Over expression of gene products that destroy or neutralize a component of the pathogen (lipase inhibition), thus reducing its growth into plant tissue.

(c) Expression of gene products that release signals regulating plant defence such as salicylic acid.

Enhancement of resistance genes involved in race specific interactions (not directly related to viruses) and hypersensitive response stands for another promising application. Notable examples to be mentioned are chitinase and ribosome inactivating protein genes (*RIP*) for fungal diseases in grapevine (Bornhoff et al., 2005) and melon (Brotman et al., 2002). Antisense technology of mRNA (Waterhouse et al., 1998; Beclin et al., 2002; Butterfield et al., 2002), cDNA coding virus satellite RNA (Zhang and Simon, 2003) and coat protein genes for resistance to viruses are some other innovative approaches **(Table 2.3)**.

Potato leafroll virus (PLRV) is a spherical RNA virus belonging to the luteovirus group and transmitted in a persistent manner by means of aphids, primarily the green peach aphid (*Myzus persicae*). The transgenic Russet Burbank NewLeaf® RBMT21-129, RBMT21-350 and RBMT22-082, were produced using recombinant DNA techniques and contained two novel genes, whose individual expression resulted in resistance to attack by Colorado Potato Beetle (CPB), and resistance to infection by PLRV. Introduction of DNA sequences corresponding to the ORF (Open Reading Frame)-1 and ORF-2 regions from PLRV conferred resistance to PLRV infection. These transgenic potato cultivars exhibit the trait of resistance to infection and subsequent disease caused by PLRV through an incompletely understood process that has been termed "replicase-mediated resistance", which may involve silencing of viral gene translation (Dunwell, 2000).

Table 2.3. Transgenic horticultural crops for resistance to diseases (fungi and viruses)

Inserted Gene	*Gene Product*	*Gene Source*	*GM Species*	*Engineered Useful Trait*	*Reference*
DAH53	Endochitinase	*Pisum sativum*	Potato, pea, bean	Resistance to *Fusarium solani*	Chang et al., 1995
STH2	STH proteins		Potato	Semi-resistance to *Phytophthora infestans*	Constabel et al., 1993
GOX	Glucose oxidase	*Aspergillus niger*	Potato, tomato	Delayed lesion due to *Phytophthora infestans*; resistance to *Alternaria solani* and *Verticillium dahliae*	Wu et al., 1995
PL3	Pectate lyase	*Erwinia carotovora* ssp. *Atroseptica*	Carrot	Resistance to *Erwinia carotovora* ssp. *atroseptica*	Wu et al., 1995
SB-37	Cercosporin		Potato	Resistance to *Erwinia carotovora*	Arce et al., 1999
HP	Hc-Pro	*Erwinia amylovora*	Potato, stone fruits	Resistance to *Phytophthora infestans*/ Semi-resistance to *Sharka* PP Virus	Li et al., 2005 Negri et al., 2005
Cf 9 Avr 9	Induce dependent hypersensitive response	*Lycopersicum esculentum*	Canola, tomato	Delayed disease development due to *Leptosphaeria maculans* Resistance against *Cladosporium fulvum*	Hennin et al., 2001 Punja, 2004
RIP	Ribosome inactivating protein gene		Grapevine, Melon	Resistance to *Oidium* spp. and *Leveillula taurica*	Bornhoff et al., 2005 Brotman, 2002
Hut	Antisense mRNA	mRNA	Sugarcane	Resistance to SrMV	Butterfield, 2002
CP SatC	Coat proteins of viruses	Viruses	Papaya, watermelon	Resistance to PRSV and CGMM viruses	Zhang and Simon, 2003 Park et al., 2005

Another example is the development of two papaya lines (55-1 and 63-1) using rDNA techniques to resist infection by Papaya Ring Spot Virus (PRSV) by inserting virus-derived sequences that encode the PRSV coat protein (CP). Furthermore, a squash line was developed using recombinant DNA techniques to resist infection by Cassava Mosaic Virus (CMV), Zucchini Yellow Mosaic Virus (ZYMV), and Watermelon Mosaic Virus (WMV)2 by inserting virus-derived sequences that encode the CPs from each of these viruses (Sharma and Lavanya, 2002). A chimeric satellite RNA (SatC) associated with Turnip crinkle virus (TCV) capable of expressing low levels of a defective CP was used for transformation. Since the TCV CP is a suppressor of RNA silencing, increased levels of resultant free CP could augment silencing suppression, thus resulting in enhanced colonization of the plant (Zhang and Simon, 2003).

The fungicidal class I endochitinases (EC) are associated with the biochemical defence of plants against potential pathogens. Chang et al. (1995) isolated and sequenced a genomic clone (DAH53), corresponding to a class I basic endochitinase gene in pea. The pea genome contains one gene corresponding to the chitinase DAH53 probe. Chitinase RNA accumulation was observed in pea pods within two to four hours after inoculation with the incompatible fungal strain *Fusarium solani* f. sp. *phaseoli* (Chang et al., 1995).

The *STH-2* gene was rapidly activated in potato leaves and tubers following elicitation or infection by *Phytophthora infestans.* In order to ascertain whether STH-2 protein is directly involved in the defence of potato against pathogens, the STH-2 coding sequence under the control of the cauliflower mosaic virus (CaMV) 35S promoter was introduced into potato plants (Constabel et al., 1993).

The tomato *Cf9* resistance gene induces an Avr9-dependent hypersensitive response (HR) in tomato and transgenic Solanaceaous species. Hennin et al. (2001) made investigations to find out whether the *Cf9* gene product responded functionally to the corresponding *Avr9* gene product when introduced in a heterologous plant species. They successfully expressed the *Cf9* gene under control of its own promoter and the *Avr9* or *Avr*9R8K genes under control of the p35S1 promoter in transgenic oilseed rape. Phytopathological analyses revealed that disease development of *Leptosphaeria maculans* was delayed when the pathogen was applied on an Avr9-mediated HR site. This research group demonstrated that the *CJ9* and *Avr*9 genes can be functionally expressed in a heterologous plant species and that the two components confer an increase in disease resistance (Hennin, 2001).

Cladosporium fulvum (syn. *Passalora fulva*) is a biotrophic fungal pathogen causing leaf mould on tomato (*Solanum esculentum*). The fungus grows exclusively in the tomato leaf apoplast where it secretes several small (<15 kDa, i.e., kilodalton) cysteine-rich proteins that are thought to play a role in disease establishment. To investigate the role of these proteins, a targeted proteomics approach was undertaken (Van Esse et al., 2006). *C. fulvum* proteins were expressed as recombinant fusion proteins carrying various affinity-tags at either their C- or N-terminus. Although these fusion proteins were correctly expressed and secreted into the leaf apoplast, detection of affinity-tagged *C. fulvum* proteins failed, and affinity purification did not result in the recovery of these proteins. However, when using *C. fulvum* effector protein-specific antibodies, specific signals were obtained for various proteins (Van Esse et al., 2006). The application of the hairpin-mediated RNA silencing technology for obtaining resistance to Plum pox virus (PPV) was applied to *Nicotiana benthamiana* and *Solanum tuberosum* plants. The transgenic plants, resistant to PPV infection, indicated that the RNA silencing of PPV suppressor P1/HCPro sequences results in an efficient and predictable PPV resistance, which may be utilized in obtaining stone fruit plants resistant to the devastating *Sharka* disease (Negri et al., 2005). Moreover, commercially important watermelon varieties were developed using a cDNA encoding the Cucumber Green Mottle Mosaic Virus (CGMMV) *CP* gene and successfully transformed a watermelon rootstock named 'gongdae'. The transformation rate was as low as 0.1-0.3%, depending on the transformation method used (ordinary co-culture vs injection, respectively). This was the first report of GM watermelons resistant to CGMMV infection (Park et al., 2005). In *Vitis vinifera* cv. Seyval blanc was transformed with *Agrobacterium tumefaciens* strain LBA 4404 carrying genes for chitinase and RIP (ribosome-inactivating protein) in an attempt to improve fungal resistance (Bornhoff et al., 2005).

Several commercially available products have shown considerable disease reduction through various pathogen and disease reduction mechanisms. Production of hydrolytic enzymes and antibiotics, competition for plant nutrients and niche colonization, induction of plant host defence mechanisms, and interference with pathogenicity factors in the pathogen stand for the most important mechanisms. Other approaches were adopted towards developing disease resistant transgenic plants to augment systemic acquired resistance (SAR) or generic immune response to many pathogens. Biotechnological techniques are becoming increasingly important towards elucidating the mechanisms of action of fungi and yeasts, and providing genetic characterization and molecular markers to monitor the spreading of these agents (Punja, 2003).

TOLERANCE TO ABIOTIC STRESSES

There are many proposed transgenic routes for the improvement of stress related responses in plants. Progress was recently made in the identification and characterization of mechanisms allowing plants to withstand high salt concentrations. The development of salt tolerant plants is crucial for enhancing the world's overall agricultural output. It is known that one third of irrigated land all over the world contains prohibitive salt levels. In order to produce a salt tolerant plant, the latter must either retain Na^+ ions which infiltrate into its cells, or find a place to store these ions where they cannot interfere with the plant's metabolism.

Understanding metabolic fluxes and the main constraints for production of compatible solutes opens up the possibility of genetically engineering entire pathways that could lead to osmoprotectant production. This, together with the identification of different sodium transporters that could provide the needed ion homeostasis during salt stress, opens up the possibility of engineering crop plants with improved salt tolerance (Apse and Blumwald, 2002).

The over expression of various glutamate dehydrogenases (GDH) that is involved in directing the transportation of Na^+ ions into vacuole, was reported by Apse et al. (1999) and would be applied in some horticultural crops. As an example, tomato plants transformed with genes from alga *Chlorella sorokiniana*, showed salt tolerance and a better performance in cultivation with improved plant growth under stress conditions (Apse and Blumwald, 2002).

Another example is the regulation of two enzymes closely related with trehalose expression. Trehalose is a substance produced by trehalose phosphate synthase (*TPS*) gene. Plant transformation with this gene activated two enzymes (trehalose phosphate and trehalose phosphate phosphatase) expressed by means of forming larger leaves, altering growth and improving tolerance to stress factors. Transgenic tomato lines with various levels of trehalose are currently being field tested (Pilon-Smits et al., 1998). An alternative approach deals with the introduction of uridine diphosphate glucose pyrophosphorylase (*UDGP*) gene from the bacterium *Acetobacter xylinum*. These transgenics were claimed to have increased growth rates and yield, and improved response to stress conditions (Dunwell, 2000).

Plant tolerance to high temperature is very important because it offers the possibility for some horticultural crops to be grown in areas up to now unavailable for horticulture. It was found that this kind of tolerance is considerably influenced by trienoic fatty acids common in chloroplast membrane. Plants grown in colder temperatures have higher content of trienoic acids. Transgenic plants of *Nicotiana tabacum* and *Lathyrus sativa*, in which the gene encoding chloroplast omega-3 fatty acid desaturase was silenced, had as a consequence a lower level of trienoic fatty acids than wild type plants and better acclimation to higher temperature climates (Murakami et al., 2000). Probably the reverse transformation process in some horticultural crops would have the stimulation of omega-3 fatty acids, which have beneficial effect on human health. Another published claim involves the introduction of a gene encoding a plant farnesyltransferase. Also inhibitors of this enzyme, when expressed in plants, will most likely enhance drought tolerance, improve resistance to senescence and modify the growth habit (Dunwell, 2000).

METABOLIC ENGINEERING AND IMPROVEMENT OF CROP CHARACTERISTICS SUCH AS YIELD AND NUTRITIVE VALUE

One of the most significant contributions of genetic engineering to agriculture is the improvement of nutritional status of horticultural crops. Several nutritional traits such as, carbohydrates, proteins, oils, fats, vitamins and amino acids constitute the main target of genetic engineering **(Table 2.4)**. An experimental approach to increase crop yield is to modify plant biochemical processes in order to introduce the C4 photosynthetic system into commonly cultivated C3 plants. Matsuoka et al. (1994) transferred the gene for PEPC (phosphoenolpypuvate carboxylase) from *Zea mays* into *Oryza sativa* using the *Agrobacterium tumefaciens* - transformation method. Physiologically, rice plants exhibited reduced O_2 inhibition and had increased photosynthetic products into their seeds. The same application in tomato has shown increased foliar sucrose/starch ratios in leaves and decreased amounts of foliar carbohydrates when grown with CO_2 enrichment (Signora et al., 1998). Another attempt was made by means of sucrose phosphate synthase (SPS), a key enzyme in the regulation of sucrose metabolism. Transgenic plants expressing SPS controlled by a special promoter were developed in some horticultural crops (Hellwege et al., 2000). Furthermore, the introduction of inorganic pyrophosphatase from *Escherichia coli* (Schaffer et al., 1997), or modification of hexokinases was used to alter the amount of sugar in developing legume seeds. This application opened new horizons for altering the chemical composition of plant food products to meet specific requirements. Transgenic modifications were carried out for altering the ratio of amylase to amylopectin in starch of potatoes and corn (Schwall et al., 2000). Modification of metabolites activity of TCA (tricarboxylic acid) cycle by reducing the amount of the NAD (nicotinamine adenine dinucleotide)-malic enzyme can also be exploited

Table 2.4. Transgenic horticultural crops for nutritive and quality traits

Inserted Gene	*Gene Product*	*Gene Source*	*GM Species*	*Engineered Useful Trait*	*Reference*
ADP	Glucose pyrophosphorylase	*Escherichia coli*	Rice, potato	Increased starch content	Schaffer and Petreikov, 1997
SBEA/B	Inhibition of SBE A/B	*Pisum sativum*	Potato	Very high amylase content of starch	Schwall et al., 2000
SST–1	Sucrose:fructose syltransferase	*Cynara scolymus*	Potato	Increased insulin content	Hellwege et al., 2000
FFT–1	Fructan:fructan 1-fructosyltransferase	*Cynara scolymus*	Potato	Enhanced fructan content	Hellwege et al., 2000
Sgt	Antisense sterol glucotransferase		Potato	Enhanced fructan content Regulated starch content	McCue et al., 2004
ACC synthase	Aminocyclopropane-1-carboxylic acid	*Pseudomonas chlororaphis*	Tomato	Delayed softening/ reduced pectin degradation	Xiong et al., 2003
PG/CR3	Antisense polygalacturonase	*Lycopersicum esculentum*	Tomato	Delayed softening/ reduced pectin degradation	Smith et al., 1990 Good et al., 1994
SF-1 gene	Sucrose fructosyl transferase		Tomato	Enhanced fructan content	Smeekens and Rook, 1997
A-Toc	Overexpression of γ-tocopherol	*Arabidopsis thaliana*	Potato, canola	Enhanced vitamin E content	Sintani and De la Penna, 1998
Δ6/Δ12	Linoleic (Δ6/) desaturate activity Fatty acid dehydrogenase		Canola	Increased γ-linoleic acid content	Lee et al., 1991
Fl 3p/5p	Flavonoid hydrolase	*Petunia hybrida/ Viola* ssp.	Eggplant, avocado, tomato carnation	Modified colour of fruits	Sompornpailin et al., 2002
Ch B2	Fat B2,thioesterase	*Cuphea hookeriana*	Canola	Greater number of medium chain fatty acids	Dehesh et al., 1996

for increasing starch concentrations in potato (McCue et al., 2004). Other reported approaches resulted in decreasing the oligosaccharide amounts (raffinose and stachyose), thus improving the digestibility in humans and animals (Signora et al., 1998).

Metabolic engineering is the modification of endogenous metabolic pathways to increase flux towards particular desirable molecules, or divert flux towards the synthesis of new molecules (Slater et al., 2003). In vegetables many useful substances were detected like phytochemicals which proved to be highly protective for the human body from several chronic diseases (e.g., diabetes). It is very important to identify the genetic mechanisms regulating the synthesis of phytochemicals such as flavonoids, glucisinolates and thiosulfides. Since each vegetable group contains a unique combination of phytochemicals, it is necessary to develop new cultivars endowed with a variety of specific substances so as to ensure a proper mixture being introduced in the human diet (Kushad et al., 1999). Furthermore, an increase in concentration of lycopene or carotenoids in species like tomato or carrot, should be only with a more sophisticated proposition for the transfer of these compounds into species where such compounds do not exist (e.g., potato and melon).

Nowadays, the main target for food-grade oils is high oleic and low linolenic acid content. Success in applying appropriate promoters and targeting expression to oil bodies was reported (Lee et al., 1991). Antisense technology was used to effectively modify the degree of fatty acid saturation. Plant varieties of low linolenic acid minimize the need for partial hydrogenation, a process routinely used towards prolonged shelf life and flavour stability in fried, baked, snack and other processed foods. More recently, the focus of metabolic engineering shifted from a single gene to multiple gene strategies and novel methods were developed with the objective of coordinated regulation of larger groups of genes, thus controlling agronomically and nutritionally industrial or pharmaceutical metabolites (Capell and Cristou, 2004).

Tocopherols (vitamin E) are lipophilic antioxidants presumed to play a key role in protecting chloroplast membranes and the photosynthetic apparatus from photooxidative damage. Additional non-antioxidant functions of tocopherols were suggested after the recent finding that *Suc exports defective1* mutant (*sxd1*) in maize (*Zea mays*) carries a defect in tocopherol cyclase (TC) and thus is devoid of tocopherols. However, the corresponding *vitamin E deficient1* Arabidopsis mutant (*vte1*) lacks a phenotype analogous to *sxd1*, suggesting differences in tocopherol function between C4 and C3 plants. Therefore, the potato (*Solanum tuberosum*) ortholog of SXD1 was isolated and functionally characterized. *StSXD1* encoded a protein with high TC activity *in vitro*, and chloroplastic localization was demonstrated by transient expression of green fluorescent protein-tagged fusion constructs. RNAi (ortholog ribonucleic acid of *SXD1*) mediated silencing of *StSXD1* in transgenic potato plants resulted in the disruption of TC activity and severe tocopherol deficiency similar to the orthologous *sxd1* and *vte1* mutants (Hofius et al., 2004). It was observed that nearly complete absence of tocopherols caused a characteristic photoassimilate export-defective phenotype comparable to *sxd1*, which appeared to be a consequence of vascular-specific callose deposition observed in plant leaves. Also, CO_2 assimilation rates and photosynthetic gene expression decreased in source leaves in close correlation with excess sugar accumulation, thus suggesting a carbohydrate-mediated feedback inhibition rather than a direct impact of tocopherol deficiency on photosynthetic capacity. These data provide evidence that tocopherol deficiency leads to impaired photoassimilate export from source leaves in both monocot and dicot plant species and suggest significant differences among C3 plants in response to tocopherol reduction (Hofius et al., 2004).

Steroidal glycoalkaloids (SGAs) are secondary metabolites of Solanaceous plants. An antisense transgene was constructed to down-regulate glycoalkaloid biosynthesis using a potato cDNA encoding a solanidine glucosyltransferase (SGT1). Introduction of this construct into potatoes resulted in an almost complete inhibition of α-solanine accumulation and elevated levels of α-chaconine, resulting in wild type total SGA levels in the transgenic lines (McCue et al., 2004).

High-amylose starch is in great demand by the starch industry because of its unique functional properties. However, very few high-amylose crop varieties are commercially available. Schwall et al. (2000) described the generation of very-high-amylose potato starch by means of genetic modification. This was achieved by simultaneously inhibiting two isoforms of starch branching enzyme to below 1% of the wild type activities. Starch granule morphology and composition were shown to be noticeably altered. The normally occurring high-molecular-weight amylopectin was absent, whereas the amylose content increased to levels comparable to the highest commercially available maize starches. Moreover, the phosphorus content of starch increased more than fivefold. This unique starch, with its high amylose, low amylopectin, and high phosphorus levels, is very promising for food and industrial applications (Schwall et al., 2000).

POSTHARVEST QUALITY AND SHELF LIFE PROLONGATION

The conservation of quality traits in fruits and vegetables by prolonging shelf life but without loss of their nutritive values, is another useful approach. Delayed softening of vegetables and fruits was obtained by inserting a truncated version of the polygalacturonase (PG) encoding gene in the sense or anti-sense orientation in order to reduce expression of the endogenous PG, and thus reducing pectin degradation. The insertion of an additional copy of PG encoding gene into tomato's genome led to similar results (Kou et al., 2004). The Flavr Savr^{TM} tomato line was genetically engineered to express delayed softening by insertion of an additional copy of the PG encoding gene in the "antisense" orientation, resulting in reduced translation of the endogenous PG mRNA. The antisense PG gene was essentially a reverse copy of part of the native tomato PG gene that effectively suppressed the expression of endogenous PG enzyme prior to the onset of fruit ripening (Smith et al., 1990). The mechanism of decreased PG activity in Flavr Savr^{TM} tomato is most likely linked to the hybridization of antisense and sense mRNA transcripts, resulting in decreased amount of free positive sense mRNA available for protein translation. Reduced PG expression decreases the breakdown of pectin, thereby resulting in fruit with slowed cell wall breakdown, better viscosity characteristics and delayed softening.

A tomato line (1345-4) was developed using recombinant DNA techniques to express the trait of delayed ripening of tomato fruit. The transgenic line contained a truncated version of the tomato 1-aminocyclopropane-1-carboxylic acid (ACC) synthase gene. This endogenous enzyme was responsible for the conversion of S-adenosylmethionine (SAM) to ACC, the immediate precursor of ethylene (Good et al., 1994). Genetic modification for a reduced accumulation of S-adenosylmethionine hydrolase reducing the ethylene biosynthesis in *Cucumis melo* L. (http://www.agbios.com). Furthermore, indirect approaches for preventing the softening and senescence of fruits are the regulation and distribution of growth regulators (GthR). As an example, the cytokinin is a plant hormone naturally preventing senescence and thereby maintaining photosynthetic activity in leaves. The process of leaf senescence can be blocked through a gene (isopentyltransferase) encoding the cytokinin synthesis enzyme. Introduction of farnesyl transferase and isopentyltransferase (IPT) genes into some horticultural plants delays senescence, resulting in higher plant vegetative growth, seed and fruit production and prolonging the shelf life of vegetables (Pei et al., 1998). As an example, GM horticultural varieties with low linolenic acid are characterized by lesser need for partial hydrogenation. Therefore, this stands for a process routinely used to increase shelf life and flavour stability in fried, baked, snack and other processed foods (Lee et al., 1991). Genetically modified cultivars, which delay ripening by either down-regulating ethylene production or disrupting ethylene perception, were developed in tomato and broccoli.

This had previously succeeded in papaya by an insertion of additional copy or copies of the native capacs1 and capacs2 genes in a sense and antisense orientation, resulting in decreased ethylene production due to inhibition of ACC (Smith et al., 1990; Yang et al., 2003). Another tentative application in horticultural plants comes from a transformation in papayas by insertion of a gene (*etr*1) from *Arabidopsis thaliana.* Plants carrying this non-functional gene would be insensitive to ethylene, thus exhibiting delayed fruit ripening and floral senescence (Kou et al., 2004). The presence of the truncated ACC synthase gene suppresses the normal expression of the native ACC synthase gene, and while not completely understood, the mechanism of "down-regulation" is considered to be linked to the coordinate suppression of transcription of both the endogenous gene and the introduced truncated ACC synthase gene. The in situ accumulation of ethylene in the transgenic tomatoes was only about 1/50 the level found in the unmodified parental line and the fruit does not fully ripen unless an external source of ethylene is applied (http://www.agbios.com). Two lines of *Cucumis melo* L. were developed through a specific genetic modification to express a reduced accumulation of SAM and consequently the trait of delayed ripening. The conversion of SAM to ACC is the first step in ethylene biosynthesis and the lack of sufficient pools of SAM results in significantly reduced synthesis of this phytohormone, which is known to play a key role in fruit ripening (Xiong et al., 2003)

MOLECULAR "FARMING" AND DEVELOPMENT OF BIOTECH VACCINES AND GM PHARMACEUTICALS

The new branch of plant biotechnology defined as "molecular farming" is mainly focused on the exploitation of plants of agronomic relevance such as factories for large-scale production of biomolecules. Plants can be used for the large-scale production of a variety of recombinant proteins destined for agro-industrial, biomedical and pharmaceutical applications **(Table 2.5)**. In 1991, the World Health Organization (WHO) challenged scientists to develop simpler, safer and less expensive methods for vaccinating the people from the "Third World" (FAO/WHO, 2001). One of the proposed axis was the use of plant vaccines. Plants produce various secondary compounds and metabolites with pharmaceutical properties. If someone extends the expression system of plants, more efficient biofactories than bacteria or yeast cells for the production of high value molecules should be considered. Plant metabolic systems offer several unique advantages. First of all, plant cells are able to produce proteins structurally and functionally equivalent to those synthesized in mammalian cells, thus guaranteeing high quality products free of human

Table 2.5. Antibodies, antigens and recombinant proteins expressed in transgenic plants

Antibody/Antigen	*Specificity/Pathogen*	*Transgenic Species*	*Reference*
scFv	Carciniembryonic antigen	*Lycopersicum esculentum, Pisum sativum*	Stoger et al., 2000
Coat protein epitope	Human hepatitis B virus	*Solanum tuberosum, Lactuca sativa*	Thanavala et al., 2005
Coat protein NV	Norwalk virus	*Musa* spp.	Tacket et al., 2005
Heat labile enterotoxin B subunit	Enterotoxic *Escherichia coli*	*Solanum tuberosum*	Tacket et al., 1998
Glycoprotein	Rabies virus	*Lycopersicum esculentum*	Mc Garvey et al., 1995
Cholera Toxin B subunit	*Vibrio cholerae*	*Solanum tuberosum*	Arakawa et al., 1998
hGAD65	Diabetes associated autoantigen	*Lycopersicum esculentum*, *Daucus carota*	Jiang et al., 2007 Avesani et al., 2003
Recombinant Protein			
HG	Human growth hormone	*Solanum tuberosum*	Sijmons et al., 1990
HSA	Human serum albumin	*Solanum tuberosum*	Sijmons et al., 1990
ACh	Human acetylcholinesterase	*Lycopersicum esculentum*	Sijmons et al., 1990
B-casein		*Solanum tuberosum*	Chong et al., 2000
hLF	Human lactoferin	*Solanum tuberosum*	Chong et al., 2000

pathogens (Tacket, 2005). In fact, blood components, hormones, enzymes, cytokines, antibodies and vaccines were successfully produced in plants **(Table 2.5)**.

The first recombinant protein to be synthesized in plants was the human growth hormone (Barta et al., 1986). The possibility of obtaining antibody expression in plants was first demonstrated by Hiatt et al. (1989). The typical structure of these glycoproteins consists of four polypeptides, two "heavy" chains and two "light" chains, linked together with disulfide bridges and non-covalent bonds. These complex proteins are involved in the specific recognition of antigens (substances extraneous to the organism able to induce the activation of the immune response). The antibody expression levels in transgenic plant systems are variable, depending on the plant species used and the tissue where it is expressed. A stable response was exploited in an attempt to express several antigens in potato (*Solonum tuberosum*), lettuce (*Lactuca sativa*), tomato (*Lycopersicum esculentum*) and carrot (*Daucus carota*) **(Table 2.5)**.

A modified CaMV 35S promoter was used to direct the expression of chimeric genes encoding human serum albumin (HSA) in transgenic potato and tobacco plants. To secrete the protein, either the human prepro-sequence or the signal sequence from the extracellular pathogenesis related to tobacco protein (PR-S) was used. Expression of the human preproHSA gene led to partial processing of the precursor and secretion of proHSA. Fusion of HSA to the plant PR-S presequence resulted in cleavage of the presequence at its natural site and secretion of correctly processed HSA that is indistinguishable from the authentic human protein (Sijmons et al., 1990).

A cDNA fragment encoding human lactoferrin (hLF) linked to a plant microsomal retention signal peptide (SEK-DEL) was stably integrated into the *Solanum tuberosum* genome by *Agrobacterium tumefaciens* transformation method (Chong and Langridge, 2000). The lactoferrin gene was expressed under control of both the auxin-inducible manopine synthase (mas) P2 promoter and the CaMV 35S tandem promoter. The presence of the hLF cDNA in the genome of regenerated transformed potato plants was detected by polymerase chain reaction (PCR) amplification methods. Antimicrobial activity against four different human pathogenic bacterial strains was detected in extracts of lactoferrin-containing potato tuber tissues. This was the first report of synthesis of full length, biologically active hLF in edible plants (Chong and Langridge, 2000).

Furthermore, transgenic potatoes were engineered to synthesize a cholera toxin B subunit (CTB) pentamer with affinity for ganglioside mimics oligosaccharide (GMI)-ganglioside. Both serum and intestinal

CTB-specific antibodies were induced in orally immunized mice. Mucosal antibody titters declined gradually after the last immunization but were restored following an oral booster of transgenic potato. The cytopathic effect of cholera holotoxin (CT) on Vero cells was neutralized by serum from mice immunized with transgenic potato tissues. Following injection with CT, the plant immunized mice showed up to a 60% reduction in diarrhoeal fluid accumulation in the small intestine. Protection against CT was based on inhibition of enterotoxin binding to the cell-surface receptor GMI-ganglioside. These results demonstrate the ability of transgenic food plants to generate protective immunity in mice against a bacterial enterotoxin (Arakawa et al., 1998).

The smaller isoform of the enzyme glutamic acid decarboxylase (GAD65) is a major islet autoantigen in autoimmune type 1 diabetes mellitus (T1DM). Transgenic plants expressing human GAD65 (hGAD65) are a potential means of direct oral administration of the islet autoantigen in order to induce tolerance and prevent clinical onset of disease. Avesani et al. (2003) reported the successful generation of transgenic tobacco and carrot that express immunoreactive, full-length hGAD65. Also, they tested the hypothesis if the expression levels of recombinant hGAD65 in transgenic plants can be increased by targeting the enzyme to the plant cell cytosol and by mediating expression through the potato virus X (PVX) vector. By substituting the NH2-terminal region of hGAD65 with a homologous region of rat GAD67, a chimeric GAD67 (1-87)/GAD65 (88-585) molecule was expressed in transgenic tobacco plants. Transient expression of wild-type, full-length hGAD65 in *Nicotiana benthamiana* mediated by PVX infection was associated with expression levels of immuno-reactive protein as high as 2.2% of total soluble protein. This substantial improvement of the expression of hGAD65 in plants paves the way for immuno-prevention studies of oral administration of GAD65-containing transgenic plant material in animal models of spontaneous autoimmune diabetes (Avesani et al., 2003).

Transgenic plants present a novel system for both production and oral delivery of vaccine antigens. Production of protein antigen in food plants is substantially cheaper than production in bacterial, fungal, insect cell, or mammalian cell culture. Edible plants themselves can also serve as the oral vaccine delivery system. Phase-1 studies have been conducted on: (1) raw transgenic potatoes expressing the B subunit of *Escherichia coli* heat labile enterotoxin (LT-B), (2) potatoes expressing Norwalk virus capsid protein, and (3) defatted corn germ meal expressing LT-B. New oral vaccines based on other transgenic plants will soon be evaluated on humans (Tacket, 2005). Mc Garvey et al. (1995) engineered tomato plants (*Lycopersicum esculentum* Mill var. UC82b) to express a gene for the glycoprotein (G-protein), which coats the outer surface of the rabies virus.

The recombinant constructs contained the G-protein gene from the ERA strain of rabies virus, including the signal peptide, under the control of the 35S promoter of CaMV. Plants were transformed by *Agrobacterium tumefaciens*-mediated transformation of cotyledons and tissue culture on selective media. The G-protein expressed in tomato appeared as two distinct bands with an apparent molecular mass of 62 and 60 kDa, as compared to the 66 kDa observed for G-protein from virus grown in BHK type of cells. These results administered the rabies G-protein from the same ERA strain elicits protective immunity in animals (Mc Garvey et al., 1995). Some examples of transgenic horticultural plants will hopefully provide a valuable tool for the development of edible oral vaccines (Falorni et al., 2005; Giddings, 2000).

GENETIC ENGINEERING IN FRUIT TREES

The application of genetic engineering in perennial horticultural trees is very difficult because of their extended biological cycle and long period for fruit production (two to five years) from the time of transplantation. The control and selection processes are time consuming and unreliable for progeny test, especially for characteristics related to quality of fruits. In this case, it is necessary to transfer together the transgene with a marker gene closely related to early selectable phenotypic characteristics. Furthermore, most of perennial fruit crops are cross-pollinated or self-incompatible, thus the presence of other cultivars used as pollinators is necessary in the same field. Usually the genetic modification for a useful trait is "one genotype event" and was applied in cross-pollinated species in case of hybrid production, but is not possible in fruit trees (female and male parents). Many times the commercial fruit cultivars are grafted on a wild species or another cultivar which influences the physiology and quality characteristics of fruits. Especially in this case, interactions among introduced genes need to be considered. It is known that repeats of some genetic elements (transposable) can trigger gene silencing. These are some difficulties for broadening the application of genetic transformation into fruit crops. Most of GM applications referred to species which were multiplied by clonal propagation (like *Musa* spp., *Vitis vinifera, Fragaria* spp. etc.) (Tacker et al., 1998). However, in these species the genetic modification is more convenient because their propagation takes place facultative by means of *in vitro* culture, which is used for some of the transformation methods. Research managed to enhance resistance in pests and viruses, nutrition and consumer-oriented commodities (Bornhoff et al., 2005; Zhao et al., 2005). The major contribution was the development of virus resistant fruit trees, like apple, citrus, plum, walnut, coffee tree and grapevine **(Table 2.6)**.

Table 2.6. Number of GM cultivars in horticultural crops which were under field testing in European countries (2002-2005)

Species	*A*	*B*	*D*	*DK*	*ES*	*FL*	*FR*	*GB*	*GR*	*IR*	*IT*	*NL*	*PR*	*S*	*Total*
Actinidia deliciosa											3				**3**
Beta vulgaris var. *cicla*			1												**1**
Brassica juncea		3													**3**
Brassica napus		50	38	4	3	2	116	105			4	15		26	**363**
Brassica oleraceae						1									**1**
B. oleraceae var. *botrytis*		5			1							2			**8**
Chicorium intybus		15					5			3	9				**32**
Citrullus lanatus											1				**1**
Citrus sinensis					1										**1**
Coffea canephora							1								**1**
Cucumis melo					4		3				1				**8**
Cucurbita pepo					2		1				3				**6**
Daucus carota												3			**3**
Lactuca sativa							7				1				**8**
Lycopersicum esculentum					16		5	1	1		46	2	2		**73**
Malus domestica		1										1		1	**3**
Olea eurorea					4		1				8				**13**
Pisum sativum			1												**1**
Prunus domestica					1										**1**
Prunus avium											3				**3**
Solanum melongena											8				**8**
Solanum tuberosum	2	2	46	10	10	3	12	36			7	50	4	24	**206**
Vitis vinifera			1				3				1				**5**
	2	**76**	**87**	**14**	**42**	**6**	**154**	**142**	**1**	**3**	**95**	**73**	**6**	**51**	**752**

A: Austria, B: Belgium, D: Deutschland, DK: Denmark, ES: Spain, FL: Finland, FR: France, GB: United Kingdom, GR: Greece, IR: Irleand, IT: Italy, NL: Netherlands, PR: Portugal, S: Switzerland

FUTURE PERSPECTIVES

Between the years 1992 to 2002, the total grown cultivated area with GMO crops has increased 30 times (James, 2000). The estimated global area of approved biotech crops for 2006 amounted to 102 million hectares (http://www.isaa.org). According to ISAAA status report (http://www.isaaa.org). GM crops were grown by approximately 10.3 million farmers in 22 countries (USA, Argentina, China, India, Canada, Paraguay, South Africa, Australia, Mexico, Uruguay, Colombia, Philippines, Brazil, Honduras, Iran, France, Germany, Czech Republic, Romania, Slovakia Portugal and Spain) (http://www.bio.org). It is worth mentioning that only four of them represent the 95% of the total GM plant production. The main producers are USA [54.6 million hectares (mh)], Argentina (18 mh), Brazil (11.5 mh), Canada (6.1 mh), India (3.8 mh), China (3.5 mh), Paraguay (2 mh) and South Africa (1.4 mh) (http://www.gmocompass.org). In 2004 the 99% per cent of the above hectarage was covered only by four commercialized GM crops (soybean, cotton, maize, and canola). The increased hectarage in five developing countries, the cultivation of GMO in Spain and countries like, Hungary, Poland, Bulgaria and Romania, which have recently joined the European Community or are expected to join in the period 2007-2010, have gathered with Appelation d'Origin Protegée (AOP) and are likely expand the use of GMO worldwide in the very near future.

However, in the case of horticultural crops, the anticipated increase will be dependent on the more conservative approach adopted. Except for GM tomatoes, potatoes and canola developed and commercialized between the years 1996-2000 (James, 2000), four new GM products (squash, carnation, melon, sweet corn and papaya) were available commercially at the USA market in 2003 (http://www.ific.org). Sixty-one different crops were under global experimentation in 2004 and 18 different traits were under study at the experimental phase. Also, in Europe, 750 new horticultural GM cultivars were under field testing during the period 2002-2005 (**Table 2.6**). It becomes evident from **Table 2.6** that most EU countries can be tentatively divided in two groups: (a) countries like France, Germany, UK, Italy, Spain and the Netherlands with a high number of experimental GM cultivars, and (b) Austria, Greece, Finland and Ireland, which could be described as GMO free zones.

The future perspective in GM horticultural crops is the pyramiding of different transgenes into the same cultivar for increasing the protective effectiveness, widening activity range and durability of resistance (Zhao et al., 2005). Furthermore, GM horticultural crops can be manipulated for increasing their resistance to insects or viruses and produce completely artificial substances from the precursors to plastics and from pharmaceutical substances to vaccines. The new GM horticultural crop generation might be accompanied with a "clean vector technology" to produce GM plants with only the gene of interest as a newly introduced gene function with a plant derived promoter and without any superfluous gene or selectable marker sequences.

Like all new technologies, several risks, both known and unknown, are involved. Controversies regarding GM crops and foods commonly focus on human and environmental safety. It is important to identify the potential hazards as well the risks and any license conditions imposed to manage the identified risks. There was a brief interlude where Monsanto experimented with introducing a new technology by using a terminator into food crops, which produced plants with sterile seeds. Monsanto claimed that this is as necessary as protecting their intellectual property rights.

Genetic modification might also be the cause for some rearrangements in plant genome or activation of silence genes and transposonable elements, with unknown consequences. These potential risks are focused on:

- Toxicity and allerginicity for human and other organisms
- Weediness and spread into environment
- Gene flow through related species
- Horizontal transfer of undesirable genes
- Gene rearrangement and activation or regulation of transposable elements
- The patentability of DNA sequences (genes, plasmids, promoters, terminators, etc.)

Furthermore, it is anticipated that issues like legislation for GM food products, respect for consumer's choice for safe foods, intellectual property rights and bioethics (discussed in Chapter 1 in this Volume), are some of the most important issues to be ensured in an attempt to avoid disastrous consequences in the future.

REFERENCES

Apse, M.P., Aharon, G.S., Snedden, W.A. and Blumwald, E. (1999). Salt tolerance conferred by over expression of a vacuolar Na^+/H^+ antiport in *Arabidopsis*. Science 285(5431): 1256-1258.

Apse, M.P. and Blumwald, E. (2002). Engineering salt tolerance in plants. Curr. Opin. Biotechnol. 13(2): 146-150.

Arakawa, T., Chong, D.K. and Langridge, W.H. (1998). Efficacy of a food plant-based oral cholera toxin B subunit vaccine. Nat. Biotechnol. 16(5): 292-297.

Arce, P., Moreno, M., Gutierrez, M., Gebauer, M., Dell'Orto, P., Torres, H., Acuna, I.., Oliger, P., Vegenas, A., Jordana, X., Kalazich, J. and Holuighe, J. (1999). Enhanced resistance to bacterial infection by *Erwinia carotovora* spp. *atroseptica* in transgenic potato plants expressing the attacin or the cercosporin *SB-37* genes. Am. J. Potato Res. 76: 169-177.

Avesani, L., Falorni, A., Tornielli, G.B., Marusic, C., Porceddu, A., Polverari, A., Faleri, C., Calcinaro, F. and Pezzotti, M. (2003). Improved *in planta* expression of the human islet autoantigen glutamic acid decarboxylase (GAD65). Transgenic Res. 12(2): 203-212.

Barta, A., Sommergruber, K., Thimpson, D., Harmuth, K., Matzke, M.E. and Matzke, A.J. (1986). The expression of nopaline synthase-human growth hormone chimeric gene in transformed tobacco and sunflower callus tissue. Plant Mol. Biol. 6: 347-357.

Barton, K., Whiteley, H. and Yang, N.S. (1987). *Bacillus thuringiensis* δ-endotoxin in transgenic *Nicotiana tabacum* provides resistance to lepitopteran insects. Plant Physiol. 85: 1103-1109.

Beclin, C., Boutet S., Waterhouse P. and Vaucheret H. (2002). A branced pathway for transgene-induced RNA silencing in plants. Curr. Biol. 112: 684-688.

Bornhoff, B.A., Harst, M., Zyptian, E. and Topfer, R. (2005). Transgenic plants of *Vitis vinifera* cv. Seyval blanc. Plant Cell Rep. 24(7): 433-438.

Brotman, Y., Silberstein, L., Kovalski, I., Perin, C., Dogimont, C., Pitrat, M., Klinger, J., Thompson, A. and Perl-Treves, R. (2002). Resistance gene homologues in melon are linked to genetic loci conferring disease and pest resistance. Theort. Appl. Genet. 104: 1055-1063.

Butterfield, K., Irvine, E., Valdez Garza, M. and Mirkov, E. (2002). Inheritance and segregation of virus and herbicide resistance transgenes in sugarcane. Theort. Appl. Genet. 104(5): 797-803.

Capell, T. and Cristou, P. (2004). Progress in plant metabolic engineering. Curr. Opin. Biotechnol. 15: 148-154.

Chang, M.M., Horovitz, D., Culley, D. and Hardwiger, L.A. (1995). Molecular cloning and characterization of pea chitinase gene expressed in response to wounding, fungal infection and the elicitor chitosan. Plant Mol. Biol. 28(1): 105-111

Cho, H.S., Cao, J., Ren, J.P. and Earle, E.D. (2001). Control of Lepidopteran insect pests in transgenic Chinese cabbage (*Brassica rapa* spp.) transformed with synthetic *Bacillus thuringiensis cryIc* gene. Plant Cell Rep. 20: 1-7.

Chong, D.K. and Langridge, W.H. (2000). Expression of full-length bioactive antimicrobial human lactoferin in potato plants. Transgenic Res. 9(1): 71-78.

Constabel, C.P., Bertrand, C. and Brisson, N. (1993). Transgenic potato plants overexpressing the pathogenesis-related *STH-2* gene show unaltered susceptibility to *Phytophora infestans* and potato virus X. Plant Mol. Biol. 22(5): 775-782.

Daily, G., Dasgupta, P., Bolin, B., Crosson, P., du Guerny, J., Ehrlich, P., Folke, C., Jansson, A.M., Jansson, B., Kautsky, N., Kinzig, A., Levin, S., Maler, K.G., Pinstrup-Andersen, P., Siniscalco, D. and Walker, B. (1998). Food production, population growth and the environment. Science 281: 1291-1292.

De Block, M., Botterman, J., Vandewiede, M. Dockx, J., Thoen, C., Movva, V., Thompson, C., Van Montagu, M. and Leemans, J. (1987). Engineering herbicide resistance in plants by expression of a detoxifying enzyme. EMBO Journal 6(9): 2513-2518.

Dehesh, K., Jones, A., Knutzon, D.S. and Voelker, T.A. (1996). Production of high levels of 8:0 and 10:0 fatty acids in transgenic canola by overexpression of Ch FatB2, a thioesterase cDNA from *Cuphea hookeriana*. Plant J. 9 (2): 167-172.

Delannay, X., La Vallee, B.J., Proksch, R.K., Fuchs, S.K., Sims, S.K., Greenplate, J.T., Marrone, P.G., Dodson, R.B., Augustine, J.J., Layton, J.G. and Fischhoff, D.A. (1989). Field performance of transgenic tomato plants expressing *Bacillus thuringiensis* var. *kurstaki* insect control protein. Biol. Technol. 7: 1265-1269.

Douches, D.S., Pett, W., Santos, F., Coombs, J., Grafius, E., Li, W., Metro, E.A., El-Din, T.N. and Madkour, M. (2004). Field and storage testing Bt potatoes for resistance to potato tuberworm (Lepidoptera: Gelichiidae). J. Econ. Entomol. 97(4): 1425-1431.

Down, R.E., Ford, L., Bedford, S.J., Gatehouse, L.N., Newell, C., Gatehouse, J.A. and Gatehouse, A.M. (2001). Influence of plant development and environment on transgene expression in potato and consequence for insect resistance. Transgenic Res. 10(3): 223-236.

Down, R.E., Ford, L., Woodhouse, S.D., Davison, G.M., Majerus, M.E., Gatehouse, J.A. and Gatehouse, A.M. (2003). Tritrophic interactions between transgenic potato expressing snowdrop lectin (GNA), an aphid pest (peach, potato) *Myzus percicae* Sulz. and a beneficial predator (2-spot ladybird; *Adalia bipunctata* L.). Transgenic Res. 12(2): 229-241.

Dunwell, J.M. (2000). Transgenic approaches to crop improvement. J. Exp. Bot. 51: 487-496.

FAO/WHO (2001). Genetically modified organisms, consumers, food safety and the environment. Food and Agriculture Organization/World Health Organization Consultation. Ethics Series 2. Rome, Italy.

Falorni, A., Avesanti, A. L., Porcedu, K. and Pezzotti, M. (2005). Expression of human glutamic acid decarboxylase, a major autoantigen in Type I diabetes in transgenic plants In: Proc. of International Congress: In the wake of the Double Helix: From Green Revolution to Gene Revolution (eds.) R. Tuberosa, R.I. Phillips and M. Gale, Bologna, Italy, pp. 596-613.

Fischhoff, D.A. (1989). Field performance of transgenic tomato plants expressing *Bacillus thuringiensis* var. *kurstaki* insect control protein. Biol. Technol. 7: 1265-1269.

Fraley, R.T., Rogers, S.G., Horsch, R.B., Sanders, P.R., Flick, J.S., Adams, S.P., Bittner, M.L., Brand, L.A., Fink, C.L., Fry, J.S., Galluppi, G.R., Goldberg, S.B., Hoffmann, N.L. and Woo, S.C. (1983). Expression of bacterial genes in plant cells. Proc. Natl. Acad. Sci., USA 80 (15): 4803-4807.

Gasson, M. and Burke, D. (2001). Scientific perspectives on regulating the safety of genetically modified foods. Nature Reviews Genetics 2: 217-222.

Giddings, G.A. (2000). Transgenic plants as factories for biopharmaceuticals. Nat. Biotechnol. 18: 1151-1155.

Good, X., Kellogg, J.A, Wagoner, W., Langhoff, D., Matsumura, W. and Bestwick, R.K. (1994). Reduced ethylene synthesis by transgenic tomatoes expressing S-adenosylmethionine hydrolase. Plant Mol. Biol. 26 (3): 781-790.

Hartl, D.L., Dykhuizen, D.E., Miller, R.D., Green, L. and De Framond, J. (1983). Transposable element IS50 improves growth rate of *Escherichia coli* cells without transposition. Cell 35: 503-510.

Hellwege, E.M, Czapla, S., Jahnke, A., Willmitzer, L. and Heyer, A.G. (2000). Transgenic potato tubers synthesize the full spectrum of inulin molecules naturally occurring in globe artichoke (*Cynara scolymus*) roots. Proc. Natl. Acad. Sci., USA 97 (15): 8699-8704.

Hennin, C., Hofte, M. and Diederichsen, E. (2001). Functional expression of *Cf*9 and *Avr*9 genes in *Brassica napus* includes enhanced resistance to *Leptosphaeria maculans*. Mol. Plant Microbe Interact. 14 (9): 1075-1085.

Herrera-Estrella, L., Block, M.D., Messens, E., Hernalsteens, J.P., Montagu, M.V. and Schell, J. (1983). Chimeric genes as dominant selectable markers in plant cells. EMBO J. 2 (6): 987-995.

Hiatt, K., Cafferkey, R. and Bowdish, K. (1989). Production of antibodies in transgenic plants. Nature 342: 76-78.

Hinder, V.A. and Boulter, D. (1999). Genetic engineering of crop plants for insect resistance: A critical review. Crop Protec. 18: 191-199.

Hofius, D., Hajirezaei, M.R., Geiger, M., Tschiersch, H., Melzer, M. and Sonnewald, U. (2004). RNA II mediated tocopherol deficiency impairs photo-assimilate export in transgenic potato plants. Plant Physiol. 135: 1256-1268.

James, C. (2000). Global status of commercialized transgenic crops, ISAAA Briefs No. 21 Ithaca, NY, USA.

Jansens, S., Cornelissen, M., Clercq, R., de Reynaerts, A. and Peferoen, M. (1995). *Pathorimaea opercullella* resistance in potato by expression of *Bacillus thuringiensis* Cry 1A(b) insecticidal crystal protein. J. Econ. Entomol. 88: 1469-1476.

Jiang, X.L., He, Z.M., Peng, Z.Q., Qi, Y., Chen, Q., and Yu, S.Y. (2007). Cholera toxin B protein in transgenic tomato fruit induces systemic immune response in mice. Transgen. Res. 16(1): 27-32.

Kou, X.H., Zhu, B.Z., Luo, Y.B., Tian, H.Q. and Yu, B.Y. (2004). Relationship between ethylene and polygalacturonase in tomato fruits. (http: // www. ncbi.nlm.nih.gov).

Kushad, M.M., Brown, A.F., Kurilich, A.C., Juvik, J.A, Klein, B.P., Walling, M.A. and Jeffery, E.H. (1999). Variation of glucosinolates in vegetable crops of *Brassica oleraceae*. J. Agric. Food Chem. 47(4): 1541-1548.

Lee, W.S., Tzen, J.T., Kridl, J.C., Radke, S.E. and Huang, A.H. (1991). Maize oleosin is correctly targeted to seed oil bodies in *Brassica napus* transformed with the maize oleosin gene. Proc. Natl. Acad. Sci., USA 88: 6181-6185.

Li, M., Shao, M., Lu, X.Z. and Wang, J.S. (2005). Biological activities of purified harpin (Xoo) and harpin (Xoo) detection in transgenic plants using its polyclonal antibody. Acta Biochem. Sin. 37(10): 713-718.

Matsuoka, M., Kyozuka, J., Shimamoto, K. and Murakami, Y. K. (1994). The promoters of two carboxylases in a C4 plant direct cell-specific, light- regulated expression in a C3 plant. Plant J. 6(3): 311-319.

McCue, K.F., Sheperd, L., Allen, P.V., Maccree, M.M., Rockhold, D.R., Corsini, D.L., Davies, H.V. and Belknap, W.R. (2004). Metabolic compensation in transgenic potato tubers. Nature Magazine 168: 267-273.

Mc Garvey, P.B., Hammond, J., Dienelt, M.M., Hooper, D.C., Fu Z.F., Dietzschold, B., Koprowski, H. and Michaels, F.H. (1995) Expression of rabies virus glycoprotein in transgenic tomatoes. Biotechnology 13(13): 1484-1487.

Melander, M., Ahman, I., Kamnert, I. and Stromdahl, A.C. (2003). Pea lectin expressed transgenically in oilseed rape reduces growth rate of pollen beetle larvae. Transgenic Res. 12(5): 555-567.

Mengistu, L.W., Christoffers, M.J. and Lym, R.G. (2005). A psbA mutation in *Kochia scoparia* L. Schard from railroad rights of way with resistance to diuron, tebuthiuron and metribuzin. Pest Manag. Sci. 61(11): 1035-1042.

Murakami, Y., Tsuyama, M., Kobayashi, Y., Kodama, Y. and Iba, K. (2000). Trienoic fatty acids and plant tolerance of high temperature. Science 287: 476-479.

Negri, N.E., Brunetti, A., Tavazza, M. and Ilardi, V. (2005). Hairpin RNA-mediated silencing of Plum Pox Virus P1 and HC-Pro genes for efficient and predictable resistance to the virus. Transgenic Res. 14(6): 989-994.

Park, S.M., Lee, J.S., Jegal, S., Jeon, B.Y., Jung, M., Park, Y.S., Han, S.L, Shin, Y.S., Her, N.H., Lee, J.H., Lee, M.Y., Ryu, K.H., Yang, S.G. and Harn, C.H. (2005). Transgenic watermelon rootstock resistant to CCMMV infection. Plant Cell Rep. 24(6): 350-356.

Pei, Z.M., Ghassemian, M., Kwak, C.M., Mc Court, P. and Schroeder, I. (1998). Role of farnesyl transferase in ABA regulation of quard cell anion channels and plant water loss. Science 282: 287-290.

Perlak, F.J., Stone, T.B., Muskopf, Y.M., Petersen, L.J., Parker, G.B., McPherson, S.A., Wyman, J., Love, S., Reed, G., Biever, D. and. Fischoff, D.A. (1993). Genetically improved potatoes: Protection from damage by Colorado potato beetle. Plant Mol. Biol. 22: 313-321.

Pilon-Smits, E.A., Terry, N., Sears, T., Kim, H., Zayed, A., Hwang, S.B., Van Dun, K., Voogd, E., Verwoerd, T.C., Krutwagen, R. W. and Gooddijn, O.J. (1998). Trehalose-producing transgenic tobacco plants show improved growth performance under drought stress. Plant Physiol. 152: 525-532.

Prescott, V.E., Campbell, P.M., Moore, A., Matters, J., Rothenberg, M.E, Forster, P.S., Higgins, T.J. and Hogan, S.P. (2005). Transgenic expression of bean alpha-amylase inhibitor in peas results in altered structure and immunogenicity. J. Agric. Food. Chem. 53(23): 9023-9030.

Punja, Z.K. and. Utkhede R.S. (2003). Using fungi and yeasts to manage vegetable crop diseases. Trends Biotechnol. 21(9): 400-407.

Punja, Z.K. (2004). Genetic engineering of plants to enhance resistance to fungal pathogens. In: Fungal Disease Resistance in Plants (ed.) Z.K. Punja, Haworth Press Inc., New York, USA, pp. 207-258.

Schaffer, A.A. and Petreikov, M. (1997). Sucrose to starch metabolism in tomato fruit undergoing transient starch accumulation. Plant Physiol. 113 (3): 739-746.

Schwall, G.P., Safford, R., Westcott, R.J., Jeffcoat, R., Tayal, A., Shi, Y.C., Gindley, M.J. and Jobling, S.A. (2000). Production of very-high amylase potato starch by inhibition of SBE A and B. Nat. Biotechnol. 18 (5): 551-554.

Sharma, K.K. and Lavanya, M. (2002). Recent developments in transgenics for abiotic stress in legumes of the semi-arid tropics. In: Genetic Engineering of Crop Plants for Abiotic Stress (ed.) M. Ivanaga, JIRCAS (Jap. Int. Res. Cent. Agric. Sci.), Japan, pp. 61-73.

Shelton, A.M., Tang, J.D., Roush, R.T., Metz, T.D. and Earle, E.D. (2000). Field tests on managing resistance to Bt-engineered plants. Nat. Biotechnol. 18 (3): 339-342.

Signora, L., Galtier, N., Skot, L., Lukas, H. and. Foyer, C.H. (1998). Over expression of phosphate synthase in *Arabidopsis thaliana* results in increased foliar sucrose/starch ratios and flavours decreased foliar carbohydrate accumulation in plants after prolonged growth with CO_2 enrichment. J. Exp. Bot. 4: 669-680.

Sijmons, P.C., Dekker, B.M., Schrammeijer, B., Verwoerd, T.C., Van den Elzen, P.J. and Hoekema, A. (1990). Production of correctly processed human serum albumin in transgenic plants. Biotechnology 8(3): 217-221.

Sintani, D. and De la Penna, D. (1998). Elevating the vitamin E content of plants through metabolic engineering. Science 282 (5396): 2098-2100.

Slater, A., Scott, N. and Fowler, M. (2003). Plant Biotechnology: The Genetic Manipulation of Plants. Oxford University Press, Oxford, UK.

Smeekens, S. and Rook, F. (1997). Sugar sensing and sugar mediated signal transduction in plants. Plant Physiol. 115(1): 7-13.

Smith, C.J., Watson, C.F., Morris, P.C., Bird, C.R., Seymour, G.B., Gray, J.E., Arnold, C., Tucker, G.A., Schuch, W. and Harding, S. (1990). Inheritance and effect on ripening of antisense polygalacturonase genes in transgenic tomatoes. Plant Mol. Biol. 14(3): 369-379.

Sompornpailin, K., Makita, Y., Yamazaki, M. and Saito, K. (2002). A WD repeat containing putative regulatory protein in anthocyanin biosynthesis in *Perilla frutescens*. Plant Mol. Biol. 50(3): 485-495.

Stalker, D.M., Malyj, L.D. and McBride, K.E.(1988). Purification and properties of a nitrilase specific for the herbicide bromoxynil and corresponding nucleotide sequence analysis of the *bxn* gene. J. Biol. Chem. 263(13): 6310-6314

Stoger, E., Vaquero, C., Torres, E., Sack, M., Nicholson, L., Drossard, J., Williams, S., Keen, D., Perrin, Y., Christou, P. and Fisher, R. (2000). Cereal crops as viable production and storage systems for pharmaceutical scFv antibodies. Plant Mol. Biol. 42(4): 583-590.

Tacket, C.O., Mason, H.S., Losonsky, G., Clements, J.D., Lavine, M.M. and Arntzen, L.J. (1998). Immunogenecity in bananas of recombinant bacterial antigen delivered in a transgenic potato. Nat. Med. 4: 607-609.

Tacket, C.O. (2005). Plant derived vaccines against diarrhoeal diseases. Vaccine 723(15): 1866-1869.

Tan, S., Evans, R. and Sinhgh, B. (2006). Herbicidal inhibitors of amino acid biosynthesis and herbicide tolerant crops. Amino Acids 30(2): 195-204.

Thanavala, Y., Mahoney, M., Pal, S., Scott, A., Richter, L., Natarajan, N., Goodwin, P., Arntzen, C.J. and Mason, H.S. (2005). Immunogenicity in humans of an edible vaccine for hepatitis B. Proc. Natl. Acad. Sci. 102 (9): 3378-3382.

Van Esse, H.P., Thomma, B.P., Vant Klooster, J.W. and de Wit, P.J. (2006). Affinity-tags are removed from *Cladosporium fulvum* effector proteins expressed in the tomato leaf apoplast. J. Exp. Bot. 57(3): 599-608.

Wang, J., Chen, Z., Du, J., Sun, Y. and Liang, A. (2005). Novel insect resistance in *Brassica napus* developed by transformation of chitinase and scorpion toxin genes. Plant Cell Rep. 24(9): 549-555.

Waterhouse, P.M., Graham M.W. and Wang M.B. (1998).Virus resistance and genesilencing in plants can be induced by simultaneous expression of sense and antisense RNA. PNAs 95: 13959-13964.

Wu, G., Shortt, B.J., Lawrence, E.B., Levine, E.B, Fitzsimmons, K.C., Shah, D.M. (1995). Disease resistance conferred by expression of a gene encoding H_2O_2 generating glucose oxidase in transgenic potato plants. Plant Cell 7(9): 1375-1368.

Xiong, A.S., Yao, Q.H., Li, X., Fan, H.Q. and Peng, R.H. (2003). Double antisense ACC oxidase and ACC synthase fusion gene introduced into tomato by *Agrobacterium*-mediated transformation and analysis of the ethylene production of transgenic plants. (http://www.ncbi.nlm.nih.gov).

Yang, C.Y., Chu, F.H., Wang, Y.T., Chen Y.T., Yang, S.F. and Shaw, J.F. (2003). Novel broccoli 1-aminocyclopropane-1-carboxylate oxidase gene (Bo-ACO3) associated with thelate stage of post harvest floret senescence. J. Agric. Food. Chem. 51(9): 2569-2575.

Yang, G.D., Zhu, Z., Li, Y.O. and Zhu, Z.J. (2002). Transformation of Bt-CpTi fusion protein gene to cabbage (*Brassica oleraceae* var. *capitata*) mediated by *Agrobacterium tumefaciens* and particle bombardment. Shi Yan Sheng Wu Xue Bao 35(2): 117-122.

Zhang, F. and Simon, A.E. (2003). Enhanced viral pathogenesis associated with a virulent mutant virus or virulent satellite RNA correlates with reduced virion accumulation and abundance of free coat protein. Virology 312: 8-13.

Zhao, J.Z., Cao, J., Collins, H.L., Bates, S.L., Roush, R.T., Earle, E.D. and Shelton, A.M. (2005). Concurrent use of transgenic plants expressing a single and two *Bt* genes speeds insect adaptation to pyramided plants. Proc. Natl. Acad. Sci. USA 102(24): 8426-8430.

Zhou, M., Xu, H., Wei, X., Ye, Z., Wie, L., Gong, W., Wang, Y. and Zhu, Z. (2006). Indentification of a glyphosate-resistant mutant of r5-enolpyruvylshikimate 3-phosphate synthase using a directed evolution strategy. Plant Physiol. 140(1): 184-195.

http://www.agbios.com

http://www.gmo-compass.org

http://www.ific.org

http://www.isaaa.org

http://www.bio.org

3

Plant Genetic Engineering: A Tip of Microbial Iceberg

Haseena Khan

INTRODUCTION

Genetic Engineering of plants is a recent development in the field of biotechnology. Ever since the origin of this technology in the 1970s, molecular biologists have found microbes very useful for producing new microbial strains, novel pharmaceuticals, etc., *via* manipulation. Bacteria have always provided a convenient reservoir for storing, multiplying and recombining genes necessary for the recombinant DNA technology. These microorganisms have provided all the tools of the trade of modern biotechnology, from cutting and splicing DNA pieces, to the vectors for ferrying these genetic material to different hosts, to the tiny factories for making proteins from the genetic information, to enabling amplification of DNA, to the production of bio-diesel, and even to cleaning up an environment contaminated by humans. Microbes are continually providing the means by which agriculturists, medical and forensic scientists, environmentalists and even industrialists are improving their skills. So vast is the scope of microbes in the field of genetic engineering, that the process is considered to be only the tip of the microbial iceberg.

In particular, there has been an exponential growth of interest and activity in the application of the techniques and ideas of modern molecular genetics and molecular biology to problems related to the improvement of plants and for the generation of transgenic plants. The previous two chapters in this volume focus on general applications, ethics, issues and legislation concerning transgenic plants. Microorganisms have contributed tremendously in this development. Among the countless contributions of tiny microbes in the field of genetic engineering, this chapter concentrates mainly on:

(i) recent developments in the understanding of bacterial- and viral-mediated transformation of plants
(ii) the role of microbes in the new and emerging field of synthetic biology.

THE MAKING OF TRANSGENIC PLANTS

Several technologies for the introduction of DNA into cells are well-known and can be divided into categories including:

(1) chemical methods (Graham and van der Eb, 1973; Zatloukal et al., 1992);
(2) physical methods such as microinjection (Capecchi, 1980), electroporation (Wong and Neumann, 1982; Fromm et al., 1985), and the gene gun (Fynan et al., 1993; Johnston and Tang, 1994)
(3) viral vectors (Eglitis and Anderson, 1988; Clapp, 1993; Lu et al., 1993).

It is possible to insert DNA into plant cells directly without using a biological vector. Such DNA may be taken up and integrated into the plant genome. Until recently, the methods employed for some monocot species included direct DNA transfer into isolated protoplasts and microprojectile-mediated DNA delivery (Shimamoto et al., 1989; Fromm et al., 1990). The protoplast methods have been widely used in rice, where DNA is delivered to the protoplasts through liposomes, polyethylene glycol (PEG), and electroporation. This delivery can lead to a high proportion of the protoplasts becoming infected. Thus nucleic acids can be taken up by protoplasts without loss of biological activity. While a large number of transgenic plants have been recovered in several laboratories (Shimamoto et al., 1989; Datta et al., 1990), the protoplast methods require the establishment of long-term embryogenic suspension cultures. Some regenerants from protoplasts are infertile and phenotypically abnormal due to the long-term suspension culture (Rhodes et al., 1988; Davey et al., 1991).

Electroporation is a process where cells are integrated with a DNA construct and then briefly exposed to pulses of high electrical voltage. Fromm et al. (1985) have developed a general method for electrically introducing DNA into plant cells. The cell membrane of the host cell is penetrable, thereby allowing foreign DNA to enter the host cell (Terzaghi and Cashmore, 1997; Prescott et al., 1999). Gene transfer occurs when a high-voltage electric pulse is applied to a solution containing protoplasts and DNA. It was shown that gene transfer efficiency increased with the DNA concentration and was affected by the amplitude and duration of the electric pulse, as well as by the composition of the electroporation medium. Some of these cells incorporate the new DNA and express the desired gene. While direct introduction of DNA into monocots by electroporation has proved successful, this technique is used in a relatively small proportion of cases in comparison to other methods.

Microinjection involves the direct injection of material into a host cell using a finely drawn micropipette needle. The procedure of direct injection of new genes using ultra-fine cannulae, which is now almost a routine with animal cells and in medicine, has so far proved very arduous with plant cells. However, microinjection would be an elegant way of avoiding marker genes. During microinjection, DNA is injected directly into the cell, or even into the cell nucleus via an inserted cannula. The process is observed and controlled under a microscope. The DNA then integrates into the plant genome - probably during the cell's own DNA repair processes.

The gene gun or *particle bombardment* or *biolistics*, uses rapidly propelled tungsten microprojectiles that have been coated with DNA (Taylor and Fauquet, 2002). These DNA-coated particles are fired into plant cells, and the DNA of interest is often incorporated into the plant DNA. Improved particle bombardment devices for transporting DNA into living cells are being routinely developed (Altpeter et al., 2005).

Since the crystallization of tobacco mosaic virus in 1935 (Stanley, 1935), developments in virology and molecular biology have gone hand in hand. Active research is being carried out on exploring the possibility of adding a foreign gene to a viral genome so that the gene could be introduced into a plant cell during the infection process, and perhaps be expressed there. Such a vector system would differ significantly from the *Agrobacterium* plasmid vector. Viruses can move from cell to cell through the plant. Thus, a viral vector would be able to introduce a gene into growing intact plants, thereby avoiding the problems associated with regenerating plants from single cells. The use of these viruses in plant transformation will be addressed in more detail later in this chapter.

However, even with all the developments and the use of different avenues through which DNA is made to enter plant cells, *Agrobacterium*-mediated transformation remains the most popular method of plant transformation. More recently, monocot species have been successfully transformed via *Agrobacterium*-mediated transformation. *Agrobacterium*-mediated transformation provides a viable alternative to all the different methods, and it also allows quick molecular analysis of transgenic lines.

AGROBACTERIUM TUMEFACIENS – A GENE JOCKEY

When the concept of using *Agrobacterium tumefaciens* (which has only recently become a very important microorganism in the field of plant genetic engineering) was first considered for ferrying novel genes in the creation of transgenic plants, many wondered if it was wishful thinking. However, since then, in the last 26 years the use of *Agrobacterium* to genetically transform plants has advanced from wishful thinking to a reality. Agricultural scientists have modified these bacteria and can use them as carriers for useful genes to incorporate novel traits of other species. Modern agricultural biotechnology is, therefore, heavily indebted to *Agrobacterium* in the creation of transgenic plants, and it is difficult to think of an area of plant science research that has not benefited from this technology. *Agrobacterium* and the genetic engineering of plants have been the subject of a significant number of review articles (Hookyas, 1998; Gelvin, 2003; Cheng et al., 2004; Jones et al., 2005; Nester et al., 2005; Opabode, 2006).

The molecular basis of genetic transformation of plant cells by *Agrobacterium* is the transfer from the bacterium and integration into the plant nuclear genome of a region of a large tumor-inducing (*Ti*), causing crown gall (Zaenen et al., 1994), or rhizogenic (*Ri*) producing hairy roots plasmids that reside in *Agrobacterium* (Nilsson and Olsson, 1997). Proteins encoded by *vir* genes of *Agrobacterium* play an essential role in this transformation process. The transferred DNA, called T-DNA, carries a set of oncogenes as well as genes for catabolizing opines (Bomhoff et al., 1976), which lead to the production of opines (amino acid derivatives), which allow the bacterium to utilize the same as an energy source. This process leads to the formation of crown gall, tumorous lesions in the host plants (Chilton et al., 1977; Zhu et al., 2000).

When integrated into the plant genome, the genes on the T-DNA code for:

- production of cytokinins
- production of indole acetic acid
- synthesis and release of novel plant metabolites – the opines and agrocinopines, which are unique phosphorylated sugar derivatives

These plant hormones upset the normal balance of cell growth, leading to the production of galls and thus to a nutrient-rich environment for the bacteria. All these compounds are used by the bacterium as the sole carbon and energy source, and because they are absent in normal plants, they provide *Agrobacterium* with a unique food source that other bacteria cannot use.

MODIFICATION PROCESS OF *Ti* PLASMIDS FOR *AGROBACTERIUM*-BASED CLONING

Ti plasmids are very large and the T-DNA region does not generally contain unique restriction endonuclease sites, thus making the cloning of desired DNA fragments within this region very difficult. Therefore, to make the cloning into *Agrobacterium* simple, scientists constructed binary vectors. In this simplification process, the use of two different vectors for introducing foreign genes into plants based on *Agrobacterium*-mediated transformation was first described by Bevan (1984). A vector molecule for the efficient transformation of higher plants has been constructed with several features that make it efficient to use. It utilizes the trans-acting functions of the *vir* region of a co-resident *Ti* plasmid in *A. tumefaciens* to transfer sequences bordered by left and right T-DNA border sequences into the nuclear genome of plants. The T-region was engineered to contain a dominant selectable marker gene that confers high levels of resistance to kanamycin, and a *lac* alpha-complementing region from M13mpl9 that contains several unique restriction sites for the positive selection of inserted DNA.

Binary vectors helped to prevent the T-DNA introduced into the *Ti* plasmid from spreading uncontrollably. In constructing these vectors, the *vir* genes are removed and the DNA of choice is integrated between the borders of the T-DNA. The binary plasmids so 'disarmed' have several advantages: for example, they are easy to propagate in the laboratory, like *Escherichia coli*. A binary plasmid is also much smaller than a normal *Ti* plasmid, and therefore, very easy to handle. The *vir* genes, needed for transfer to the plant, are arranged on a second plasmid, from which the T-DNA and the two borders of T-DNA are removed. With a binary vector, the target gene (T-DNA) and virulence gene, which are natural residents of one plasmid, are split between two. One transfers the target gene, while the other helps with the transformation through the activity of the *vir* genes, without the information for the transfer itself being integrated into the plant as well (**Fig. 3.1**).

These T-DNA binary vectors have brought about a revolution in the use of *Agrobacterium* for heterologous transfer of genes into plants. Researchers with very little training in microbial genetics can now easily manipulate *Agrobacterium* to create transgenic plants. These plasmids are

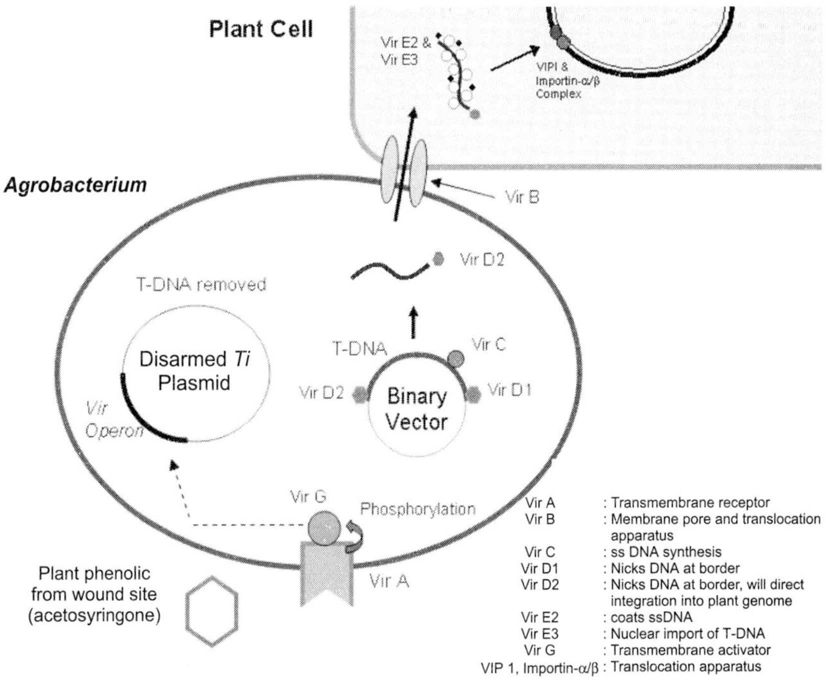

Fig. 3.1. Binary vectors of *Agrobacterium* and the different bacterial and plant proteins necessary for T-DNA transfer

now being produced in a small size and, therefore, are easy to manipulate in both *E. coli* and *Agrobacterium*. These plasmids now contain multiple unique restriction endonuclease sites within the T-region into which genes of interest can be cloned. The binary vectors have been further improved to contain super-linker regions carrying unique sites for more than 20 restriction enzymes (Hennegan and Danna, 1998). Other researchers have designed these vectors for specialized purposes, containing different plant selectable markers, promoters, and poly (A) addition signals, between which genes of interest can be inserted, including translational enhancers to boost the expression of transgenes, and protein-targeting signals to direct the transgene-encoded protein to particular locations within the plant cell (Indrani et al., 1992; Vlasak and Ondrej, 1992; Taylor and Fauquet, 2002). Vectors are being refined for solving problems with transgene expression that begin with the lack of control over the region where the transgene is to be inserted. Solutions have been sought for the random nature of the integration and the possibility of multiple insertions. The solutions have been provided by the *Cre/lox* system from bacteria (Albert et al., 1995) and a chloroplast transformation method (Odell and Russell, 1994).

The vectors contain marker genes, which can be both selected and scored. These are functional genetic components, as they yield a product that serves a function in the identification of a transformed plant, or a product of desired utility. The DNA, that serves as a selection device, functions in a regenerable plant tissue to produce a compound that would confer upon the plant tissue resistance to an otherwise toxic compound. Genes of interest for use as a selectable, screenable, or scorable marker would include, but are not limited to, β-glucuronidase (GUS), green fluorescent protein (GFP), luciferase (LUX) and antibiotic or herbicide tolerance genes. Examples of transposons and associated antibiotic resistance genes include the transposons Tns (*bla*), Tn5 (*nptII*), Tn7 (*dhfr*); penicillins; kanamycin (and neomycin, G418, bleomycin); methotrexate (and trimethoprim); chloramphenicol; and tetracycline.

Characteristics which have been found useful for selectable markers in plants include:

- stringent selection with minimum number of non-transformed tissues
- large numbers of independent transformation events with no significant interference with the regeneration
- application to a large number of species
- availability of an assay to score the tissues for presence or absence of the marker.

Several antibiotic resistance markers satisfy these criteria, including those resistant to kanamycin (nptII), hygromycin B (aph IV), and gentamycin. A number of selectable marker genes are known. Particularly preferred selectable marker genes for use include genes that confer resistance to compounds such as antibiotics like kanamycin (Dekeyser et al., 1989), and herbicides like glyphosate (Della-Cioppa et al., 1987).

REGULATION OF TRANSGENE EXPRESSION

A number of promoters that are active in plant cells have been used for the construction of *Ti* plasmid vectors. Such promoters include: nopaline synthase (NOS) and octopine synthase (OCS) promoters, which are carried on tumor-inducing plasmids of *A. tumefaciens* Caulimovirus promoters, such as:

(i) cauliflower mosaic virus (CaMV) 19S and 35S promoters
(ii) figwort mosaic virus (FMV) 35S promoter;
(iii) enhanced CaMV35S promoter (e35S)

(iv) light-inducible promoter from the small subunit of ribulose bisphosphate carboxylase (ssRUBISCO, a very abundant plant polypeptide).

All these promoters have been utilized to create various DNA constructs used for expression in plants.

Promoter hybrids have also been constructed to enhance transcriptional activity, or to combine desired transcriptional activity, inducibility, and tissue or developmental specificity. Promoters that function in plants are promoters that are inducible, viral, synthetic, constitutive (Odell et al., 1985), and temporally-, spatially- and spatio-temporally-regulated (Benfey and Chau, 1989). Other promoters that are tissue-enhanced, tissue-specific, or developmentally regulated are also in use. Promoters may be obtained from a variety of sources such as plants and plant DNA viruses, and include the CaMV35S and FMV35S promoters, and promoters isolated from plant genes such as *ssRUBISCO* genes. It is preferred that the particular promoter selected should be capable of causing sufficient expression to result in the production of an effective amount of the gene product of interest.

The promoters used in such constructs (i.e., chimeric/recombinant plant genes) are modified, if desired, to affect their control characteristics. Promoters can be derived by means of ligation with operator regions, random or controlled mutagenesis, etc. Furthermore, the promoters may be altered to contain multiple "enhancer sequences" to assist in elevating gene expression. Examples of such enhancer sequences have been reported by Kay et al. (1987).

The mRNA produced by a DNA construct may also contain a 52 non-translated leader sequence. This sequence can be derived from the promoter selected to express the gene and can be specifically modified so as to increase translation of the mRNA. The 5′ non-translated regions can also be obtained from viral RNAs, from suitable eukaryotic genes, or from a synthetic gene sequence. Such "enhancer" sequences may be desirable to increase or alter the translational efficiency of the resultant mRNA. The 3′ non-translated region of the chimeric constructs should contain a transcriptional terminator, or an element having equivalent function, and a polyadenylation signal, which functions in plants to cause the addition of polyadenylated nucleotides to the 3′ end of the RNA. Examples of suitable 3′ regions are: (1) the 3′ transcribed, non-translated regions containing the polyadenylation signal of *Agrobacterium* tumor-inducing (*Ti*) plasmid genes, such as the nopaline synthase (*NOS*) gene, and (2) plant genes such as the soybean storage protein genes and the small subunit of the ribulose-1,5-bisphosphate carboxylase (*ssRUBISCO*) gene.

PROCESS FOR REMOVAL OF THE REPORTER/MARKER GENE(S) FROM TRANSGENIC PLANTS

It has been mentioned earlier that for reasons of selection and scoring, it is important to introduce marker genes with the vector DNA. However, proteins produced by the marker genes may be toxic or may cause allergies. It is also possible that the antibiotic resistance gene could get transferred to pathogens. Moreover, the continued expression of selectable markers in field grown plants is not only unnecessary but also highly undesirable. Also, in order to appease the concerns of the general public it is essential to remove the useless marker gene(s). This will help in overcoming public, concerns. Several strategies have been developed to remove selectable markers after they have served their function. Ongoing refinements in the technology of precisely excising foreign genetic material will, therefore, continue to form an integral part of strategies to improve transformation rates and to safeguard against the unintended spread of genes encoding novel traits.

This problem has been overcome through the strategy of co-transformation of plant cells using two vectors: (i) one vector is manipulated to contain a reporter gene, and (ii) the other contains the target gene. Some cells with the reporter will also contain the target gene, but will be integrated at different chromosomal locations (the two genes are initially unlinked). The reporter gene is then eliminated from the plant by traditional breeding methods. This is followed by crossing transgenic plants with non-transgenic plants. The progeny that has the target gene, but not the reporter gene, is then identified. The transgenic plants lacking the reporter are next propagated.

A recent suggestion (Kumar and Fladung, 2001) made for achieving the removal of selectable marker genes is with the use of both a site-specific recombinase and a site-specific endonuclease that could be modified to ensure the simultaneous elimination of components not required any longer, leaving only the recombinase recognition site with the gene of choice. Kumar and Fladung (2001) describe how the marker gene can be eliminated by placing it between two directly orientated recognition sites of a recombinase (cre-*lox* or FLP/FRT system) followed by the expression of recombinase resulting in recombination between flanking recognition sites, thus deleting the marker gene. The FLP-FRT system is similar to the cre-*lox* system and is becoming more frequently used in mouse-based research. It involves using flippase (FLP) recombinase, derived from the yeast *Saccharomyces cerevisiae* (Sadowski, 1995). FLP recognizes a pair of FLP recombinase target (FRT) sequences that flank a genomic region of interest. A similar but simpler method, based on intra-chromosomal recombination between bacteriophage λ attachment (attP) regions, has recently been developed to remove selectable marker genes from tobacco transgenes (Zubko et al., 2000). Unlike the recombinase-based methods this method does not require the transfer of an additional recombinase gene.

A new strategy called the "clean gene" technology has been developed to generate transgenic rice plants free of undesirable antibiotic marker genes and containing simple transgenic locus (Afolabi et al., 2004). The technology involves the use of vectors that are normally used for *A. tumefaciens*. The system calls for the use of multiple binary plasmids (*pGreen/pSoup* system). *pGreen* is a small binary plasmid unable to replicate in *Agrobacterium* without the presence of another binary plasmid, *pSoup*, in the same strain. Both *pGreen* and *pSoup* can carry a T-DNA with different genes of interest. When co-transferred into the rice genome, the transgenes carried by *pGreen* and *pSoup*, can integrate at unlinked loci allowing the recovery of rice plant progeny containing only the gene of interest (on *pGreen* T-DNA) but without any selectable marker gene (on *pSoup* T-DNA).

ENTRY PROCESS OF THE T-DNA INTO THE HOST NUCLEUS AND ITS INTEGRATION INTO THE HOST GENOME

Even with the widespread use of *Agrobacterium*-mediated transformation as a genetic tool, the mechanisms of T-DNA transfer into the plant cell and subsequent integration into the host genome are not well understood. There are some species of plants that are not amenable to genetic transformation by *Agrobacterium*. A few species of dicotyledonous plants and most species of monocotyledonous still resist transformation by *A. tumefaciens*. Detailed knowledge of bacterial and plant proteins involved in the regulation of *Agrobacterium*-mediated transformation will lead to improvements in engineering of the plant transformation systems and will enable the use of this valuable tool for recalcitrant species as well. Thus, transformation of many crop species, which has been relatively inefficient in the past years, may soon change due to recent advances made in the transformation technology.

Initiation of transformation by *Agrobacterium* and the subsequent tumor induction has been the subject of several reviews (Gelvin, 2003; Nester et al., 2005). In order to further enhance the role of *Agrobacterium* as a vector, it is important to have a better understanding of T-DNA transfer from the bacteria to the host genome. Such a finding will also shed light on the fundamental mechanism of genetic transformation of eukaryotic cells. Inside the host the T-strand exists as a protein–DNA complex (T-complex) that is said to have a well-defined structure (Abu-Arish et al., 2004). However, what is not understood very clearly is the intracellular movement of the *Agrobacterium* T-complex through the dense cytoplasm before its importation into the nucleus. It is apparent that the diameter and the extended size of the T-complex demands active transport to achieve entry into the nucleus.

It is noted that the importation requires the combined action of several bacterial virulence proteins (VirE3, VirE2 and VirD2) and host receptors VIP1 and importin-α (Tzfira et al., 2005; Lacroix et al., 2006). Importin-α, also known as karyopherin-α, is a component of the nuclear import machinery that can recognize and bind to classical nuclear localization signal (NLS) sequences composed mainly of basic amino acid residues. Usually importin-α forms a heterodimer with importin-β, another nuclear transport receptor, which then helps in the entry into the nucleus of the cell through the formation of the protein-importin α/β complex.

VirE2, the most abundant protein of the T-complex, is sufficient for nuclear import of artificially produced VirE2–ssDNA complexes in plant cells. VirE2 interacts with VIP1. VirE2 is distinguished by its nuclear targeting, occurring only in plant cells but not in animal cells, and is facilitated by the cellular VIP1 protein. VIP1 is a basic leucine zipper motif protein. It shows no significant homology to known animal or yeast proteins, and it is suggested that VIP1 is a cellular factor responsible, at least partially, for plant-specific VirE2 nuclear import. Mature T-complex is formed by the association of the single-stranded T-DNA, covalently linked at its 5′ end to VirD2, and coated cooperatively with VirE2 molecules. VIP1, which has a conventional NLS, interacts with importin-α and might help in the transfer of VirE2 into the nucleus.

It is interesting to note that the two putative NLS regions in VirE2 are specific for plants. They do not allow the entry of VirE2 or VirE2–ssDNA complexes into non-plant systems (Tzfira et al., 2005; Lacroix et al., 2006) but mutations in any of the two nuclear localization signal regions of VirE2 have been shown to permit transport into animals. As for VirD2, Ballas and Citovsky (1997) have shown that it interacts directly with importin-α of *Arabidopsis thaliana*. The entire T-complex is thus targeted to the nucleus through the nuclear pore with the help of importin-α that binds to the nuclear localization sequences of VirD2, VIP1 and VirE2.

Salman et al. (2005) have addressed the question of how protein products, bearing nuclear localization signals, reach the nuclear membrane before gaining entry into the nucleus, i.e., by simple diffusion or perhaps with the assistance of cytoskeletal elements or cytoskeleton-associated motor proteins. They have shown that the presence of nuclear localization signals invokes active transport along microtubules in a cell-free *Xenopus* egg extract. Chemical and antibody inhibition of minus-end directed cytoplasmic dynein blocks this active movement. It has, therefore, been hypothesized that in the intact cell, where microtubules project radially from the centrosome, such an interaction would effectively deliver nuclear-targeted cargo to the nuclear envelope in preparation for entry. Tzfira and his group (Tzfira, 2006) have developed a similar motility assay using the VirE2 NLS mutants, and with an automated tracking method they showed that the microtubule network is used for the movement of the DNA in these mutants. However, as the author (Tzfira, 2006) has pointed out, plant microtubule associated motor proteins are yet to be identified and, therefore, a mechanical model cannot be proposed for the transport of DNA protein complexes such as the T-complex along the microtubule networks in plants for entry into the nucleus.

In a recent review, Lacroix et al. (2006) have speculated on the fate of the T-DNA in the host nucleus. It is clear that first of all the proteins coating the T-DNA will have to be removed and the T-complex has to be targeted to that site on the chromatin where it will later be integrated. VirF may play a role in the removal of the coating proteins (Tzfira et al., 2004). Inside the nucleus VirF sets up an association with the T-complex through it's binding with VIP1. This interaction involves the recruitment of a host protein called ASK1, which is a plant homologue of the yeast Skp1 protein involved in targeted proteolysis. This results in the proteasomal degradation of VIP1 and VirE2, leading to the uncoating of the T-DNA.

It has also been shown that when a proteasome inhibitor was applied to *Agrobacterium*-inoculated plant tissues, the expression of T-DNA was greatly reduced with no reduction in host cell gene expression (Tzfira et al., 2004). This was a clear demonstration of the involvement of proteasomal degradation in *Agrobacterium*-mediated genetic transformation. The importance of targeted proteolysis has also been stressed from the finding that VirE2, VirE3, VirD2 and VirD5 and not VirF proteins have sequences enriched for proline (P), glutamic acid (E), serine (S) and threonine (T) residues forming the PEST motif. These PEST motifs act as proteolytic signals and are, therefore, present only in proteins destined for proteolysis (Rechsteiner and Rogers, 1996).

Chromosomal breaks are repaired by homologous recombination (HR) or non-homologous end joining (NHEJ) mechanisms. Several observations (Leskov et al., 2001; Windels et al., 2003) suggest that double-stranded T-DNA integration at double strand break (DSB) sites in the host genome is the mode of T-DNA integration. DSBs or DSB repair enzymes are hypothesized to act as baits in attracting the invading T-DNA molecules to the sites where their integration will occur (Lacroix et al., 2006).

Non-host species, like yeast, have also been used to identify the function of the host and cellular systems in the mechanism for T-DNA integration (van Attikum and Hooykaas, 2003). Yeast mutants with knocked-out enzymes of the DNA repair machinery were used for the determination of specific regulators of the integration event. Two genes identified in this process are Rad52 (a single-stranded DNA binding protein) and Ku70 (a double-stranded DNA binding protein). Rad52 is required for integration via non-homologous recombination. Ku70 works in association with Ku80, a protein involved in the repair of DNA double-strand breaks. Ku70/Ku80 forms a heterodimer that is required for integration *via* the homologous recombination pathway. The integrating double-stranded DNA molecules are ligated to the ends of the double-strand breaks through the activity of plant DNA ligases. Based on a trial conducted by Oleinick et al. (1994), in which it has been shown that decondensed chromatin of mammalian cells are more susceptible to DSB-inducing agents, Lacroix et al. (2006), suggest that by analogy plant DNA may have just as high frequency as that of DSB in active decondensed chromatin, explaining their high susceptibility to T-DNA integration.

USE OF *AGROBACTERIUM* IN GENETIC MANIPULATION OF DIVERSE SPECIES OF ORGANISMS

Recent reports of *Agrobacterium*-mediated transformation of several fungal species (Chen et al., 2000; Lima et al., 2006) suggest that *Agrobacterium* may be a useful "gene-jockeying tool" for more than just plant species. This offers new prospects for the use of *Agrobacterium* in the genetic manipulation of diverse species. It appears that *A. tumefaciens,* the only known natural example with the capability of trans-kingdom DNA transfer, may have a long list of organisms that it can transform.

Kunik et al. (2001) have shown that T-DNA can be transferred to the chromosomes of human cancer cells. Under laboratory conditions the T-DNA has also been transferred to human cells, demonstrating the diversity of insertion application. In fact, *Agrobacterium* gets attached to, and genetically transforms, several types of human cells. They showed that integration occurred at the right border of the *Ti* plasmid's T-DNA of transformed HeLa cells, an event similar to transformation of the plant cell genome. This suggests that *Agrobacterium* transforms human cells by a mechanism similar to what it uses for transformation of plants cells. They have shown that human cancer cells, along with neuron and kidney cells, could be transformed with the *Agrobacterium* T-DNA.

PATENT RIGHTS OF *AGROBACTERIUM*

As is evident from the discussions above, a lot of the basic research that led to *Agrobacterium*-mediated transformation was done in public institutions. However, it is the private sector that holds the keys to the patent rights. Monsanto and a few companies hold the patents on the use of *Agrobacterium* and, therefore, make it difficult for any transgenic variety to be developed without infringing on these patents. Researchers wishing to transform plants, not only have to seek permission from the owner of the patents on the transformation methods, but also have to contend with the patents on the promoter and marker genes as well as patents on the actual gene of interest that is being introduced into the plant (e.g., a *Bt* gene or a herbicide tolerance gene). Compounding this further is the fact that a novel plant variety may also be patent protected (Pray and Naseem, 2005).

However, the good news is that the stranglehold on the use of *Agrobacterium* is about to change now with the discovery of the ability of gene transfer to plants by diverse species of bacteria (Broothaerts et al., 2005) by scientists working at Center for the Application of Molecular Biology to International Agriculture (CAMBIA), Canberra, Australia. Biotechnology tool kits developed by CAMBIA have been introduced under an open source license. The technologies include TransBacter, a new method for transferring genes to plants, using bacterial species outside the genus *Agrobacterium*, developed as an alternative to *Agrobacterium*-mediated transformation. TransBacter is designed to be a work-around to the many patents covering *Agrobacterium* transformation, and thus aims at overcoming the current Intellectual Property (IP) restrictions on the commercialization of products created by using bacteria-mediated gene transformation in plants.

PLANT VIRUSES AS VECTORS FOR PLANT TRANSFORMATION

Viral vectors are extremely promising for the expression of recombinant proteins in plants. Development of plant virus gene vectors for expression of foreign genes in plants provides attractive biotechnological tools to complement conventional breeding and transgenic methodology. Plants viruses are versatile in allowing the rapid and convenient production of recombinant proteins in plants.

Howell et al. (1980) showed that copies of the viral DNA, propagated in *E. coli* using a plasmid vector, were infectious when inoculated into host plants. This opened up the avenues for performing recombinant DNA manipulations on the viral genome in *E. coli* and for testing their effects in plants. There are several sites where foreign DNA might be inserted into the viral DNA without destroying infectivity. There is one large and one small intergenic region. In addition, two non-essential coding regions have been identified. Small insertions of bacterial DNA at these non-essential sites were successfully propagated through a cycle of infection in the plant. However, these experiments revealed two serious problems. Firstly, there appears to be a stringent limitation on the amount of extra DNA, which can be inserted without causing deletions. This is about 250 base pairs. The second limitation is perhaps more serious. During several cycles of propagation of an engineered virus in plants the inserted DNA tends to be lost through recombination (Howell et al., 1980).

Findings that a full length DNA copy of the RNA is infectious for appropriate cells in culture paved the way for making the RNA genomes amenable to the various manipulations that can be carried out *in vitro* on double-stranded DNA. At present, various laboratories are exploring the use of plant viruses with RNA genomes. Many plant viruses have their genomes divided into two or three separate pieces of RNA. Some of these multipartite viruses are particularly attractive as potential vectors. For example, tobacco rattle virus has a large and a small RNA housed in two separate rod-shaped particles. The small RNA contains the coat protein gene, which is not essential for the replication of the larger RNA as a naked RNA. The possibility, therefore, exists of introducing foreign RNA (via a dsDNA intermediate) with either the long or the short RNA and achieving replication of the engineered molecule. Furthermore, there would be no packaging problem with a naked RNA. Some isolates of several RNA plant viruses contain satellite RNAs which are not essential for replication of the virus. They are templates for their own replication, but require functions of the helper virus for that replication. It appears possible that additional genetic material could be inserted into the satellite RNA without affecting replication of the virus.

The virus-based transient RNA and DNA replicons are better than transgenic gene expression because they allow rapid and convenient engineering, and provide flexibility for expeditious application in various plant species. Viral vectors are very useful when very high levels of gene expression are desired within a short time. The strategies that have been tested for foreign gene expression in various virus-based vectors include gene replacement, gene insertion, epitope presentation, use of virus controlled gene expression cassettes, and complementation. Virus vectors are being utilized for foreign gene expression in fundamental research and biotechnology applications (Scholthof et al., 1996).

Researchers are working on the development of new viral vectors (Gleba et al., 2004) that are not simply carbon copies of wild type viruses carrying heterologous coding sequences. Instead, new vectors are being designed to function as integrated expression systems, with transgenic host plants pre-engineered to provide some of the functions that are normally provided by the vector. Such integrated systems are expected to provide a more efficient and controlled gene expression, and improve safety by preventing any escape of infectious viral particles outside the host plant.

Compared with production systems based on transgenic plants, viral vectors are easier to manipulate and recombinant proteins can be produced not only very quickly but also in greater yields. In the last few years, a great deal of interest has been evinced in the development of plant viruses as vectors for the production of vaccines, either as whole polypeptides or epitopes displayed on the surface of chimeric viral particles. Quite a few viruses have been extensively developed for vaccine production, including tobacco mosaic virus, potato virus X and cowpea mosaic virus. Vaccine candidates have been produced against a range of human and animal diseases, and in many cases have shown immunogenic activity and protection in the face of disease challenge. In a review, Yusibov et al. (2006) discuss the advantages of plant virus vectors, the development of different viruses as vector systems, and the immunological experiments that have demonstrated the principle of plant virus-derived vaccines.

FEASIBILITY OF ENGINEERING PLANTS FOR DRUG PRODUCTION

Only a handful of small-molecule-drugs used are natural products obtained from plants, or are synthetic compounds based on molecules found naturally in plants. However, the vast majority of the protein therapeutics (or biopharmaceuticals) used today are from animal or human source. The possibility of producing huge quantities of proteins of therapeutic importance has added fuel to our desire to use plants as bioreactors for such production (Larrick and Thomas, 2001). Over the last few years, it has become clear that plants have great potential for the production of human proteins and other protein-based therapeutic entities. Plants offer the prospect of inexpensive biopharmaceutical production without sacrificing product quality or safety, and following the success of several plant-derived technical proteins, the first therapeutic products are now approaching the market. In a recent review, Twyman et al. (2005) have discussed the different plant-based pharmaceutical production systems evaluating the merits of transgenic plants in comparison with other means of making these therapeutics. A detailed discussion is provided of the development issues that remain to be addressed before plants become an acceptable mainstream production technology (Twyman et al., 2005). The many different proteins that have already been produced using plants are described, and a sketch of the current market and the activities of the key players are provided. Even without a clear regulatory framework, the benefits of plant-derived pharmaceuticals have brought the prospect of inexpensive veterinary and human medicines closer than ever before.

The future production of recombinant proteins in plants is, therefore, becoming increasingly promising. It is expected that the number of applications will increase as more therapeutic proteins are required to be made in quantities that cannot be attained economically in cell culture through the use of fermenters. Antibodies have long been recognized for their diagnostic and therapeutic potential. The number of monoclonal antibodies approved for immunotherapy is rapidly increasing. This has paved the way for an even greater demand for these molecules. In order to meet this growing demand and to enhance the production capacity, alternative systems based on antibody production in transgenic plants are being actively explored and many plant species have been successfully used in the production of biologically active antibodies.

A common problem with plant-produced proteins is that they are often rejected by the human body's immune system, which treats them as foreign. Previous attempts, at using plants to make human proteins, have failed due to the body recognizing the products as foreign substances, and triggering unwanted immune reactions Plant-specific *N*-glycosylation is said to represent an important limitation for the use of recombinant glycoproteins of mammalian origin produced by transgenic plants. In order to reduce immunogenic potential of plant-based therapeutics, efforts have been made to modify glycosylation, which represents a major step forward in the engineering of the *N*-glycosylation of recombinant proteins. To reduce such rejection of plant-made proteins the 'humanization' of such proteins is under active research. Bakker et al. (2001) provide a major effort in the *in planta* engineering of the *N*-glycosylation of recombinant antibodies.

Joshi's research team (Shah et al., 2003) has also uncovered a pathway in plants that brings scientists significantly closer to the generation of biocompatible proteins via plants. Plants have, therefore, been engineered to generate human-like proteins. This team of researchers has found a method of attaching the needed sugar groups (sialic acids) to the proteins to prevent undesired removal of proteins from the body. Presently, the group is focusing on "metabolically engineering" plants to enhance the levels of sugars put on the proteins to get the maximum yield. This work is of great importance for large-scale transgenic plant production systems of recombinant human protein therapeutics, which often require complex glycosylation, such as sialylation, for stability, solubility, folding, pharmacokinetics/pharmacodynamics and/or activity (Seveno et al., 2004).

Gomord et al. (2004) also focus their attention on transgenic plants as a promising system for scale-up and processing of plant-made pharmaceuticals. The advantages and limitations induced by glycosylation of plant-made antibodies for human therapy are also the focus of attention of these authors.

Control of post-transcriptional gene silencing in plants will also enhance the value of this important technology. An increasing number of transgenic plant-derived proteins is entering clinical testing. The initial success of these development programs suggests an important role for cost effective and large-scale production technologies.

EMERGING SCIENCE OF SYNTHETIC BIOLOGY

Scientists have played around for decades with genes and whole biochemical and regulatory pathways to create products such as insulin-producing bacteria, plants that synthesize antibodies and other therapeutic molecules. They have been producing a number of proteins of therapeutic importance by simply slipping an extra gene into *E. coli*. This was genetic engineering to them. However, to a new generation of researchers, genetic engineering is not just moving a preexisting gene from one organism to another. To them , engineering is designing what one has in mind, making sure the approach is rationale, and eventually building biological systems simply from scratch. This is the emerging science of synthetic biology.

In this system one is able to specify all the DNA that enters the organism so that it is possible to determine the function in a way that can be predicted. Some compare this to "etching a microprocessor or building a bridge". The goal is to create genetically custom-made organisms for energy production and the production of environment friendly industrial chemicals, or also for engineering organisms for mass production of drugs.

Synthetic biology aims at designing and constructing new enzymes, gene control circuits and even cells. This emerging science is also directed at redesigning existing biological systems with the help of advances made in molecular, cell and systems biology. It is seeking to change biology in the same way as the integrated circuit design transformed computing. In a manner similar to engineers designing integrated circuits, through fabrication of processors, synthetic biologists will design and build biological systems. It is important to know the physical details of gene sequences, protein properties and biological systems in order to recast the same with a set design rules. The construction of *de-novo* genetic circuits begins with the assembly and characterization of genetic parts, the cellular building blocks. Scientists are discovering useful engineering principles in biochemical circuitry (Blake and Isaacs, 2004).

At the heart of synthetic biology is the pioneering work of French researchers, Francois Jacob and Jacques Monod. Their outstanding work on how genes can be regulated paved the path of this new discipline of science. The studies of gene regulation began with a small number of bacterial systems and it led to the celebrated operon model of Jacob and Monod, in 1961. This model, a paradigm for gene regulation in all organisms, introduced concepts such as regulator gene and transcriptional repression (Jacob and Monod, 1961). Since then the model has been elaborated extensively (Reznikoff, 1972) as different regulatory mechanisms, such as transcriptional activation, (Englesberg et al., 1965) were discovered. These analyses of gene expression patterns observed over a range of conditions and time points have become widely used in modern biology to discover relationships between different genes in a genome.

ENGINEERING CELLS

Knowledge of the existence of many circuitries for the regulation of their genes in natural organisms is paving the pathway for manufacturing of functional artificial biological circuitry. Such a circuit was built when three genes: *tetR, lacI* and *λcI* encoding for the proteins TetR, LacI and λcI were placed on a plasmid (Baker et al., 2006). The arrangement was such that the protein products of the three genes in the circuit were bound selectively to one another's promoter sequences. LacI protein binds to the *tetR* gene's promoter, whereas the λcI protein binds to the *lacI* gene's promoter, and TetR binds the promoter of *λcI* gene. Such interrelations allow the protein product of one gene to block RNA polymerase from binding to the promoter of another gene. Thus, synthesis of the three proteins occurs through an oscillatory cycle: abundance of LacI protein represses *tetR* gene activity; absence of TetR protein will then enable *λcI* gene to be turned on. The latter will then repress the synthesis of LacI and so forth. If one of the protein products in the circuitry was hooked on to the gene for green fluorescent protein and the whole system introduced into bacteria, then the bacteria either fluoresced or did not fluoresce, depending on the gene activation. It is expected that such genetic switches would allow the use of bacteria for a number of purposes, e.g., detection of cellular DNA damage, etc.

Synthetic biology researchers are hoping they will be able to create bacteria that can produce drugs, which presently have to be extracted from rare plants. It involves creating new biochemical pathways by inserting plant and yeast genes into the bacterial genome and placing them under the tight control of regulatory networks. This is also known as metabolic engineering, which is the modification of existing, or the introduction of entirely new, metabolic pathways in organisms. Introduction of new pathways enables the use of nature's diversity to meet the needs of humans in a sustainable manner. Bacteria and yeast are already in use in numerous biotechnology applications such as fermentation and drug synthesis. The ability to engineer multicellular systems will trigger vast improvements in existing applications as well as opening up a wide variety of new possibilities.

RETOOLING A MICOORGANISM'S METABOLIC PATHWAY

At present synthetic metabolic networks make use of transcriptional and translational control elements to regulate the expression of enzymes that synthesize and breakdown metabolites. In these systems, metabolite concentration acts as an input for other control elements. Researchers have retooled a microorganism's metabolism to make a chemical precursor of artemisinin - the best drug available today to cure malaria. They made all the elements necessary for the production of the anti-malarial drug work as a coordinated circuit. Artemisinins are currently produced using a Chinese herb, wormwood (*Artemesia annua*).

An entire metabolic pathway, the mevalonate isoprenoid pathway for synthesizing isopentyl pyrophosphate from *Saccharomyces cerevisiae,* was first successfully transplanted into *E. coli* (Martin et al., 2003). In combination with an inserted synthetic amorpha-4,11-diene synthase, this pathway produced large amounts of a precursor to the anti-malarial drug artemisinin. cDNA libraries from *A. annua* were screened and those genes were identified which encoded artemisinic acid biosynthetic enzymes. Genes so identified were then expressed in the amorphadiene-producing *E. coli* host and optimized to maximize production of artemisinic acid. This optimized strain was then cultivated in large bioreactors to produce great quantities of artemisinic acid inexpensively.

Recently the same group has achieved a similar feat in yeast (Ro et al., 2006). They have engineered an artemisinic-acid-producing yeast in three steps, by (a) engineering the farnesyl pyrophosphate (FPP) biosynthetic pathway to increase FPP production and decrease its use for sterols (b) introducing the amorphadiene synthase gene (ADS) from *A. annua* into the high FPP producer to convert FPP to amorphadiene, and (c) cloning a novel cytochrome P450 that performs a three-step oxidation of amorphadiene to artemisinic acid from *A. annua* and expressing it in the amorphadiene producer.

By manipulating the different gene regulatory elements, the output of one of its products, farnesyl pyrophosphate, was maximized. Then they added a gene from a single wormwood enzyme, amorphadiene synthase, which converted the pyrophosphate to amorphadiene. Finally, they introduced a novel cytochrome P450 that performs a three-step oxidation of amorphadiene to artemisinic acid, a precursor to artemisinin. Thus, the goal of engineering the production of artemisinin, a plant product, in yeast was complete and was only one simple chemical alteration away.

CONCLUSION

"Modern Biotechnology" came to be applied to those applications of biological science that incorporated DNA science and this gave birth to the technology of "genetic engineering" or "recombinant DNA technology". Classical biotechnology industry has been literally transformed because of the tools of DNA science. This "modern" part of biotechnology shows a high potential for solving various problems of our modern world. The methodology of the earliest agriculture was the first concept of biotechnology. However, in recent years, biotechnological applications for the improvement of plants has progressed tremendously. Genetic engineering of plants was found to be much easier than that of animals. The main reason for this was the presence of a natural transformation system for plants in the bacterium, *A. tumefaciens*. Detailed knowledge of the mechanism of T-DNA transfer and integration in the host genomes has fine-tuned our capability of using this microorganism for plant transformation.

Genetic engineering has embraced various methods adopting biological organisms, or parts thereof, to make useful products for solving some problems. Synthetic biology is another innovation that will make possible the building of living machines from off-the-shelf chemical ingredients. Scientists are engineering microbes to perform complex multi-step syntheses of natural products by assembling animal or plant genes that encode for almost all the enzymes in a synthetic pathway. This new and emerging branch of biology is all about redesigning natural biological systems for greater efficiency, as well as for the construction of functional "genetic circuits" and metabolic pathways for practical purposes, like the synthesis of anti-malarial drugs. Synthetic-biology techniques have been used to program yeast cells to manufacture the immediate precursor of the drug artemisinin.

REFERENCES

Abu-Arish, A., Frenkiel-Krispin, D., Fricke, T., Tzfira, T., Citovsky, V., Wolf, S.G. and Michael, E. (2004). Three-dimensional reconstruction of *Agrobacterium* VirE2 protein with single-stranded DNA. J. Biol. Chem. 279: 25359-25363.

Afolabi, A.S., Worland, B., Snape, J.W. and Vain, P. (2004). A large-scale study of rice plants transformed with different T-DNAs provides new insights into locus composition and T-DNA linkage configurations. Theort. Appl. Genet. 109: 815-826.

Albert, H., Dale, E.C., Lee, E. and Ow, D.W. (1995). Site-specific integration of DNA into wild-type and mutant lox sites placed in the plant genome. Plant J. 7: 649–659.

Altpeter, F., Baisakh, N., Beachy, R., Bock, R., Capell, T., Christou, H., Datta, K., Datta, S., Dix, P.J., Fauquet, C., Huang, N., Kohli, A., Mooibroek, H., Nicholson, L., Nguyen, T.T., Nugent, G., Raemakers, K., Romano, A., Somers D.A., Stoger, E., Taylor, N. and Visser, R. (2005). Particle bombardment and the genetic enhancement of crops: myths and realities. Molecular Breeding 15(3): 305-327.

Baker, D., Church, G., Collins, J., Endy, D., Jacobson, J., Keasling, J., Mordich, P., Smolke, C. and Weiss, R. (2006). Engineering life: Building a FAB for biology. Scientific American 294(6): 44-51.

Bakker, H., Bardor, M., Molthoff, J.W., Gomord, V., Elbers, I., Stevens, L.H., Jordi, W., Lommen, A., Faye, L., Lerouge, P. and Bosch, D. (2001). Galactose-extended glycans of antibodies produced by transgenic plants. Proc. Natl. Acad. Sci. USA 98(5): 2899-2904.

Ballas, N. and Citovsky, V. (1997). Nuclear localization signal binding protein from Arabidopsis mediates nuclear import of Agrobacterium VirD2 protein. Proc. Natl. Acad. Sci., USA 94: 10723-10728.

Benfey, P.N. and Chau, N.H. (1989). Regulated genes in transgenic plants. Science 244: 174-181.

Bevan, M. (1984). Binary *Agrobacterium* vectors for plant transformation. Nucleic Acid Res. 12 (22): 8711-8721.

Blake, W.J. and Isaacs, F.J. (2004). Synthetic biology evolves. Trends Biotechnol. 22(7): 321-324.

Bomhoff, G., Klapwijk, P.M., Kester, H.C.M., Schilperoort, R.A., Hernalsteens, J.P. and Schell, J. (1976). Octopine and nopaline synthesis and breakdown genetically controlled by a plasmid of Agrobacterium tumefaciens. Mol. Gen. Genet. 145: 177-181.

Broothaerts, W., Mitchell, H.J., Weir, B., Kaines, S., Smith, L.M.A., Yang, W., Mayer, J.E., Roa-Rodriguez, C. and Jefferson, R.A. (2005). Gene transfer to plants by diverse species of bacteria. Nature 433: 629-633.

Capecchi, M.R. (1980). High efficiency transformation by direct microinjection of DNA into cultured mammalian cells. Cell 22: 479-488.

Chen, X., Stone, M., Schlagnhaufer, C. and Romaine, C.P. (2000). A fruiting body tissue method for efficient *Agrobacterium*-mediated transformation of *Agaricus bisporus*. Appl. Environ. Microbiol. 66(10): 4510–4513.

Cheng, M., Lowe, B.A., Spencer, T.M., Ye, X.D. and Armstrong, C.L. (2004). Factors mediating transformation of monocotyledonous species. In vitro Cell Dev. Biol. Plant. 40: 31-45.

Chilton, M.D., Drummond, M.H., Merio, D.J., Sciaky, D., Montoya, A.L., Gordon, M.P. and Nester, E.W. (1977). Stable incorporation of plasmid DNA into higher plant cells: the molecular bases of crown gall tumorigenesis. Cell 11(2): 263-271.

Clapp, D.W. (1993). Somatic gene therapy into hematopoietic cells: Current status and future implications. Clin. Perinatol. 20(1): 155-168.

Datta, S.K., Peterhans, A., Datta, K. and Potrykus, I. (1990). Genetically engineered fertile *indica* rice recovered from protoplasts. Bio/Technology 8: 736-740.

Davey, M.R., Kothari, S.L., Zhang, H., Rech, E.L., Cocking, E.C. and Lynch, P.T. (1991). Transgenic rice: Characterization of protoplast-derived plants and their seed progeny. J. Exp. Bot. 42: 1129-1169.

Dekeyser, R., Claes, B., Marichal, M., Montagu, M. and Caplan, A. (1989). Evaluation of selectable markers for rice transformation. Plant Physiol. 90: 217-223.

Della-Cioppa, G., Bauer, S.C., Taylor, M.L., Rochester, D.E., Klein, B.K., Shah, D.M., Fraley, R.T. and Kishore, G.M. (1987). Targeting a herbicide-resistant enzyme from *Escherichia coli* to chloroplasts of higher plants. Bio/Technology 5: 579-584.

Eglitis, M.A. and Anderson, W.R.(1988). Retroviral vectors for introduction of genes into mammalian cells. Biotechniques 6(7): 608-614.

Englesberg, E., Irr, J., Power, J. and Lee, N. (1965). Positive control of enzyme synthesis by gene C in the L-arabinose system. J. Bacteriol. 90: 946-957.

Fromm, M., Loverine, M.F., Taylor, P. and Virginia, W. (1985). Expression of genes transferred into monocot and dicot plant cells by electroporation. Proc. Natl. Acad. Sci., USA 82(17): 5824-5828.

Fromm, M.E., Morrish, F., Armstrong, C., Williams, R., Thomas, J. and Klein, T.M. (1990). Inheritance and expression of chimeric genes in the progeny of transgenic maize plants. Bio/Technology 8: 833-844.

Fynan, E.F., Webster, R.G., Fuller, D.H., Haynes, J.R., Santoro, J.C. and Robinson, H.L. (1993). DNA vaccines: Protective immunizations by parenteral, mucosal, and gene-gun inoculations. Proc. Natl. Acad. Sci., USA 90(24): 11478-11482.

Gelvin, S.B. (2003). *Agrobacterium*-mediated plant transformation: The biology behind the "gene-jockeying" tool. Microbiol. Mol. Biol. Rev. 67: 16–37.

Gleba, Y., Marillonnet, S. and Klimyu, V. (2004). Engineering viral expression vectors for plants: the 'full virus' and the 'deconstructed virus' strategies. Curr. Opin. Plant Biol. 7(2): 182-188.

Gomord, V., Sourrouille, C., Fitchette, A.C., Bardor, M., Pagny, S., Lerouge, P. and Faye, L. (2004). Production and glycosylation of plant-made pharmaceuticals: The antibodies as a challenge. Plant Biotechnol. J. 2(2): 83–100.

Graham, F.L. and van der Eb, A.J. (1973). A new technique for the assay of infectivity of human adenovirus 5 DNA. Virology 54(2): 536-539.

Hennegan, K.P. and Danna, K.J. (1998). pBIN20: an improved binary vector for *Agrobacterium*-mediated transformation. Plant Mol. Biol. Rep. 16: 129-131.

Hookyaas, P.J.J. (1998). *Agrobacterium* molecular genetics. In: Plant Molecular Biology Manual (eds.) S.D. Gelvin and R.A. Schilperoort, Kluwer Academic Press, Dordrecht, The Netherlands.

Howell, S.H., Walker, L.L. and Dudley, R.K. (1980). Cloned cauliflower mosaic-virus DNA infects turnips (*Brassica rapa*). Science 208(4449): 1265-1267.

Irdani, T., Bogani, P., Mengoni, A., Mastromei, G. and Buiatti, M. (1997). Construction of a new vector conferring methotrexate resistance in *Nicotiana tabacum* plants. Plant Mol. Biol. 37(6): 1079-1084.

Jacob, F. and Monod, J. (1961). Genetic regulatory mechanisms in the synthesis of proteins. J. Mol. Biol. 3: 318-356.

Johnston, S.A. and Tang, D. (1994). Gene gun transfection of animal cells and genetic immunization. Methods Cell Biol. 43: 353-365.

Jones, H.D., Doherty, A. and Wu, H. (2005). Review of methodologies and a protocol for the *Agrobacterium*-mediated transformation of wheat. Plant Methods 1: 5-9.

Kay, R., Chan, A., Daly, M. and McPherson, J. (1987). Duplication of CaMV 35S promoter sequences creates a strong enhancer for plant genes. Science 236: 1299-1302.

Kumar, S. and Fladung, M. (2001). Controlling transgene integration in plants. Trends Plant Sci. 6(4): 155-159.

Kunik, T., Tzfira, T., Kapulnik, Y., Gafni, Y., Dingwall, C. and Citovsky, V. (2001). Genetic transformation of HeLa cells by *Agrobacterium*. Proc. Natl. Acad. Sci., USA 98: 1871-1876.

Lacroix, B., Tzfira, T., Vainstein, A. and Citovsky, V. (2006). A Case of promiscuity: *Agrobacterium*'s endless hunt for new partners. Trends Genet. 22(1): 29-37.

Larrick, J.W. and Thomas, D.W. (2001). Producing proteins in transgenic plants and animals. Curr. Opin. Biotechnol. 12(4): 411-418.

Leskov, K.S., Criswell, T., Antonio, S., Li, J., Yang, C.R., Kinsella, T.J. and Boothman, D.A. (2001). When X-ray-inducible proteins meet DNA double strand break repair. Semin Radiat Oncol. 11: 352–372.

Lima, I.G.P., Duarte, R.T.D., Furlaneto, L., Baroni, C.H., Fungaro, M.H.P. and Furlaneto, M.C. (2006). Transformation of the entomopathogenic fungus *Paecilomyces fumosoroseus* with *Agrobacterium tumefaciens*. Lett. Appl. Microbiol. 42: 631-636.

Lu, L., Xiao, M., Clapp, D.W., Li, Z.H. and Broxmeyer, H.E. (1993). High efficiency retroviral mediated gene transduction into single isolated immature and replatable CD34 (3+) hematopoietic stem/progenitor cells from human umbilical cord blood. J. Exp. Med. 178: 2089-2096.

Martin, V.J.J., Pitera, D.J., Withers, S.T., Newman, J.D. and Keasling, J.D. (2003). Engineering the mevalonate pathway in *Escherichia coli* for production of terpenoids. Nat. Biotechnol. 21(7): 796–802.

Nester, E., Gordon, M.P. and Kerr, A. (editors) (2005). *Agrobacterium tumefaciens*: from Plant Pathology to Biotechnology. American Phytopathological Society Press, St. Paul, MN, USA.

Nilsson, O. and Olsson, O. (1997). Getting to the root: The role of *Agrobacterium rhizogenes rol* genes in the formation of hairy roots. Physiol. Plant. 100: 463-473.

Odell, J.T., Nagy, F. and Chua, N.H. (1985). Identification of DNA sequences required for activity of the cauliflower mosaic virus 35S promoter. Nature 313: 810-812.

Odell, J.T. and Russell, R.H. (1994). Use of site-specific recombination systems in plants. In: Homologous Recombination in Plant. (ed.). J. Paszkowski. Kluwer Academic Publishers, Dordrecht. Netherlands, pp. 219-270.

Oleinick, N.L., Balasubramaniam, U., Xue, L. and Chiu, S. (1994). Nuclear structure and the microdistribution of radiation damage in DNA. Int. J. Radiat. Biol. 66(5): 523-529.

Opabode, J.T. (2006). *Agrobacterium*-mediated transformation of plants: emerging factors that influence efficiency. Biotechnol. Mol. Biol. Rev. 1(1): 12-20.

Pray, C.E. and Naseem, A. (2005). Intellectual property rights on research tools: Incentives or barriers to innovation? Case studies of rice genomics and plant transformation technologies. AgBioForum, 8(2&3): 108-117. Available at: http://www.agbioforum.

Prescott, M., Harley, J. and Klein, D. (1999). Microbiology (4th ed.), WCB McGraw-Hill Publishing, International Edition, USA, pp. 326-237.

Rechsteiner, M. and Rogers, S.W. (1996). PEST sequences and regulation by proteolysis. Trends Biochem. Sci. 21: 267-271.

Reznikoff, W.S. (1972). The Operon Revisited. Annu. Rev. Genet. 6: 133-156.

Rhodes, C.A., Pierce, D.A., Mettler, I.J., Mascarendas, D. and Detmer, J.J. (1988). Genetically transformed maize plants from protoplasts. Science 240: 204-207.

Ro, D-K., Paradise, E.M., Ouellet, O., Fisher, K.J., Newman, K.L. and Keasling, J.D. (2006). Production of the antimalarial drug precursor artemisinic acid in engineered yeast. Nature 440 (7086): 940-943.

Sadowski, P. (1995). The Flp recombinase of the 2-μm plasmid of *Saccharomyces cerevisiae*. Prog. Nuclear Acid Res. Mol. Biol. 51: 53-91.

Salman, H., Abu-Arish, A., Oliel, S., Loyter, A., Klafter, J., Granek, R. and Elbaum, M. (2005). Nuclear localization signal peptides induce molecular delivery along microtubules. Biophys. J. 89: 2134-2145.

Scholthof, H.B. and Scholthof, K-B.G. and Jackson, A.O. (1996). Plant virus gene vectors for transient expression of foreign proteins. Annu. Rev. Phytopathol. 34: 299-323.

Séveno, M., Bardor, M., Paccalet, T., Gomord, V., Lerouge, P. and Faye, L. (2004). Glycoprotein sialylation in plants? Nat. Biotechnol. 22: 1351-1352.

Shah, M.M., Fujiyama, K., Flynn, C.R. and Joshi, L. (2003). Sialyated endogenous glycoconjugates in plant cells. Nat. Biotechnol. 21(12): 2470-2471.

Shimamoto, K., Terada, R., Izawa, T. and Fujimoto, H. (1989). Fertile transgenic rice plants regenerated from transformed protoplasts. Nature 338: 274-276.

Stanley, W.M. (1935). Isolation of a crystalline protein possessing the properties of tobacco-mosaic virus. Science 81(2113): 644-645.

Taylor, N.J. and Fauquet, C.M. (2002). Microparticle bombardment as a tool in plant science and agricultural biotechnology. DNA and Cell Biol. 21(12): 963-977.

Terzaghi, W.B. and Cashmore, A.R. (1997). Plant cell transfection by electroporation. Methods Mol. Biol. 62: 453-462.

Twyman, R.M., Schillberg, S. and Fischer, R. (2005). Transgenic plants in the biopharmaceutical market. Expert Opinion on Emerging Drugs 10(1): 185-218.

Tzfira, T. (2006). On tracks and locomotives: The long route of DNA to the nucleus. Trends Microbiol. 14(2): 61-63.

Tzfira, T., Vaidya, M. and Citovsky, V. (2004). Involvement of targeted proteolysis in plant genetic transformation by *Agrobacterium*. Nature 431: 87-92.

Tzfira, T., Lacroix, B. and Citovsky, V. (2005). Nuclear import of *Agrobacterium* T-DNA. In: Nuclear Import and Export in Plants and Animals (eds.) T. Tzfira and V. Citovsky, Landes Bioscience and Kluwer Academic Publishers, Dordrecht. The Netherlands, pp. 83-99.

Van Attikum, H. and Hooykaas, P.J.J. (2003). Genetic requirements for the targeted integration of *Agrobacterium* T-DNA in *Saccharomyces cerevisiae*. Nucleic Acids Res. 31: 826-832.

Vlasak, J. and Ondrej, M. (1992). Construction and use of *Agrobacterium tumefaciens* binary vectors with *A. tumefaciens* C58T-DNA genes. Folia Microbiol. (Praha) 37(3): 227-230.

Windels, P., De Buck, S., Van Bockstaele, E., De Loose, M. and Depicker, A. (2003). T-DNA integration in *Arabidopsis* chromosomes. Presence and origin of filler DNA sequences. Plant Physiol. 133: 2061-2068.

Wong, T.K. and Neumann, E. (1982). Electric field mediated gene transfer. Biochim. Biophys. Res. Commun. 107(2): 584-587.

Yusibov, V., Rabindran, S., Commandeur, U., Twyman, R.M. and Fischer, R. (2006). The potential of plant virus vectors for vaccine production. Drugs in R & D. 7(4): 203-217.

Zaenen, I., Van Larebeke, N., Van Montagu, M. and Schell, J. (1994). Supercoiled circular DNA in crown gall inducing *Agrobacterium* strains. J. Mol. Biol. 86(1): 109-127.

Zatloukal, K., Wagner, E., Cotton, M., Phillips, S., Plank, C., Steinlein, P., Curiel, D.T. and Birnstiel, M.L. (1992). Transferrinfection: a highly efficient way to express gene constructs in eukaryotic cells. Ann. N.Y. Acad. Sci., USA 660: 136-153.

Zhu, J., Oger, P.M., Schrammeijer, B., Hooykaas, P.J.J., Farrand, S.K . and Winans, S.C. (2000). The bases of crown gall tumorigenesis. J. Bacteriol. 182 (4): 3885-3895.

Zubko, E., Scutt, C. and Meyer, P. (2000). Intrachromosomal recombination between *attP* regions as a tool to remove selectable marker genes from tobacco transgenes. Nat. Biotechnol. 18: 442–445.

4

Plant Growth Promoting Rhizobacteria: Significance in Horticulture

Cheryl L. Patten and Bernard R. Glick*

INTRODUCTION

Plant growth promoting rhizobacteria (PGPR) are free-living soil bacteria that colonize seeds and roots, and provide some benefit to the host plant. These bacteria are distinct from symbiotic bacteria such as Rhizobia, that form nitrogen-fixing nodules on host legumes and therein provide the reduced nitrogen necessary for their host plants to grow in nitrogen poor soils, and from biocontrol bacteria that benefit host plants indirectly by inhibiting phytopathogenic bacteria (discussed in Chapters 4 and 5 in Volume 1 of this series (Microbial Biotechnology in Horticulture). Under consideration here are the associative PGPR that benefit their host plant by producing compounds or enzymes that directly contribute to plant growth. PGPR strains that are known to enhance plant growth belong to a variety of genera; the most thoroughly studied are *Pseudomonas*, *Enterobacter*, *Bacillus*, *Azospirillum*, *Burkholderia*, and *Kluyvera* (Glick et al., 1999; Reed and Glick, 2004). Many of these are available as commercial formulations under such trade names such as Azo-Green and Foundation RGP (*Azospirillum brasilense*); Zea-Nit (*Azospirillum brasilense* and *Azospirillum lipoferum*); Subtilex, Kodiak, Epic, Companion, and Serenade (all *Bacillus subtilis*); Deny and Intercept (*Burkholderia cepacia*); and Conquer and Victus (*Pseudomonas fluorescens*) (Fravel, 2005). Most of

*Corresponding Author

them are described as effective biopesticides to combat root and foliar pathogens; however, a few of them, such as those containing *Azospirillum* spp., are marketed as biofertilizers for accelerated seed germination and increased root biomass (e.g., http://www.encoretechllc.com).

Bacteria, both beneficial and pathogenic, colonize plant roots because plants release substantial quantities of metabolites from their roots; it is estimated that 5-21% of total photosynthates are exuded from plant roots (Marschner, 1995). Root exudates contain low molecular weight organic compounds such as amino acids, organic acids including tricarboxylic acid cycle intermediates, sugars, and phenolics like flavonoids and phenylpropanoids, a variety of inorganic compounds such as phosphate, and high molecular weight polymers such as polysaccharides and proteins, that can be used as nitrogen, carbon and/or energy sources by bacteria (Neumann and Romheld, 2001). Thus, plant roots are a rich source of nutrition for colonizing bacteria. Why do plants expend energy and resources to synthesize metabolites only to lose much of these through their roots? One answer is that they do so to attract bacteria from which they derive some benefit. This may be especially important when plants attempt to grow under conditions that are less than ideal. Indeed, the composition of exudates is altered when plants encounter sub-optimal environments (Walker et al., 2003).

In exchange for, and even in response to root exudates, PGPR benefit their host plant by increasing germination rate (Reed et al., 2005; Lytvynenko et al., 2006), stimulating growth of seedling roots (Patten and Glick, 2002a), increasing lateral and adventitious root formation (Patten and Glick, 2002a; Mayak et al., 1999) and promoting higher seed yields (Dey et al., 2004). These bacteria help plants survive less than ideal growing conditions such as salt stress (Mayak et al., 2004a), cold temperatures (Sun et al., 1995), flooding (Grichko and Glick, 2001), drought (Mayak et al., 2004b), and toxic environments (Burd et al., 2000; Huang et al., 2004; Reed and Glick, 2005). Plants of horticultural interest that are known to respond to PGPR include ornamentals such as French marigolds (*Tagetes patula* L.), *Amaranthus paniculatus*, and carnations (Pandey et al., 1999; Li et al., 2005; Lytvynenko et al., 2006); and a variety of vegetable and fruit crops like as tomato, potato, cucumber, radish, lettuce and pepper (van Peer and Schippers, 1988; Hall et al., 1996; Polyanskaya et al., 2000; Mayak et al., 2001; Kokalis-Burelle et al., 2002).

BENEFICIAL TRAITS OF PGPR

Traits that Facilitate Plant Colonization

In order to promote plant growth, rhizosphere bacteria must be able to colonize and proliferate on plant roots and seeds. This is a highly competitive environment, and inconsistencies in the performance of PGPR in field applications may be due to the inability of introduced strains to compete effectively with the indigenous microflora. Many plant and bacterial factors contribute to the successful establishment of PGPR in the rhizosphere. As previously mentioned, bacteria are attracted to seed and root exudates because they contain an abundance of nutrients compared to the bulk soil. Highly effective PGPR are generally fast-growing strains that can use a variety of substrates for growth and are, therefore, good competitors for root colonization. The chemotactic response to root exudates and the role of flagella in this response has been shown to be important for colonization of tomato, potato, and wheat roots by *Pseudomonas fluorescens* (de Weger et al., 1987; Turnbull et al., 2001; de Weert et al., 2002). Mutants deficient in the protein flagellin, which makes up the flagella filament, are unable to colonize seeds (DeFlaun et al., 1994).

The production of an extracellular polysaccharide slime layer and protein appendages such as fimbriae or pili for attachment to host cells is common among both phytopathogens and PGPR (Amellal et al., 1998; Dörr et al., 1998; Burdman et al., 2000; Espinosa-Urgel, 2004). Although direct evidence for the involvement of exopolysaccharides in root colonization by free-living PGPR is limited, exopolysaccharide deficient mutants of the PGPR *Azospirillum lipoferum* were impaired in their ability to colonize roots of pearl millet (Garg et al., 1996), and the root nodule symbionts *Rhizobium leguminosarum* and *Sinorhizobium meliloti* were impaired in their ability to colonize infection threads and invade root nodules, respectively (Laus et al., 2004; Leigh et al., 1985). Similar mutants of the plant pathogen *Pseudomonas syringae* could not colonize tomato leaves (Yu et al., 1999). In animal pathogens, pili or fimbriae are responsible for specific adherence to host cells. Pili have also been implicated in the recognition and attachment of bacteria to plant cells. Type IV pili are involved in attachment of the plant pathogen *P. syringae* to bean (Romantschuk and Bamford, 1986), the biocontrol strain *P. fluorescens* to corn roots (Vesper, 1987), and the diazotroph *Azoarcus* sp. to rice (Dörr et al., 1998).

Sensing and attachment of bacteria to a surface via an exopolysaccharide layer and pili triggers events that lead to formation of biofilms (O'Toole et al., 2000). Biofilms are structured bacterial communities, often comprised of several different species, which form on surfaces. PGPR such as *Pseudomonas* spp. and *Azospirillum* spp. have been shown to colonize roots as microcolonies or biofilms (Fukui et al., 1994; Bloemberg et al., 2000). Bacteria within a biofilm are protected from environmental insults such as antibiotics secreted by other bacteria and predation by protozoa by a thick exopolysaccharide matrix thus promoting the survival of the bacteria on roots (Morris and Monier, 2003).

While it is known that PGPR are present on root surfaces, primarily in the upper regions, and that they generally migrate to roots following seed treatment (e.g., Ma et al., 2001; Wang et al., 2004), some strains may also colonize the interior of plant root tissues. Forty percent of the endophytic bacteria isolated from stems of potato plants promoted growth of potato tissues (Sessitsch et al., 2004). These included strains of *Arthrobacter* spp., *Pseudomonas* spp., and *Brevundimonas* spp. The PGPR strain *Burkholderia* sp. PsJN, that promotes growth of potatoes, vegetables and grapevines, initially colonizes the root surface, and then enters the interior of the root at sites of lateral root emergence and through the root tip (Compant et al., 2005). This colonization pattern is similar to that described for *Herbaspirillum seropedicae* (James et al., 2002) and *Bacillus megaterium* (Liu et al., 2006). Following penetration of the root vascular tissue, *B. megaterium* was observed to migrate to the stems and leaves of rice and maize (Liu et al., 2006). Like bacteria in biofilms, endophytic bacteria are protected from environmental stresses.

Plant exudates not only contain a variety of important bacterial nutrients, but roots secrete chemical signals that may regulate the bacterial community (Walker et al., 2003). Classical examples include the secretion of flavonoids by legumes that induce expression of Rhizobiaceae lipopolysaccharide Nod factors required for nodulation (Van Rhijn and Vanderleyden, 1995), and secretion of phenolic compounds from wounded plants that induce virulence genes required for tumor formation by *Agrobacterium tumefaciens* (Lee et al., 1996). Many bacterial genes have proved to be specifically activated in the rhizosphere. In the PGPR *Pseudomonas putida*, more than 100 genes that were specifically up-regulated during colonization of maize plants were identified using a promoter trapping technique (Ramos-Gonzalez et al., 2005). These include genes involved in nutrient acquisition, stress response, attachment and colonization, and protein secretion. The nature of the regulating signal(s) has not yet been elucidated.

Plants have also been shown to secrete compounds that affect bacterial quorum sensing (Gao et al., 2003). Quorum sensing is a microbial cell-to-cell signaling mechanism that allows bacteria to sense other bacteria in the environment and to modify their behavior in accordance with cell density. The communication system consists of a small diffusible signal molecule, either N-acyl homoserine lactones in gram-negative bacteria or a small peptide in gram-positive bacteria, that accumulates in the environment in a cell density-dependent manner. When signal concentrations reach a threshold level, a response is triggered in neighboring bacteria that results in transcriptional changes in a number of genes (Waters and Bassler, 2005). Many bacterial processes are regulated in this way, including pathogenesis, biofilm formation, root colonization and biocontrol activity (Fray, 2002; Arevalo-Ferro et al., 2005). Substances secreted by plants, as yet unidentified, bind to bacterial quorum sensing signal receptors and either promote degradation of the receptor, thereby disrupting signal transduction, or activate expression of bacterial genes that respond to quorum sensing signals (Bauer and Mathesius, 2004; Keshavan et al., 2005).

Type III protein secretion systems have been well characterized in plant pathogens and are comprised of a protein channel that provides a direct route for passage of bacterial proteins from the bacterial cytoplasm to host plant cytoplasm (Alfano and Collmer, 1997). Mutants lacking this system are unable to colonize host plants. Hundreds of effector proteins that are secreted through the channel into host cells have been identified, and it is estimated that a given pathogen secretes about 50 effector proteins into a host plant cell; many of these function to suppress the plant's attempts to defend itself against the invader (Mudgett, 2005). Homologous systems have been found in non-pathogenic *Pseudomonas* strains, and although the effector proteins that are secreted through this system have not yet been identified, the Type III secretion system of *P. fluorescens* was functional and able to secrete recombinant proteins into host plants (Preston et al., 2001).

Traits that Alter Plant Hormone Levels

Some of the studies that look at the early effects of PGPR on plant growth utilize sterile growth pouches containing only water. Exogenous nutrients are not provided, therefore, factors responsible for any observed seedling growth enhancement must be provided directly by the inoculated bacterium. Known PGPR products include the plant hormones auxin or cytokinin that increase the host plant's endogenous pool of these hormones, and an enzyme that decreases plant ethylene levels.

Ethylene Reduction

Ethylene is an important plant hormone required for normal development of vegetative and reproductive tissues (Deikman, 1997). It coordinates a variety of developmental processes including seed germination, floral senescence, leaf and petal abscission, and fruit ripening, stages during which ethylene biosynthesis increases (Abeles et al., 1992; Smalle and Van Der Straetem, 1997). Ethylene production also increases in response to a variety of environmental stresses such as pathogen attack, wounding, low temperature, increased levels of salt or heavy metals in soil, drought and flooding (Abeles et al., 1992). A consequence of increased levels of "stress ethylene" is often premature senescence; it seems that the plant protects itself from environmental insults at the expense of further growth. From a horticultural perspective, ethylene-induced senescence can lead to huge economic losses from premature ripening of fruits and vegetables, seed loss, lower plant biomass, and reduced shelf life (Child et al., 1998; Huxtable et al., 1998). Attempts to reduce ethylene levels postharvest include application of chemicals that inhibit ethylene biosynthesis such as aminoethoxyvinylglycine (AVG) or ethylene perception like 1-methylcyclopropene (Sisler et al., 1999; Abeles et al., 1992). Deleterious ethylene effects can also be controlled by cultivating transgenic plants with reduced ethylene production or sensitivity (Stearns and Glick, 2003; Czarny et al., 2006), and by inoculating plants with PGPR that lower ethylene levels in host plants (Glick, 2005).

In plants, ethylene is synthesized from L-methionine. Three enzymes convert methionine to ethylene: S-adenosyl-L-methionine (AdoMet) synthase catalyzes the conversion of methionine to AdoMet which is then converted to 1-aminocyclopropane-1-carboxylic acid (ACC) by ACC synthase, and finally to ethylene by ACC oxidase. Several isoforms of both ACC synthase and ACC oxidase exist within plant tissues and are differentially expressed in response to developmental and environmental clues (Fluhr and Mattoo, 1996; Kende and Zeevaart, 1997). The auxin, indole-3-acetic acid, has been shown to increase expression of some ACC synthase genes in a tissue specific manner (Kende, 1993; Tsuchisaka and Theologis, 2004).

Many bacteria that colonize plant roots synthesize the enzyme ACC deaminase that cleaves ACC, the immediate precursor for ethylene biosynthesis in plants, to ammonia and a ketobutyrate (Glick et al., 1998; Hontzeas et al., 2006). These include several *Pseudomonas* spp., *Enterobacter cloacae, Kluyvera acorbata, Burkholderia* spp., *Bacillus* spp., *Methylobacterium fujisanaense* and *Rhizobium* spp. (Glick et al., 1995; Burd et al., 1998; Babalola et al., 1999; Belimov et al., 2001; Ghosh et al., 2003; Ma et al., 2004; Hontzeas et al, 2005; Sessitsch et al., 2005; Madhaiyan et al., 2006; Shaharoona et al., 2006; Safronova et al., 2006). A great deal of evidence suggests that the production of ACC deaminase is a major mechanism by which bacteria promote plant growth, especially under stressful conditions, by ameliorating the effects of ethylene overproduction in plants (Glick, 2005). Typically this effect is measured as an increase in seed germination or as an increase in the length or biomass of seedling roots within days of germination (Glick et al., 1995). Levels of

ACC are reduced in seeds and roots of plants treated with these bacteria (Penrose and Glick, 2001; Penrose et al., 2001). Bacteria exhibiting ACC deaminase activity enhanced growth of seedling roots of several plants of agricultural and horticultural importance including tomato, canola, various grasses, and maize relative to uninoculated seeds under normal conditions and in response to environmental stresses (Sun et al., 1995; Li et al., 2000; Grichko and Glick, 2001; Mayak et al., 2004a, b; Reed and Glick, 2005). Isogenic mutants in which ACC deaminase activity had been abolished no longer promoted root growth (Hall et al., 1996; Li et al., 2000). The effect of inoculating seeds with PGPR with ACC deaminase activity is similar to that following treatment of seeds with the ethylene inhibitor AVG (Hall et al., 1996).

Auxin Production

Auxins are an important class of plant hormones that are responsible for many aspects of plant growth and development including apical dominance (Tamas, 1995), vascular tissue differentiation (Aloni, 1995), initiation of lateral and adventitious root formation (Gaspar et al., 1996; Malamy and Benfey, 1997), elongation growth in stems and roots (Yang et al., 1993; Kende and Zeevaart, 1997), cell division (Kende and Zeevaart, 1997) and tropic responses to gravity and light (Kaufman et al., 1995). The production of auxin, notably indole-3-acetic acid (IAA), is widespread among both beneficial and pathogenic rhizosphere bacteria and can contribute to a plant's endogenous pool of auxin (Patten and Glick, 1996). Several IAA biosynthetic routes have been identified in bacteria, mostly proceeding from the precursor tryptophan through distinct intermediates (Patten and Glick, 1996). Although both pathogenic and beneficial rhizobacteria have been found to produce IAA, they appear to do so predominantly by different pathways, with different modes of regulation, and with different consequences for the host plant (Patten and Glick, 1996). Genes encoding IAA production via the indoleacetamide pathway, involving oxidative decarboxylation of tryptophan to indole-3-acetamide followed by conversion to IAA by indoleacetamide hydrolase, are present on the T-DNA (transferred DNA) region of the tumor-inducing (Ti) plasmid of *Agrobacterium tumefaciens* that produces crown gall tumors in a variety of plants (Schröder et al., 1984; Thomashow et al., 1984; Van Onckelen et al., 1986). Induction of plant cell division following transfer of the T-DNA into the chromosome of plant cells and expression of the IAA biosynthetic genes allows *A. tumefaciens* to increase its factory of plant cells harnessed to produce opines, a group of unusual amino acid derivatives for which *A. tumefaciens* possesses catabolic genes (Nilsson and Olsson, 1997). Incorporation of IAA biosynthesis genes directly into the plant genome is not the only mechanism for IAA-induced tumor formation. IAA-producing strains of *Pseudomonas syringae* and *Erwinia herbicola* that carry genes encoding the indoleacetamide pathway on a plasmid that is not transferred into plants also induce gall formation on plants such as olive, oleander, privet, and gypsophilia (Kuo and Kosuge, 1970; Comai et al., 1982; Manulis et al., 1991). A role for IAA in gall formation is supported by the failure of tumors to develop in plants infected with *P. syringae* and *E. herbicola* strains that had lost these plasmids, or in which the IAA biosynthetic genes were deleted. On the other hand, tumor-inducing functions could be transferred to nonpathogenic strains with plasmid transfer (Smidt and Kosuge, 1978; Comai and Kosuge, 1980; Comai et al., 1982; Surico et al., 1984; Yamada, 1993).

IAA is also known to be produced by a variety of PGPR including species of *Pseudomonas, Enterobacter, Azospirillum, Bacillus, Azotobacter,* and *Rhizobium* (Patten and Glick, 1996). Synthesis of IAA in plant beneficial bacteria is catalyzed by different enzymes encoded by genes that are distinct from those of pathogens. In known and well-characterized PGPR, IAA is synthesized via the intermediate indole-3-pyruvate. A key enzyme in this pathway is indole-3-pyruvate decarboxylase, which catalyzes the conversion of indolepyruvate, produced from tryptophan by tryptophan aminotransferase, to indole-3-acetylaldehyde, which is finally converted to IAA either spontaneously or by indole-3-acetaldehyde oxidase. In contrast to the function of IAA produced by plant tissues, this compound does not appear to be essential for bacterial cells; bacterial mutants in which indolepyruvate decarboxylase activity has been abolished are not compromised in growth or plant colonization, but are no longer able to promote plant growth to the same extent as the wild type strain (Patten and Glick, 2002a).

Why is IAA production common among rhizobacteria? The most obvious explanation for the prevalence of bacterial production of a phytohormone is that it provides bacteria with a mechanism to stimulate plant growth. In doing so, bacteria can increase production of plant metabolites that the bacteria can utilize for their own growth. In support of the hypothesis that plants secrete compounds that attract bacteria to their roots to benefit the plant, it appears that exogenous tryptophan is required for IAA production in many plant-associated bacteria. Tryptophan activates expression of the gene encoding indole pyruvate decarboxylase, and therefore, IAA synthesis, in the plant growth promoting bacteria *Pseudomonas putida, Enterobacter cloacae, Rhizobium phaseoli,* and *Bradyrhizobium japonicum* (Kaneshiro et al., 1983; Ernstsen et al., 1987; Koga et al., 1991; Patten and Glick, 2002b). Although these bacteria are able to synthesize tryptophan, endogenous levels are not sufficient to activate IAA biosynthesis. The most likely source of exogenous tryptophan in the rhizosphere is root exudates. In other PGPR strains, notably *Azospirillum brasilense*, IAA expression is activated by a positive feedback loop by IAA itself; an auxin-response sequence element is present in the promoter of the indole pyruvate decarboxylase gene similar to that found in the promoters of some auxin-regulated plant genes (Lambrecht et al., 1999; Vande Broek et al., 1999).

Many root-promoting bacteria produce IAA, and their effect on plants mimics that observed following application of exogenous IAA. *Pseudomonas putida*, a bacterium that produces relatively low levels of IAA (Xie et al., 1996), stimulated a two- to three-fold increase in the length of primary roots compared to uninoculated controls (Caron et al., 1995). IAA-overproducing mutant of this bacterium stimulated extensive lateral root development in canola (Xie et al., 1996) and adventitious roots on mung bean cuttings (Mayak et al., 1997). Loss of IAA production following disruption of the gene encoding indole pyruvate decarboxylase significantly reduced the ability of *Enterobacter cloacae* to promote root growth (Patten and Glick, 2002a). Inoculation of wheat and pearl millet with *Azospirillum brasilense*, which naturally produces high levels of IAA, increased the number and length of lateral roots similar to the application of exogenous IAA (Tien et al., 1979; Barbieri and Galli, 1993). Application of diluted culture extracts or low density inocula of bacteria that produce high levels of IAA stimulated growth of primary roots (Harari et al., 1988; Selvadurai et al., 1991; Beyerler et al., 1997). Interestingly, in a recent study, Liu and Nester (2006) showed that at higher concentrations, IAA inhibits expression of genes required for tumor induction by *A. tumefaciens* and prevents proliferation of *A. tumefaciens* and several other beneficial and pathogenic plant-associated bacteria.

Glick et al. (1998) have proposed a model to explain the similar effects of bacterial IAA production and ACC deaminase activity on seedling roots. In this model, IAA and ACC deaminase work in concert to stimulate root elongation. IAA secreted by PGPR is taken up by the host plant and increases transcription and activity of ACC synthase, which as described above, converts AdoMet to ACC in plant tissues. Some of this ACC is released in root and seed exudates, and is utilized by PGPR with ACC deaminase as a unique source of nitrogen. In this manner, PGPR act as a sink for ACC, a consequence of which is the lowering of host plant ethylene levels. Recent studies indicate that plants (*Arabidopsis* and canola) that have been inoculated with PGPR, up-regulate auxin-regulated plant genes and down-regulate ethylene-responsive genes (Wang et al., 2005; Hontzeas et al., 2004).

Cytokinin Production

Cytokinins play an important role in plant cell proliferation and differentiation (Sakakibara, 2006). Some of the effects of cytokinins are a consequence of its interaction with other plant hormones. For example, the relative concentrations of auxin and cytokinin in the medium of callus cultures influence the development of shoots and roots, with a high cytokinin to auxin ratio inducing shoot formation and the converse inducing root formation (Skoog and Miller, 1957). Cytokinins are derivatives of adenine with variability in the side chain at the N^6 position (Kakimoto, 2003). Their presence has been detected in many soil bacteria including PGPR strains of the genera *Pseudomonas* (De Salamone et al., 1997), *Azotobacter* (Neito and Frankenberger, 1989; Taller and Wong, 1989), *Azospirillum* (Horesman et al., 1986), and *Rhizobium* and *Bradyrhizobium* (Sturtevant and Taller, 1989).

Despite their widespread production, few studies have been undertaken to systematically evaluate the role of cytokinin in plant growth promotion by PGPR, partly because the biosynthetic genes have not been identified in these bacteria, hence making mutant construction difficult. Genes encoding the key cytokinin biosynthetic enzyme isopentenyltransferase, which catalyzes the addition of the side chain to AMP, have been reported in tumor-inducing phytopathogens, including in the T-DNA region of the Ti plasmid of the gall-forming *A. tumefaciens* (Akiyoshi et al., 1984) and on a plasmid required for pathogenicity of *Erwinia herbicola* (Manulis et al., 1991). Cytokinins have been implicated in the induction of tumors by these phytopathogens (Morris, 1986; Gaudin et al., 1994). Only a few studies provide evidence for the involvement of bacterial cytokinins in beneficial interactions with host plants. Whereas wildtype *Pseudomonas fluorescens* increased the fresh weight of tobacco callus almost five-fold, transposon mutants with reduced capacity to produce cytokinin did not promote callus growth to the same extent (De Salamone and Nelson, 2000). A mutant of *Rhizobium leguminosarum* deficient in production of the cytokinin precursor adenosine did not promote growth of canola seedling roots to the same extent as the wildtype strain (Noel et al., 1996), although it is difficult to attribute these effects directly to cytokinin.

Traits that Increase Nutrient Availability to Plants

Elements essential for plant growth are often plentiful in the environment, but are largely in a form not directly utilizable by the plant. Bacteria are key players in the transformation of nutrients through ecosystems. In the soil, bacteria are important for the conversion of nitrogenous, phosphorus and ferrous compounds into forms that are available for higher organisms.

Nitrogen Fixation

In natural soils, nitrogen is often the growth-limiting element for plants. Much of the nitrogen available on Earth exists as atmospheric dinitrogen, a form that is not useful to most organisms. However, many soil bacteria, known as diazotrophs, are able to reduce gaseous N_2 to ammonia, which can be readily assimilated into organic nitrogen compounds such as protein and nucleic acids by plants and other organisms. The most thoroughly studied example of this aspect of the nitrogen cycle is the provision of fixed nitrogen to legumes by symbiotic Rhizobiaceae residing in root nodules. The specificity of this interaction is determined by the flavonoid signal secreted by the host legume, to which the bacterium responds by producing Nod factors that initiate the formation of root nodules (Van Rhijn and Vanderleyden, 1995). Within the root nodules, bacteria are supplied with photosynthates that provide the resources to drive the energetically-expensive N_2 reduction reaction.

Free-living PGPR do not form nodules on host plant roots; however, PGPR such as *Azotobacter, Azospirillum, Pseudomonas,* and *Enterobacter,* are capable of fixing atmospheric nitrogen. Although diazotroph activity is higher in the rhizosphere than in the bulk soil (Bürgmann et al., 2005), plant growth promotion by diazotrophs can occur in the absence of nitrogen fixation (Dobbelaere et al., 2003). Evidence for this comes from inoculation of plants with mutants of *Pseudomonas putida* and *Azoarcus,* that were no longer able to fix atmospheric nitrogen, but retained their ability to stimulate growth of canola and rice, respectively (Lifshitz et al., 1987; Hurek et al., 1994). Only in two cases was the loss of nitrogen (N_2) fixation associated with a reduction in the transfer of nitrogen to the host plant and a concomitant reduction in plant growth: in the interaction between *Azoarcus* sp. and Kallar grass (Hurek et al., 2002), and between *Acetobacter diazotrophicus* and sugarcane (Sevilla et al., 2001). Although N_2 fixation may not be a primary mechanism of plant growth promotion, a diazotrophic metabolism may give some rhizobacteria a competitive advantage in soil environments where alternate forms of nitrogen are limited (Bürgmann et al., 2005). Enrichment of diazotrophic PGPR on plant surfaces that promote plant growth by a mechanism other than N_2 fixation may, therefore, indirectly contribute to plant growth in soils (Dobbelaere et al., 2003).

Most of the fixed nitrogen produced by associative diazotrophs is not secreted by the bacterium, which can account for the low levels transferred to the host plant (Rao et al., 1998). An exception to this is *Acetobacter diazotrophicus*, described above, which was shown to secrete almost half of its fixed nitrogen (Cojho et al., 1993). Promising strategies to improve the transfer of fixed nitrogen to the host plant have been identified. Ammonia-secreting mutants of *A. brasilense* were developed that provided more fixed nitrogen to host plants compared to the wild type strain (Christiansen-Weniger and Van Veen, 1991; Christiansen-Weniger and Vanderleyden, 1994). Induction of nodule-like structures, or para-nodules, on plant roots or stems by application of synthetic auxin, which can then be colonized by diazotrophs, may provide an environment more suitable for N_2 fixation by PGPR (Christiansen-Weniger, 1998). Inoculation of plants with endophytic diazotrophs such as *Azotobacter diazotrophicus*, *Herbispirillum seropedicae*, *Azoarcus* spp., and others, may also enhance N_2 fixation by providing a reduced oxygen environment, which is optimal for the oxygen-sensitive enzyme nitrogenase that catalyzes the reaction, and by reducing competition with other microbes (James and Olivares, 1997).

Phosphate Solubilization

Plants obtain phosphorus from soil as inorganic orthophosphate (P_i), and use this for synthesis of important compounds such as membrane phospholipids, nucleotides and nucleic acids. Phosphorus is abundant in many soils, but its availability to plants is limited due to reaction with soil minerals (e.g., calcium, aluminum and iron) forming insoluble compounds. To solubilize phosphates, plants secrete organic acids such as malate, citrate, and succinate, which are also good substrates for bacterial growth.

There is some evidence that PGPR may improve phosphorus uptake by host plants: peanut seedlings treated with plant growth promoting *Pseudomonas* sp. had higher levels of phosphorus in shoots and kernels relative to uninoculated controls (Dey et al., 2004); phosphate-solubilizing *Rhizobium leguminosarum* increased the concentration of phosphorus in lettuce and maize (Chabot et al., 1996); and radiolabeled phosphate was taken up in larger quantities by canola treated with *Pseudomonas putida* (Lifshitz et al., 1987). Two possible mechanisms have been proposed to explain the increased phosphate uptake in plants inoculated with PGPR. Rhizobacteria enhance solubilization of phosphate minerals through acidification of the rhizosphere by secreting organic acids, especially gluconic acid (Illmer and Schinner, 1992; Liu et al., 1992). Species of *Pseudomonas, Bacillus, Rhizobium,* and *Burkholderia* capable of doing so are abundant in the rhizosphere (Rodríguez and Fraga, 1999). Rhizobacteria also produce enzymes that hydrolyze organic phosphates. These include phosphatases that remove phosphate from a variety of substrates, and phytases that hydrolyze phytate, an abundant form of phosphate in soils, generating phosphate compounds that can be taken up by plants. Due to the acidic nature of many soils, bacterial acid phosphatases are expected to have a significant impact on phosphate availability in the rhizosphere. Many common rhizosphere bacteria have acid phosphatase activity (Rodríguez and Fraga, 1999; Rengel and Marschner, 2005), and almost half of all culturable soil bacteria are able to utilize phytate (Greaves and Webley, 1965; Richardson and Hadobas, 1997). Although it has been demonstrated that these bacteria solubilize phosphate under ideal laboratory conditions, it has not yet been shown whether these strains are effective in providing plants with phosphate under field conditions. Soil bacteria might take up and metabolize phosphate efficiently; however, the resulting phosphate compounds might not be transferred to the host plant.

Iron Acquisition

Most organisms require iron for important physiological processes such as respiration, photosynthesis, and N_2 fixation. Despite the abundance of this element on Earth, it is usually present in soils as insoluble ferric hydroxides, a form that is not readily utilizable by plants and microbes. To obtain iron, bacteria, fungi and some grasses secrete low molecular weight molecules called siderophores that bind iron with a high affinity. The iron-siderophore complex is then taken up into the cell *via* specific membrane-bound receptors, and iron is released from the siderophore following reduction to the ferrous state or cleavage of the siderophore. Many different bacteria produce siderophores, and in animal and plant pathogens siderophore production contributes to their virulence. PGPR that produce siderophores with a high affinity for iron or that can recognize iron-siderophore complexes produced by other microbes may promote plant growth indirectly by competing more effectively for iron in soils, thereby making it unavailable for phytopathogens, a mechanism of biocontrol, or by directly supplying iron for plant utilization.

Microbial siderophores are structurally diverse. Most contain three iron-binding groups positioned optimally for iron-binding by a flexible backbone. Hydroxamate iron-binding groups are typically found on fungal siderophores, while catecholates, which bind iron more tightly than hydroxamates, are common functional groups on bacterial siderophores (Matzanke, 1991). Backbones are usually comprised of single amino acids, or small cyclic or linear peptides. Although iron is essential for bacterial cells, high levels are toxic and, therefore, iron acquisition through siderophore production is tightly regulated; the many genes involved in the biosynthesis of siderophores and their cognate receptors are activated only when iron availability is low (Loper and Lindow, 1994; LeVier and Guerinot, 1996; Loper and Henkels, 1997; Marschner and Crowley, 1997).

The ability to obtain iron efficiently and at the expense of other microbes provides a means for PGPR to compete for limited resources in the rhizosphere. In doing so, the bacterium may benefit a host plant by providing it with soluble iron or some other growth promoting factor that it produces once established in the rhizosphere. Competition for iron occurs on two levels: (1) competition for ferric iron, which is dependent on the properties of the siderophore produced by the PGPR, and (2) competition for iron complexed with a siderophore, which is a function of the ferric-siderophore receptors produced. Several factors contribute to the ability of siderophores to scavenge iron from the rhizosphere. Because the binding of siderophores to iron is stoichiometric, bacteria that secrete more siderophores can obtain more iron. Certainly high levels of siderophores are produced by PGPR under iron-deficient conditions *in vitro*, but in order to contribute to plant growth promotion, these must also be produced *in planta*. Using ice-nucleation as a measure of promoter activity, transcription of the pyoverdin siderophore biosynthesis genes was shown to occur when *Pseudomonas fluorescens* was associated with lupine, barley and bean, albeit at a lower level than *in vitro* (Marschner and Crowley, 1997; Loper and Henkels, 1997). Siderophores have different affinities for ferric iron and different rates of formation of iron-siderophore complexes, therefore, the type of siderophore produced also contributes to the establishment of PGPR in the rhizosphere. Enterobactin, a tricatecholate siderophore produced by the PGPR *Enterobacter cloacae*, has the highest affinity for ferric iron among known siderophores, with an association constant of 10^{52} M^{-1} (Harris et al., 1979).

Production of large amounts of high affinity siderophores seems not to be as important for bacterial competition and root colonization as the capacity to produce receptors that recognize a variety of ferric-siderophore complexes, some of which may be produced by other bacteria in the rhizosphere. Whereas mutants unable to produce siderophores can colonize the rhizosphere as efficiently as wild type strains (Bakker et al., 1987; Loper, 1988; Höfte et al., 1991, 1992), bacteria that produce a receptor that can bind a range of ferric-siderophores, or that produce several different siderophore receptors, are more competitive in the rhizosphere (Jurkevitch et al., 1992). *Pseudomonas putida* WCS358 that is able to take up several different iron-siderophore complexes was able to outcompete *P. fluorescens* WCS374, which has a more limited capacity (Bakker et al., 1986;

Raaijmakers et al., 1995). Extending the spectrum of siderophores that *P. fluorescens* WCS374 can recognize through genetic engineering, rendered it competitive with *P. putida* WCS358 (Raaijmakers et al., 1995).

Plants may directly benefit from the acquisition of iron bound to siderophores produced by PGPR. The ability of plants to take up radiolabeled iron complexed to bacterial siderophores was demonstrated for sorghum, oats, peanut, cotton, cucumber and sunflower (Cline et al., 1984; Jurkevitch et al., 1986; Crowley et al., 1988; Bar-ness et al., 1991; Wang et al., 1993), although there seems to be some specificity for the siderophores utilized, and the efficiency is low (Becker et al., 1985). In the presence of microbial siderophores, cucumber plants increased in biomass and chlorophyll content (Ismande, 1998). The latter is a good indicator of plant iron health as iron is important for the biosynthesis of chlorophyll.

APPLICATIONS OF PGPR IN HORTICULTURE

Most of the studies undertaken to assess the efficacy of PGPR in increasing crop yields have been carried out on plants of agricultural importance such as rice, wheat and maize. However, the results of these studies combined with those that have specifically inoculated plants of horticultural interest, indicate that PGPR are effective in enhancing propagation and yields of a wide range of economically important plants.

Plant Propagation

PGPR may be used to enhance plant propagation in two ways: (1) to promote seed germination, and (2) to stimulate rooting in cuttings. Some studies have demonstrated that inoculation of seeds with PGPR enhances seed germination. Five days following inoculation of lodgepole pine seeds with *Bacillus* sp. L6, the emergence of seedlings was 50% more than that of the uninoculated controls (Chanway et al., 1991). A significant increase in germination rate was also seen following soaking of beet, barley, tomato, and cucumber seeds in suspensions of *Bacillus mobilis* and *Clostridium* sp. (Polyanskaya et al., 2000). ACC deaminase was found to contribute to the increased seed germination in the common reed *Phragmites australism* by *Pseudomonas asplenii* AC-1 (Reed et al., 2005). It is not clear how ACC deaminase activity, and presumably lowering of seed ethylene levels, contributes to increased seed germination as, in many plant species, ethylene has been implicated in promotion of seed germination, and a spike of ethylene production coincides with germination. Certainly, ACC levels are lower in seeds treated with PGPR (Penrose and Glick, 2001).

Treatment of cuttings with PGPR to promote root initiation may present a viable alternative to synthetic auxins for propagation of plants in horticulture. High concentrations of auxins, such as naphthalene acetic acid (NAA) and indolebutyric acid (IBA) in commercial rooting powders, are often used to stimulate adventitious roots. High levels of bacterial IAA, whether from IAA-overproducing mutants or from strains that naturally secrete high levels, or from high density inocula, also stimulate the formation of lateral and adventitious roots (Barbieri et al., 1986; Loper and Schroth, 1986; Barbieri and Galli, 1993; Sawar and Kremmer, 1995; Xie et al., 1996; Beyerler et al., 1997; Mayak et al., 1997). *P. putida* cells that produce wild type levels of IAA stimulated formation of many short adventitious roots on mung bean cuttings, and an IAA-overproducing mutant induced even more adventitious roots than the wild type strain (Mayak et al., 1997), while an IAA-deficient mutant of *E. cloacae* stimulated fewer adventitious roots compared to the wild type strain (Patten and Glick, 2002a). Root initiation was also enhanced following treatment of carnation cuttings with *Azospirillum brasilense*, which naturally produces high levels of IAA (Li et al., 2005). Although more roots were initiated following treatment with rooting powder containing IBA, the roots that developed from PGPR treatment were more than twice the length of those chemically treated and those treated with *A. brasilense* transformed to express ACC deaminase, were nearly three times as long (Li et al., 2005). Exogenous IAA is known to increase transcription and activity of ACC synthase, which catalyzes the production of ACC from S-adenosylmethionine (SAM) (Peck and Kende, 1995). Initiation of adventitious and lateral roots may, therefore, be mediated by IAA-induced ethylene production. In support of this, an ACC deaminase deficient mutant of *P. putida* that is no longer able to reduce ethylene levels in plants, stimulated more small adventitious roots than the wild type strain (Mayak et al., 1999).

Plant Biomass Enhancement

Numerous studies have demonstrated that many different bacteria enhance plant growth (Reed and Glick, 2005). The benefit is seen mainly in the enhanced development of plant root systems; although in many cases increases in shoot height and/or biomass have been observed. The root-enhancing effect is clearly observed in early seedling development, which is important as young plants that establish roots quickly, whether by elongation of primary roots or by proliferation of lateral or adventitious roots, have a better chance for survival. Seedlings with a good root system are well anchored in the soil and can rapidly obtain water and nutrients from their environment.

Much of the effect of PGPR on roots has been attributed to bacterial modification of plant auxin and ethylene levels. The benefit of applying low levels of IAA (usually in the range of 10^{-9} to 10^{-12} M) to plants has long been recognized in horticulture as a mechanism to stimulate primary root elongation (Thimann and Lane, 1938; Alvarez 1989; Meuwley and Pilet, 1991; Gaspar et al., 1996; Malamy and Benfey, 1997). The effect on root morphology that is seen following the application of low concentrations of exogenous IAA, is also seen following the inoculation of plants with PGPR that produce low levels of IAA. The fact that changes to roots were due to bacterial production of IAA was confirmed by inoculating canola seeds with wild type and an isogenic IAA-deficient mutant of *Enterobacter cloacae* (Patten and Glick, 2002a). While the wild type bacterium stimulated a 35-50% increase in the length of canola seedling primary roots, roots from seeds treated with the IAA-mutant were not significantly different from uninoculated controls (Patten and Glick, 2002a).

IAA secreted by PGPR may promote root growth directly by stimulating plant cell division or elongation, or indirectly by stimulating plant production of ACC and consequently bacterial ACC deaminase activity. The role of ACC deaninase produced by *P. putida* Gr12-2 in root elongation was demonstrated by an ACC deaminase-deficient mutant that was no longer able to stimulate root growth (Li et al., 2000). Indeed, introducing the gene encoding ACC deaminase into bacteria that do not normally produce this enzyme, resulted in a two-fold increase in the dry weight of tomato seedling roots by *Azospirillum brasilense* (Holguin and Glick, 2003) and an increase in canola root length by *Pseudomonas fluorescens* CHA0 (Wang et al., 2000), compared to bacteria that do not synthesize ACC deaminase.

In addition to tomato and canola, an increase in root and/or shoot biomass as a consequence of PGPR treatment has been demonstrated for numerous plants of horticultural interest. Legumes such as beans and peanuts treated with *Azospirillum brasilense* and *Bacillus subtilis*, respectively, yielded significant increases in plant biomass (Vedder-Weiss et al., 1999; Kishore et al., 2005). The success of PGPR treatment of many different cereals such as wheat, maize, rice, millet, sorghum, oats, and barley has been well documented (Kloepper et al., 1989; Hall et al., 1996; Mehnaz et al., 2001; Khalid et al., 2004; Reed and Glick, 2004). Garden crops that respond positively to PGPR treatment include red pepper, sugar beet, onion, lettuce, potato, and cucumber (Reddy and Rahe, 1989; Frommel et al., 1991; Noel et al., 1996; Van Peer and Schippers, 1998; Uthede et al., 1999; Cakmakci et al., 2001; Joo et al., 2004). Yield increases of up to 200% have been achieved, although increases in the range of 15-60% are more common, depending on the host plant. For example, in a recent study, PGPR isolated from the rhizosphere of wheat increased wheat root length and dry weight by 17 and 14%, respectively, and shoot height and dry weight by 38 and 36%, respectively, when applied as a seed treatment (Khalid et al., 2004). Growth promotion has also been observed when PGPR are applied directly to the growth medium, rather than being inoculated as a seed treatment (Yan et al., 2003). In many cases, positive effects were sustained over several growing seasons (e.g., Okon and Labandera-Gonzalez, 1994). Not only were increases in root and/or shoot biomass observed, but inoculation with PGPR increased grain yields in wheat and rice (Boddey and Dobereiner, 1988; Alam et al., 2001), and increased fruit yields in tomato and cucumber (McCullagh et al., 1996; Uthede et al., 1999; Gagné et al., 1993).

Stress Tolerance Improvement

Beneficial rhizosphere bacteria have been shown to enhance plant growth under various stresses. A psychrotolerant PGPR isolated from the Canadian Arctic, *Pseudomonas putida* GR12-2, increased root length of both spring canola (*Brassica campestris* cv. Tobin) and winter canola (*Brassica napus* cv. Ceres) by 30 and 65%, respectively, at 5°C (Sun et al., 1995). This bacterium produces an antifreeze protein that inhibits formation of large ice crystals facilitating its survival at freezing temperatures (Kawahara et al., 2001; Muryoi et al., 2004). The ability to promote plant growth at cold temperatures is of significant practical value because many regions of the world experience below optimal temperatures in agricultural soils, usually early in the growing season. Soils in hot arid regions are also poor environments for growing plants due partly to accumulation of salt from irrigation practices. Mayak et al. (2004a) isolated a bacterium, *Achromobacter piechaudii,* from the dry, saline soils of the Arava region in Israel that ameliorated the growth inhibiting effect of high salt levels on tomato seedlings. Irrigating uninoculated tomato seedlings with 172 mM of NaCl, resulted in tomato seedlings with approximately half the biomass of those that did not receive salt treatment; in contrast, seedlings inoculated with *A. piechaudii* and irrigated with high salt were not significantly different from uninoculated plants grown in the absence of salt. In addition, this bacterium protected plants against drought stress (Mayak et al., 2004b). Such bacteria have enormous potential as inoculants to extend horticultural efforts to relatively unproductive regions. Tomato seedlings were also able to resist the growth-inhibiting impact of flooding stress following treatment with the PGPR strains *Pseudomonas putida* UW4 and *Enterobacter cloacae* CAL3 (Grichko and Glick, 2001).

A common element amongst all of these environmental stresses is the elicitation of growth-inhibiting levels of ethylene. For example, during flooding, roots are deprived of oxygen. Consequently, ACC cannot be oxidized and is instead transported to the shoots where it is converted to ethylene (Grichko and Glick, 2001). The impact of increased ethylene levels is seen as leaf epinasty, chlorosis, and abscission. Some ACC is also secreted into the rhizosphere where it can be used as a carbon and energy source for root colonizing bacteria that further act as a sink for ACC (Grichko and Glick, 2001). By decreasing plant levels of ACC, and hence, ethylene, bacteria that possess ACC deaminase activity ameliorate the deleterious consequences of stress ethylene. All the bacterial species described above possess ACC deaminase activity, and tomato seedlings treated with these bacteria show a marked decrease in ethylene levels under stress conditions in comparison to untreated seedlings (Grichko and Glick, 2001; Mayak et al., 2004a).

PGPR can be used to enhance tolerance of plants to toxic soil contaminants. In a process known as phytoremediation, selected plants take up and either accumulate or breakdown in their tissues toxic environmental metals, inorganic compounds like arsenate, or organic compounds like chlorinated solvents and polyaromatic hydrocarbons, and are later harvested to remove the contaminant (Glick, 2003). In practice, the growth of plants is compromised in such soils, thus plants remain small in size, which limits the amount of contaminant they can take up. A solution to enhance plant growth and hence, toxin uptake, is inoculation with PGPR. The efficacy of this synergistic approach has been demonstrated for the uptake of nickel by *Brassica juncea* (Indian mustard) in conjunction with the nickel-resistant PGPR *Kluyvera ascorbata* (Burd et al., 1998). While the bacterium did not increase the amount of nickel taken up by the plant per unit weight, the overall weight of the plant, and thus the total amount of nickel per plant, increased. Similar effects were observed for remediation of other heavy metals such as copper, and polycyclic aromatic hydrocarbons like creosote using PGPR (Huang et al., 2004; Reed and Glick, 2005). Two mechanisms have been proposed to account for this phenomenon: (1) ACC deaminase-carrying bacteria such as *Kluyvera ascorbata* lower growth-inhibiting stress ethylene that is induced in plants exposed to toxic compounds, and (2) siderophores produced by the bacterium ameliorate iron deficiency in plants that is often associated with heavy metal toxicity (Burd et al., 2000; Glick, 2003).

FUTURE PERSPECTIVES

Natural soils are teeming with microbial life and present a rich resource for useful products. However, only 1% of soil bacteria have been cultured in the laboratory. The unculturable 99% have enormous potential to produce important and interesting products. Although many beneficial plant-associated bacteria have been isolated from soils, most belong to a few genera, mainly *Pseudomonas, Bacillus, Enterobacter* and *Azospirillum,* which is probably a reflection of their abundance in the rhizosphere and their ease of cultivation in the laboratory. Thus, a considerable quantity of the rich microbial diversity of the rhizosphere remains uncharacterized. A more comprehensive understanding of growth promoting mechanisms used by soil bacteria is required. Despite the inability to cultivate the majority of rhizobacteria in the laboratory, it may be possible to use metagenomic approaches to identify bacterial genes encoding plant growth promoting traits in soil DNA. This could entail screening DNA extracted directly from soil for sequences homologous to known genes encoding plant growth promoting functions to assess their diversity and abundance. Hontzeas et al. (2005) found evidence for the horizontal transfer of the gene encoding ACC deaminase among soil bacteria, suggesting that there may be some selective pressure for this function in the rhizosphere. Indeed, it has been noted that horizontal gene transfer is enhanced in the rhizosphere and that the host plant may select for certain bacterial traits (Berg et al., 2005). Technically more difficult is the identification of bacterial genes that encode novel plant-beneficial traits using a functional metagenomic approach. If a gram of soil contains at least 2,000 distinct genomes, each containing on an average 5,000 genes, screening of metagenomic libraries of soil DNA for expressed genes that confer some benefit to host plants is a daunting task. There is some evidence to suggest that the host plant may activate plant-beneficial bacterial genes. The indolepyruvate decarboxylase gene encoding a key enzyme in the bacterial IAA biosynthesis pathway is expressed only in the presence of exogenous tryptophan (Patten and Glick, 2002b), and ACC deaminase genes appear to be induced by ACC. The most likely source of both of these inducers in the rhizosphere is host plant exudates. Thus, it may be possible to isolate novel plant growth promoting genes by screening for genes that are specifically activated by the host plant.

In particular, a broader understanding of the mechanisms used by PGPR to facilitate plant growth in suboptimal soils is required. Although a great deal is known about some of the mechanisms used by a few rhizosphere bacteria to promote plant growth, the composition of the microbial communities on the roots of plants grown under optimal conditions is different from that on roots in poor conditions (Marschner et al., 2004). The nature of the communities is perhaps influenced by the differential chemical composition of exudates secreted by plants under these conditions. Many of the commercial inoculants currently available do not work effectively in harsh environments, for example, in acidic or saline soils. Following the approach of Mayak et al. (2004a) to identify PGPR in saline soils and Burd et al. (1998) to isolate PGPR in metal-contaminated soils, it is imperative to develop inoculants that are effective under a broader range of suboptimal conditions and understand the mechanisms that they employ.

REFERENCES

Abeles, F.B., Morgan, P.W. and Saltveit, M.E. Jr. (1992). Ethylene in Plant Biology. Academic Press, New York, USA.

Akiyoshi, D.E., Klee, H., Amasino, R.M., Nester, E.W. and Gordon, M.P. (1984). T-DNA of *Agrobacterium tumefaciens* encodes an enzyme of cytokinin biosynthesis. Proc. Natl. Acad. Sci., USA 81: 5994-5998.

Alam, M.S., Cui, Z.J., Yamagishi, T. and Ishii, R. (2001). Grain yield and related physiological characteristics of rice plants (*Oryza sativa* L.) inoculated with free-living rhizobacteria. Plant Prod. Sci. 4: 125-130.

Alfano, J.R. and Collmer, A. (1997). The type III (Hrp) secretion pathway of plant pathogenic bacteria: trafficking harpins, Avr proteins, and death. J. Bacteriol. 179: 5655-5662.

Aloni, R. (1995). The induction of vascular tissues by auxin and cytokinin. In: Plant Hormones: Physiology, Biochemistry and Molecular Biology (ed.) P.J. Davies. Kluwer Academic Publishers, Dordrecht, The Netherlands, pp. 531-546.

Alvarez, R., Nissen, S.J. and Sutter, E.G. (1989). Relationship between indole-3-acetic acid levels in apple (*Malus pumila* Mill.) rootstocks cultured in vitro and adventitious root formation in the presence of indole-3-butyric acid. Plant Physiol. 89: 439-443.

Amellal, N., Burtin, G., Bartoli, F. and Heulin, T. (1998). Colonization of wheat roots by an exopolysaccharide-producing *Pantoea agglomerans* strain and its effect on rhizosphere soil aggregation. Appl. Environ. Microbiol. 64: 3740-3747.

Arevalo-Ferro, C., Reil, G., Gorg, A., Eberl, L. and Riedel, K. (2005). Biofilm formation of *Pseudomonas putida* IsoF: the role of quorum sensing as assessed by proteomics. Syst. Appl. Microbiol. 28: 87-114.

Babalola, O.O., Osir, E.O., Sanni, A.I., Odhiambo, G.D. and Bulimo, W.D. (1999). Amplification of 1-aminocyclopropane-1-carboxylic acid (ACC) deaminase from plant growth-promoting rhizobacteria in Striga-infested soil. Afr. J. Biotechnol. 2: 157-160.

Bakker, P.A.H.M., Lamers, J.G., Bakker, A.W., Marugg, J.D., Weisbeek, P.J. and Schippers, B. (1986). The role of siderophores in potato tuber yield increase by *Pseudomonas putida* in a short rotation of potato. Neth. J. Plant Pathol. 92: 249-256.

Bakker, P.A.H.M., Bakker, A.W., Marugg, J.D., Weisbeek, P.J. and Schippers, B. (1987). Bioassay for studying the role of siderophores in potato growth stimulation by *Pseudomonas* spp. in short potato rotations. Soil Biol. Biochem. 19: 443-449.

Barbieri, P., Zanelli, T., Galli, E. and Zanetti, G. (1986). Wheat inoculation with *Azospirillum brasilense* Sp6 and some mutants altered in nitrogen fixation and indole-3-acetic acid production. FEMS Microbiol. Lett. 36: 87-90.

Barbieri, P. and Galli, E. (1993). Effect on wheat root development of inoculation with an *Azospirillum brasilense* mutant with altered indole-3-acetic acid production. Res. Microbiol. 144: 69-75.

Bar-ness, E., Chen, Y., Hadar, Y., Marschner, H. and Römheld, V. (1991). Siderophores of *Pseudomonas putida* as an iron source for dicot and monocot plants. In: Iron Nutrition and

Interactions in Plants (eds.) Y. Chen and Y. Hadar. Kluwer Academic Publishers, Dordrecht, The Netherlands, pp. 271-281.

Bauer, W.D. and Mathesius, U. (2004). Plant responses to bacterial quorum sensing signals. Curr. Opin. Plant Biol. 7: 429-433.

Becker, J.O., Hedges, R.W. and Messens, E. (1985). Inhibitory effect of pseudobactin on the uptake of iron by higher plants. Appl. Environ. Microbiol. 49: 1090-1093.

Belimov, A.A., Safronova, V.I., Sergeyeva, T.A., Egorova, T.N., Matveyeva, V.A., Tsyganov, V.E., Borisov, A.Y., Tikhonovich, I.A., Kluge, C., Preisfeld, A., Dietz, K.J. and Stepanok, V.V. (2001). Characterization of plant growth promoting rhizobacteria isolated from polluted soils and containing 1-aminocyclopropane-1-carboxylate deaminase. Can. J. Microbiol. 47: 642-652.

Berg, G., Eberl, L. and Hartmann, A. (2005). The rhizosphere as a reservoir for opportunistic human pathogenic bacteria. Environ. Microbiol. 7: 1673-1685.

Beyerler, M., Michaux, P., Keel, C. and Haas, D. (1997). Effect of enhanced production of indole-3-acetic acid by the biological control agent *Pseudomonas fluorescens* CHA0 on plant growth. In: Plant Growth-Promoting Rhizobacteria: Present Status and Future Prospects (eds.) A. Ogoshi, K. Kobayashi, Y. Homma, F. Kodama, N. Kondo and S. Akino. OECD, Paris, France, pp. 310-312.

Bloemberg, G.V., Wijfjes, A.H., Lamers, G.E., Stuurman, N. and Lugtenberg, B.J. (2000). Simultaneous imaging of *Pseudomonas fluorescens* WCS365 populations expressing three different autofluorescent proteins in the rhizosphere: New perspectives for studying microbial communities. Mol. Plant Microbe Interact. 13: 1170-1176.

Boddey, R.M. and Dobereiner, J. (1988). Nitrogen fixation associated with grasses and cereals: recent results and perspectives for future research. Plant Soil 108: 53-65.

Burd, G.I., Dixon, D.G. and Glick, B.R. (1998). A plant growth-promoting bacterium that decreases nickel toxicity in seedlings. Appl. Environ. Microbiol. 64: 3663-3668.

Burd, G.I., Dixon, D.G. and Glick, B.R. (2000). Plant growth-promoting bacteria that decrease heavy metal toxicity in plants. Can. J. Microbiol. 46: 237-245.

Burdman, S., Okon, Y. and Jurkevitch, E. (2000). Surface characteristics of *Azospirillum brasilense* in relation to cell aggregation and attachment to plant roots. Crit. Rev. Microbiol. 26: 91-110.

Burgmann, H., Meier, S., Bunge, M., Widmer, F. and Zeyer, J. (2005). Effects of model root exudates on structure and activity of a soil diazotroph community. Environ. Microbiol. 7: 1711-1724.

Cakmakci, R., Kantar, F. and Sahin, F. (2001). Effect of N_2-fixing bacterial inoculations on yield of sugar beet and barley. J. Plant. Nutr. Soil. Sci. 164: 527-531.

Caron, M., Patten, C.L., Ghosh, S. and Glick, B.R. (1995). Effects of the plant growth-promoting rhizobacterium *Pseudomonas putida* GR12-2 on the physiology of canola roots. Plant Growth Regulators Society of America Quarterly 23: 297-302.

Chabot, R., Antoun, H., Kloepper, J.W. and Beauchamp, C.J. (1996). Root colonization of maize and lettuce by bioluminescent *Rhizobium leguminosarum* biovar *phaseoli*. Appl. Environ. Microbiol. 62: 2767-2772.

Chanway, C.P., Radley, R.A. and Holl, F.B. (1991). Inoculation of conifer seed with plant growth promoting *Bacillus* strains causes increased seedling emergence and biomass. Soil Biol. Biochem. 23: 575-580.

Child, R.D., Chauvaux, K.J., Ulvskov, P. and Van Onckelen, H.A. (1998). Ethylene biosynthesis in oil rape pods in relation to pod shatter. J. Exp. Bot. 49: 829-838.

Christiansen-Weniger, C. and Van Veen, J.A. (1991). NH_4^+-excreting *Azospirillum brasilense* mutants enhance the nitrogen supply of a wheat host. Appl. Environ. Microbiol. 57: 3006-3012.

Christiansen-Weniger, C. and Vanderleyden, J. (1994). Ammonia excreting *Azospirillum* sp. become intracellularly established in maize (*Zea mays*) para-nodules. Biol. Fertil. Soils 17: 1-8.

Christiansen-Weniger, C. (1998). Endophytic establishment of diazotrophic bacteria in auxin-induced tumors or cereal crops. Crit. Rev. Plant Sci. 17: 55-76.

Cline, G.R., Reid, C.P.P. and Szaniszlo, P.J. (1984). Effects of hydroxamate siderophore on iron absorption by sunflower and sorghum. Plant Physiol. 76: 36-39.

Cojho, E.H., Reis, V.M, Schenberg, A.C.G. and Döbereiner, J. (1993). Interactions of *Acetobacter diazotrophicus* with an amylolytic yeast in nitrogen-free batch culture. FEMS Microbiol. Lett. 106: 341-346.

Comai, L. and Kosuge, T. (1980). Involvement of plasmid deoxyribonucleic acid in indoleacetic acid synthesis in *Pseudomonas savastanoi*. J. Bacteriol. 143: 950-957.

Comai, L., Surico, G. and Kosuge, T. (1982). Relation of plasmid DNA to indoleacetic acid production in different strains of *Pseudomonas syringae* pv. *savastanoi*. J. Gen. Microbiol. 128: 2157-2163.

Compant, S., Reiter, B., Sessitsch, A., Nowak, J., Clement, C. and Ait, B. E. (2005). Endophytic colonization of *Vitis vinifera* L. by plant growth-promoting bacterium *Burkholderia* sp. strain PsJN. Appl. Environ. Microbiol. 71: 1685-1693.

Crowley, D.E., Reid, C.P.P. and Szaniszlo, P.J. (1988). Utilization of microbial siderophores in iron acquisition by oat. Plant Physiol. 87: 680-685.

Czarny, J.C., Grichko, V.P. and Glick, B.R. (2006). Genetic modulation of ethylene biosynthesis and signaling in plants. Biotechnol. Adv. 24: 410-419.

De Salamone, I.E.G., Nelson, L. and Brown, G. (1997). Plant growth promotion by *Pseudomonas* PGPR cytokinin producers. In: Plant Growth-Promoting Rhizobacteria: Present Status and Future Prospects (eds.) A. Ogoshi, K. Kobayashi, Y. Homma, F. Kodama, N. Kondo and S. Akino. OECD, Paris, France, pp. 316-319.

De Salamone, I.E.G. and Nelson, L. (2000). Effect of cytokinin-producing *Pseudomonas* PGPR strains on tobacco callus growth. Auburn University, Auburn, Alabama, http://www.ag.auburn.edu/argentina.

de Weert, S., Vermeiren, H., Mulders, I.H., Kuiper, I., Hendrickx, N., Bloemberg, G.V., Vanderleyden, J., De Mot, R. and Lugtenberg, B.J. (2002). Flagella-driven chemotaxis towards exudate components is an important trait for tomato root colonization by *Pseudomonas fluorescens*. Mol. Plant Microbe Interact. 15: 1173-1180.

De Weger, L.A., van der Vlugt, C.I., Wijfjes, A.H., Bakker, P.A., Schippers, B. and Lugtenberg, B. (1987). Flagella of a plant-growth-stimulating *Pseudomonas fluorescens* strain are required for colonization of potato roots. J. Bacteriol. 169: 2769-2773.

DeFlaun, M.F., Marshall, B.M., Kulle, E.P. and Levy, S.B. (1994). Tn5 insertion mutants of *Pseudomonas fluorescens* defective in adhesion to soil and seeds. Appl. Environ. Microbiol. 60: 2637-2642.

Deikman, J. (1997). Molecular mechanisms of ethylene regulation of gene transcription. Plant Physiol. 100: 561-566.

Dey, R., Pal, K.K., Bhatt, D.M. and Chauhan, S.M. (2004). Growth promotion and yield enhancement of peanut (*Arachis hypogaea* L.) by application of plant growth-promoting bacteria. Microbiol. Res. 159: 371-394.

Dobbelaere, S., Vanderleyden, J. and Okon, Y. (2003). Plant growth-promoting effects of diazotrophs in the rhizosphere. Crit. Rev. Plant Sci. 22: 107-149.

Dörr, J., Hurek, T. and Reinhold-Hurek, B. (1998). Type IV pili are involved in plant-microbe and fungus-microbe interactions. Mol. Microbiol. 30: 7-17.

Ernstsen, A., Sandberg, G., Crozier, A. and Wheeler, C.T. (1987). Endogenous indoles and the biosynthesis and metabolism of indole-3-acetic acid in cultures of *Rhizobium phaseoli* Planta 171: 422-428.

Espinosa-Urgel, M. (2004). Plant-associated *Pseudomonas* populations: molecular biology, DNA dynamics, and gene transfer. Plasmid 52: 139-150.

Fluhr, R. and Mattoo, A.K. (1996). Ethylene-biosynthesis and perception. Crit. Rev. Plant Sci. 15: 479-523.

Fravel, D.R. (2005). Commercialization and implementation of biocontrol. Annu. Rev. Phytopathol. 43: 337-359.

Fray, R.G. (2002). Altering plant-microbe interaction through artificially manipulating bacterial quorum sensing. Ann. Bot. (Lond) 89: 245-253.

Frommel, M.I., Nowak, J. and Lazarovitis, G. (1991). Growth enhancement and developmental modifications of in vitro grown potato (*Solanum tuberosum* ssp. *tuberosum*). Plant Physiol. 96: 928-936.

Fukui, R., Poinar, E.I., Bauer, P.H., Schroth, M.N., Hendson, M., Wang, X.L. and Hancock, J.G. (1994). Spatial colonization patterns and interaction of bacteria on inoculated sugar beet seed. Phytopathol. 84: 1338-1345.

Gagne, S., Dehbi, L., Le Quere, D., Cayer, F., Morin, J., Lemay, R. and Fournier, N. (1993). Increase of greenhouse tomato fruit yields by plant growth-promoting rhizobacteria (PGPR) inoculated into the peat-based growing media. Soil Biol. Biochem. 25: 269-272.

Gao, M., Teplitski, M., Robinson, J.B. and Bauer, W.D. (2003). Production of substances by *Medicago truncatula* that affect bacterial quorum sensing. Mol. Plant Microbe Interact. 16: 827-834.

Garg, R.K., Sharma, P.K. and Kundu, B.S. (1996). Role of *Azospirillum* exopolysaccharides in root colonization of pearl millet (*Pennisetum americanum* L.). Ind. J. Microbiol. 36: 193-196.

Gaspar, T., Kevers, C., Penel, C., Greppin, H., Reid, D.M. and Thorpe, T.A. (1996). Plant hormones and plant growth regulators in plant tissue culture. In vitro cell development. Biol. Plant 32: 272-289.

Gaudin, V., Vrain T. and Jouanin, L. (1994). Bacterial genes modifying hormonal balances in plants. Plant Physiol. Biochem. 32: 11-29.

Ghosh, S., Penterman, J.N., Little, R.D., Chavez, R. and Glick, B.R. (2003). Three newly isolated plant growth-promoting bacilli facilitate the growth of canola seedlings. Plant Physiol. Biochem. 41: 277-281.

Glick, B.R., Karaturovic, D.M. and Newell, P.C. (1995). A novel procedure for rapid isolation of plant growth promoting pseudomonads. Can. J. Microbiol. 41: 533-536.

Glick, B.R., Penrose, D.M. and Li, J. (1998). A model for the lowering of plant ethylene concentrations by plant growth-promoting bacteria. J. Theort. Biol. 190: 63-68.

Glick, B.R., Patten, C.L., Holguin, G. and Penrose, D.M. (1999). Biochemical and Genetic Mechanisms Used by Plant Growth Promoting Bacteria. Imperial College Press, London, UK, pp. 267.

Glick, B.R. (2003). Phytoremediation: synergistic use of plants and bacteria to clean up the environment. Biotechnol. Adv. 21: 383-393.

Glick, B.R. (2005). Modulation of plant ethylene levels by the bacterial enzyme ACC deaminase. FEMS Microbiol. Lett. 251: 1-7.

Greaves, M.P. and Webley, D.M. (1965). A study of the breakdown of organic phosphates by microorganisms from the root region of certain pasture grasses. J. Appl. Bacteriol. 28: 454-465.

Grichko, V.P. and Glick, B.R. (2001). Amelioration of flooding stress by ACC deaminase-containing plant growth-promoting bacteria. Plant Physiol. Biochem. 39: 11-17.

Hall, J.A., Peirson, D., Ghosh, S. and Glick, S. (1996). Root elongation in various agronomic crops by the plant growth promoting rhizobacterium *Pseudomonas putida* GR12-2. Isr. J. Plant Sci. 44: 37-42.

Harari, A., Kige, J. and Okon, Y. (1988). Involvement of IAA in the interaction between *Azospirillum brasilense* and *Panicium miliaceum* roots. Plant Soil 110: 275-282.

Harris, W.R., Carrano, D.J., Cooper, S.R., Sofen, S.R., Avdeef, A.E., McArdle, J.V. and Raymond, K.N. (1979). Coordination chemistry of microbial iron transport compounds. J. Am. Chem. Soc. 101: 6097-6104.

Höfte, M., Seong, K.Y., Jurkevitch, E. and Verstraete, W. (1991). Pyoverdin production by the plant growth beneficial *Pseudomonas* strain 7NSK2: ecological significance in soil. Plant Soil 130: 249-257.

Höfte, M., Boelens, J. and Verstraete, W. (1992). Survival and root colonization of mutants of plant growth-promoting pseudomonads affected in siderophore biosynthesis or regulation of siderophore production. J. Plant Nutr. 15: 2253-2262.

Holguin, G. and Glick, B.R. (2003). Transformation of *Azospirillum brasilense* Cd with an ACC deaminase gene from *Enterobacter cloacae* UW4 fused to the Tet^r gene promoter improves its fitness and plant growth promoting ability. Microb. Ecol. 46: 122-133.

Hontzeas, N., Saleh, S.S. and Glick, B.R. (2004). Changes in gene expression in canola roots induced by ACC-deaminase-containing plant-growth-promoting bacteria. Mol. Plant Microbe Interact. 17: 865-871.

Hontzeas, N., Richardson, A.O., Belimov, A., Safronova, V., Abu-Omar, M.M. and Glick, B.R. (2005). Evidence for horizontal transfer of 1-aminocyclopropane-1-carboxylate deaminase genes. Appl. Environ. Microbiol. 71: 7556-7558.

Hontzeas, N., Hontzeas, C.E. and Glick, B.R. (2006). Reaction mechanisms of the bacterial enzyme 1-aminocyclopropane-1-carboxylate deaminase. Biotechnol. Adv. 24: 420-426.

Horesman, S., Koninck, K.D., Neuray, J., Herman, R. and Vlassak, K. (1986). Production of plant growth substances by *Azospirillum* sp. and other rhizobacteria. Symbiosis 2: 341-346.

Huang, X.D., El-Alawi, Y., Penrose, D.M., Glick, B.R. and Greenberg, B.M. (2004). Responses of three grass species to creosote during phytoremediation. Environ. Pollut. 130: 453-463.

Hurek, T., Reinhold-Hurek, B., Van Montagu, M. and Kellenberger, E. (1994). Root colonization and systemic spreading of *Azoarcus* sp. strain BH72 in grasses. J. Bacteriol. 176: 1913-1923.

Hurek, T., Handley, L.L., Reinhold-Hurek, B. and Piche, Y. (2002). *Azoarcus* grass endophytes contribute fixed nitrogen to the plant in an unculturable state. Mol. Plant Microbe Interact. 15: 233-242.

Huxtable, S., Zhou, H., Wong, S. and Li. N. (1998). Renaturation of 1-aminocyclopropane-1-carboxylate synthase expressed in *Escherichia coli* in the form of inclusion bodies into a dimeric and catalytically active enzyme. Protein Express. Purif. 12: 305-314.

Illmer, P. and Schinner, F. (1992). Solubilization of inorganic phosphates by microorganisms isolated from forest soils. Soil Biol. Biochem. 24: 389-395.

Ismande, J. (1998). Iron, sulfur and chlorophyll deficiencies; a need for an integrative approach in plant physiology. Physiol. Plant 103: 139-144.

James, E.K. and Olivares, F.L. (1997). Infection of sugar cane and other graminaceous plants by endophytic diazotrophs. Crit. Rev. Plant Sci. 17: 77-119.

James, E.K., Gyaneshwar, P., Mathan, N., Barraquio, W.L., Reddy, P.M., Iannetta, P.P., Olivares, F.L. and Ladha, J.K. (2002). Infection and colonization of rice seedlings by the plant growth-promoting bacterium *Herbaspirillum seropedicae* Z67. Mol. Plant Microbe Interact. 15: 894-906.

Joo, G.J., Kim, Y.M., Lee, I.J., Song, K.S. and Rhee, I.K. (2004). Growth promotion of red pepper plug seedlings and the production of gibberellins by *Bacillus cereus*, *Bacillus macroides* and *Bacillus pumilus*. Biotechnol. Lett. 26: 487-491.

Jurkevitch, E., Hadar, Y. and Chen, Y. (1986). The remedy of line-induced chlorosis in peanuts by *Pseudomonas* sp. siderophores. J. Plant Nutr. 9: 535-545.

Jurkevitch, E., Hadar, Y. and Chen, Y. (1992). Differential siderophore utilization and iron uptake by soil and rhizosphere bacteria. Appl. Environ. Microbiol. 58: 119-124.

Kakimoto, T. (2003). Biosynthesis of cytokinins. J. Plant Res. 116: 233-239.

Kaneshiro, T., Slodki, M.E. and Plattner, R.D. (1983). Tryptophan catabolism to indolepyruvic and indoleacetic acid by *Rhizobium japonicum* L-259 mutants. Curr. Microbiol. 8: 301-308.

Kaufman, P.B., Wu, L., Brock, T.G. and Kim, D. (1995). Hormones and their orientation of growth. In: Plant Hormones. (ed.) P.J. Davies, Kluwer Academic Publishers, Dordrecht, The Netherlands, pp. 547-571.

Kawahara, H., Li, J., Griffith, M. and Glick, B.R. (2001). Relationship between antifreeze protein and freezing resistance in *Pseudomonas putida* GR12-2. Curr. Microbiol. 43: 365-370.

Kende, H. (1993). Ethylene biosynthesis. Annu. Rev. Plant Physiol. Plant Mol. Biol. 44: 283-307.

Kende, H. and Zeevaart, J.A.D. (1997). The five "classical" plant hormones. Plant Cell 9: 1197-1210.

Keshavan, N.D., Chowdhary, P.K., Haines, D.C. and Gonzalez, J.E. (2005). L-Canavanine made by *Medicago sativa* interferes with quorum sensing in *Sinorhizobium meliloti*. J. Bacteriol. 187: 8427-8436.

Khalid, A., Arshad, M. and Zahir, Z.A. (2004). Screening plant growth-promoting rhizobacteria for improving growth and yield of wheat. J. Appl. Microbiol. 96: 473-480.

Kishore, G.K., Pande, S. and Podile, A.R. (2005). Phylloplane bacteria increase seedling emergence, growth and yield of field-grown groundnut (*Arachis hypogaea* L.). Lett. Appl. Microbiol. 40: 260-268.

Kloepper, J.W., Lifshitz, R. and Zablotowicz, R.M. (1989). Free-living bacterial inocula for enhancing crop productivity. Trends Biotechnol. 7: 39-43.

Koga, J., Adachi, T. and Hidaka, H. (1991). Molecular cloning of the gene for indolepyruvate decarboxylase from *Enterobacter cloacae*. Mol. Gen. Genet. 226: 10-16.

Kokalis-Burelle, N., Vavrina, E.N., Rosskopf, E.N. and Shelby, R.A. (2002). Field evaluation of plant growth-promoting rhizobacteria amended transplant mixes and soil solarization for tomato and pepper production in Florida. Plant Soil 238: 257-266.

Kuo, T. and Kosuge, T. (1970). Role of aminotransferase and indole-3-pyruvic acid in the synthesis of indole-3-acetic acid in *Pseudomonas savastanoi*. J. Gen. Appl. Microbiol. 16: 191-204.

Lambrecht, M., Vande Broek, A., Dosselaere, F. and Vanderleyden, J. (1999). The ipdC promoter auxin-responsive element of *Azospirillum brasilense*, a prokaryotic ancestral form of the plant AuxRE? Mol. Microbiol. 32: 889-891.

Laus, M.C., Logman, T.J., Van Brussel, A.A., Carlson, R.W., Azadi, P., Gao, M.Y. and Kijne, J.W. (2004). Involvement of exo5 in production of surface polysaccharides in *Rhizobium leguminosarum* and its role in nodulation of *Vicia sativa* subsp. *nigra*. J. Bacteriol. 186: 6617-6625.

Lee, Y.W., Jin, S., Sim, W.S. and Nester, E.W. (1996). The sensing of plant signal molecules by *Agrobacterium*: genetic evidence for direct recognition of phenolic inducers by the VirA protein. Gene 179: 83-88.

Leigh, J.A., Signer, E.R. and Walker, G.C. (1985). Exopolysaccharide-deficient mutants of *Rhizobium meliloti* that form ineffective nodules. Proc. Natl. Acad. Sci., USA 82: 6231-6235.

LeVier, K. and Guerinot, M.L. (1996). The *Bradyrhizobium japonicum fegA* gene encodes an iron-regulated outer membrane protein with similarity to hydroxamate-type siderophore receptors. J. Bacteriol. 178: 7265-7275.

Li, J., Ovakim, D.H., Charles, T.C. and Glick, B.R. (2000). An ACC deaminase minus mutant of *Enterobacter cloacae* UW4 no longer promotes root elongation. Curr. Microbiol. 41: 101-105.

Li, Q., Saleh-Lakha, S. and Glick, B.R. (2005). The effect of native and ACC deaminase-containing *Azospirillum brasilense* Cd1843 on the rooting of carnation cuttings. Can. J. Microbiol. 51: 511-514.

Lifshitz, R., Kloepper, J.W., Kozlowski, M., Simonson, C., Carlson, J., Tipping, E.M. and Zaleska, I. (1987). Growth promotion of canola (rapeseed) seedlings by a strain of *Pseudomonas putida* under gnotobiotic conditions. Can. J. Microbiol. 33: 390-395.

Liu, P. and Nester, E.W. (2006). Indoleacetic acid, a product of transferred DNA, inhibits *vir* gene expression and growth of *Agrobacterium tumefaciens* C58. Proc. Natl. Acad. Sci., USA 103: 4658-4662.

Liu, S.T., Lee, L.Y., Tai, C.Y., Hung, C.H., Chang, Y.S., Wolfram, J.H., Rogers, R. and Goldstein, A.H. (1992). Cloning of an *Erwinia herbicola* gene necessary for gluconic acid production and enhanced mineral phosphate solubilization in *Escherichia coli* HB101: nucleotide sequence and probable involvement in biosynthesis of the coenzyme pyrroloquinoline quinone. J. Bacteriol. 174: 5814-5819.

Liu, X., Zhao, H. and Chen, S. (2006). Colonization of maize and rice plants by strain *Bacillus megaterium* C4. Curr. Microbiol. 52: 186-190.

Loper, J.E. and Schroth, M.N. (1986). Influence of bacterial sources of indole-3-acetic acid on root elongation of sugar beet. Phytopathol. 76: 386-389.

Loper, J.E. (1988). Role of fluorescent siderophore production in biological control of *Pythium ultimum* by a *Pseudomonas fluorescens* strain Phytopathol. 78: 166-172.

Loper, J.E. and Lindow, S.E. (1994). A biological sensor for iron available to bacteria in their habitats on plant surfaces. Appl. Environ. Microbiol. 60: 1934-1941.

Loper, J.E. and Henkels, M.D. (1997). Availability of iron to *Pseudomonas fluorescens* in rhizosphere and bulk soil evaluated with an ice nucleation reporter gene. Appl. Environ. Microbiol. 63: 99-105.

Lytvynenko, T., Zaetz, I., Voznyuk, T., Kovalchuk, M., Rogutskyy, I., Mytrokhyn, O., Lukashov, D., Estrella-Liopis, V., Borodinova, T., Mashkovska, S., Foing, B., Kordyum, V. and Kozyrovska, N. (2006). A rationally assembled microbial community for growing *Tagetes patula* L. in a lunar greenhouse. Res. Microbiol. 157: 87-92.

Ma, W., Zalec, K. and Glick, B.R. (2001). Biological activity and colonization pattern of the bioluminescence-labeled plant growth-promoting bacterium *Kluyvera ascorbata* SUD165/26. FEMS Microbiol. Ecol. 35: 137-144.

Ma, W., Charles, T.C. and Glick, B.R. (2004). Expression of an exogenous 1-aminocyclopropane-1-carboxylate deaminase gene in *Sinorhizobium meliloti* increases its ability to nodulate alfalfa. Appl. Environ. Microbiol. 70: 5891-5897.

Madhaiyan, M., Poonguzhali, S., Ryu, J. and Sa, T. (2006). Regulation of ethylene levels in canola (*Brassica campestris*) by 1-aminocyclopropane-1-carboxylate deaminase-containing *Methylobacterium fujisawaense*. Planta. (In press).

Malamy, J.E. and Benfey, P.N. (1997). Down and out in *Arabidopsis*: the formation of lateral roots. Trends Plant Sci. 2: 390-396.

Manulis, S., Valinski, L., Gafni, Y. and Hershenhorn, J. (1991). Indole-3-acetic acid biosynthetic pathways in *Erwinia herbicola* in relation to pathogenicity on *Gypsophila paniculata*. Physiol. Mol. Plant Pathol. 39: 161-171.

Marschner, H. (1995). Mineral Nutrition in Higher Plants. Academic Press, London, U.K.

Marschner, P. and Crowley, D.E. (1997). Iron stress and pyoverdin production by a fluorescent pseudomonad in the rhizosphere of white lupine (*Lupinus albus* L.) and barley (*Hordeum vulgare* L.). Appl. Environ. Microbiol. 63: 277-281.

Marschner, P., Crowley, D.E. and Yang, C.H. (2004). Development of specific rhizosphere bacterial communities in relation to plant species, nutrition and soil type. Plant Soil 261: 199-208.

Matzanke, B.F. (1991). Structures, coordination chemistry and functions of microbial iron chelates. In: CRC Handbook of Microbial Iron Chelates (ed.) G. Winkelmann. CRC Press, Boca Raton, Florida, USA, pp. 15-64.

Mayak, S., Tirosh, T. and Glick, B.R. (1997). The influence of plant growth promoting rhizobacterium *Pseudomonas putida* GR12-2 on the rooting of mung bean cuttings. In: Plant Growth-Promoting Rhizobacteria: Present Status and Future Prospects (eds.) A. Ogoshi, K. Kobayashi, Y. Homma, F. Kodama, N. Kondo and S. Akino. OECD, Paris, France, pp. 313-315.

Mayak, S., Tirosh, T. and Glick, B.R. (1999). Effect of wild-type and mutant plant growth-promoting rhizobacteria on the rooting of mung bean cuttings. J. Plant Growth Regulat. 18: 49-53.

Mayak, S., Tirosh, T. and Glick, B.R. (2001). Stimulation of the growth of tomato, pepper and mung bean plants by the plant growth-promoting bacterium *Enterobacter cloacae* CAL3. Biol. Agr. Hort. 19: 261-274.

Mayak, S., Tirosh, T. and Glick, B.R. (2004a). Plant growth-promoting bacteria that confer resistance in tomato plants to salt stress. Plant Physiol. Biochem. 42: 565-572.

Mayak, S., Tirosh, T. and Glick, B.R. (2004b). Plant growth-promoting bacteria that confer resistance to water stress in tomato and pepper. Plant Sci. 166: 525-530.

McCullagh, M., Utkhede, R., Menzies, J.G., Punja, Z.K. and Paulits, T.C. (1996). Evaluation of plant growth promoting rhizobacteria for biological control of *Pythium* root rot of cucumbers grown in rockwool and effects on yield. Europ. J. Plant Pathol. 102: 747-755.

Mehnaz, S., Mirza, M.S., Haurat, J., Bally, R., Normand, P., Bano, A. and Malik, K.A. (2001). Isolation and 16S rRNA sequence analysis of the beneficial bacteria from the rhizosphere of rice. Can. J. Microbiol. 47: 110-117.

Meuwley, P. and Pilet, P-E. (1991). Local treatment with indole-3-acetic acid induces differential growth responses in *Zea mays* L. roots. Planta 185: 58-64.

Morris, C.E. and Monier, J.M. (2003). The ecological significance of biofilm formation by plant-associated bacteria. Annu. Rev. Phytopathol. 41: 429-453.

Morris, R.O. (1986). Genes specifying auxin and cytokinin biosynthesis in phytopathogens. Annu. Rev. Plant Physiol. 37: 509-538.

Mudgett, M.B. (2005). New insights to the function of phytopathogenic bacterial type III effectors in plants. Annu. Rev. Plant. Biol. 56: 509-531.

Muryoi, N., Sato, M., Kaneko, S., Kawahara, H., Obata, H., Yaish, M.W., Griffith, M. and Glick, B.R. (2004). Cloning and expression of *afpA*, a gene encoding an antifreeze protein from the arctic plant growth-promoting rhizobacterium *Pseudomonas putida* GR12-2. J. Bacteriol. 186: 5661-5671.

Neito, K.F. and Frankenberger, W.T. (1989). Biosynthesis of cytokinins by *Azotobacter chroococcum*. Soil Biol. Biochem. 21: 967-972.

Neumann, G. and Römheld, V. (2001). The release of root exudates as affected by the plant's physiological status. In: The Rhizosphere – Biochemistry and Organic Substances at Soil-Plant Interface (eds.) R. Pinton, Z. Varanini and P. Nannipieri. Marcel Dekker, New York, USA, pp. 41-94.

Nilsson, O. and Olsson, O. (1997). Getting to the root: the role of the *Agrobacterium rhizogenes rol* genes in the formation of hairy roots. Plant Physiol. 100: 463-473.

Noel, T.C., Sheng, C., Yost, C.K., Pharis, R.P. and Hynes, M.F. (1996). *Rhizobium leguminosarum* as a plant growth-promoting rhizobacterium: direct growth promotion of canola and lettuce. Can. J. Microbiol. 42: 279-283.

Okon, Y. and Labandera-Gonzalez, C.A. (1994). Agronomic applications of *Azospirillum*: an evaluation of 20 years worldwide field inoculation. Soil Biol. Biochem. 26: 1591-1601.

O'Toole, G., Kaplan, H.B. and Kolter, R. (2000). Biofilm formation as microbial development. Annu. Rev. Microbiol. 54: 49-79.

Pandey A., Durgapal A., Joshi M. and Palni, L.M.S. (1999). Influence of *Pseudomonas corrugate* inoculation on root colonization and growth promotion of two important hill crops. Microbiol. Res. 154: 259-266.

Patten, C.L. and Glick, B.R. (1996). Bacterial biosynthesis of indole-3-acetic acid. Can. J. Microbiol. 42: 207-220.

Patten, C.L. and Glick, B.R. (2002a). Role of *Pseudomonas putida* indoleacetic acid in development of the host plant root system. Appl. Environ. Microbiol. 68: 3795-3801.

Patten, C.L. and Glick, B.R. (2002b). Regulation of indoleacetic acid production in *Pseudomonas putida* GR12-2 by tryptophan and the stationary-phase sigma factor RpoS. Can. J. Microbiol. 48: 635-642.

Peck, S.C. and Kende, H. (1995). Sequential induction of the ethylene biosynthetic enzymes by indole-3-acetic acid in etiolated peas. Plant Mol. Biol. 28: 293-301.

Penrose, D.M. and Glick, B.R. (2001). Levels of ACC and related compounds in exudate and extracts of canola seeds treated with ACC deaminase-containing plant growth-promoting bacteria. Can. J. Microbiol. 47: 368-372.

Penrose, D.M., Moffatt, B.A. and Glick, B.R. (2001). Determination of 1-aminocycopropane-1-carboxylic acid (ACC) to assess the effects of ACC deaminase-containing bacteria on roots of canola seedlings. Can. J. Microbiol. 47: 77-80.

Polyanskaya, L.M., Vedina, O.T., Lysak, L.V. and Zvyagintev, D.G. (2000). The growth-promoting effect of *Beijerinckia mobilis* and *Clostridium* sp. cultures on some agricultural crops. Microbiol 71: 109-115.

Preston, G.M., Bertrand, N. and Rainey, P.B. (2001). Type III secretion in plant growth-promoting *Pseudomonas fluorescens* SBW25. Mol. Microbiol. 41: 999-1014.

Raaijmakers, J.M., van der Sluis, I., Koster, M., Bakker, P.A.H.M., Weisbeek, P.J. and Schippers, B. (1995). Utilization of heterologous siderophores and rhizosphere competence of fluorescent *Pseudomonas* spp. Can. J. Microbiol. 41: 126-135.

Ramos-Gonzalez, M.I., Campos, M.J. and Ramos, J.L. (2005). Analysis of *Pseudomonas putida* KT2440 gene expression in the maize rhizosphere: in vivo expression technology capture and identification of root-activated promoters. J. Bacteriol. 187: 4033-4041.

Rao, V.R., Ramakrishnan, B., Adhya, T.K., Kanungo, P.K. and Nayak, D.N. (1998). Current status and future prospects of associative nitrogen fixation in rice. World J. Microbiol. Biotechnol. 14: 621-633.

Reddy, M.S. and Rahe, J.E. (1989). Growth effects associated with seed bacterization not correlated with populations of *Bacillus subtilis* inoculant in onion seedling rhizospheres . Soil Biol. Biochem. 21: 373-378.

Reed, M.L. and Glick, B.R. (2004). Applications of free-living plant growth-promoting rhizobacteria. Antonie Van Leeuwenhoek 86: 1-25.

Reed, M.L. and Glick, B.R. (2005). Growth of canola (*Brassica napus*) in the presence of plant growth-promoting bacteria and either copper or polycyclic aromatic hydrocarbons. Can. J. Microbiol. 51: 1061-1069.

Reed, M.L., Warner, B.G. and Glick, B.R. (2005). Plant growth-promoting bacteria facilitate the growth of the common reed *Phragmites australisin* in the presence of copper or polycyclic aromatic hydrocarbons. Curr. Microbiol. 51: 425-429.

Rengel, Z. and Marschner, P. (2005). Nutrient availability and management in the rhizosphere: exploiting genotypic differences. New Phytol. 168: 305-312.

Richardson, A.E. and Hadobas, P.A. (1997). Soil isolates of *Pseudomonas* spp. that utilize inositol phosphates. Can. J. Microbiol. 43: 509-516.

Rodriguez, H. and Fraga, R. (1999). Phosphate solubilizing bacteria and their role in plant growth promotion. Biotechnol. Adv. 17: 319-339.

Romantschuk, M. and Bamford, D.H. (1986). The causal agent of halo blight in bean, *Pseudomonas syringae* pv. *phaseolicola,* attaches to stomata via its pili. Microb. Pathog. 1: 139-148.

Safronova, V.I., Stepanok, V.V., Engqvist, G.L., Alekseyev, Y.V. and Belimov, A.A. (2006). Root-associated bacteria containing 1-aminocyclopropane-1-carboxylate deaminase improve growth and nutrient uptake by pea genotypes cultivated in cadmium supplemented soil. Biol. Fertil. Soils 42: 267-272.

Sakakibara, H. (2006). Cytokinins: activity, biosynthesis, and translocation. Annu. Rev. Plant. Biol. 57: 431-449.

Sawar, M. and Kremmer, R.J. (1995). Enhanced suppression of plant growth through production of L-tryptophan compounds by deleterious rhizobacteria. Plant Soil 172: 261-269.

Schröder, G., Waffenschmidt, S., Weiler, E.W. and Schröder, J. (1984). The T-region of Ti plasmids codes for an enzyme synthesizing indole-3-acetic acid. Eur. J. Biochem. 138: 387-391.

Selvadurai, E.L., Brown, A.E. and Hamilton, J.T.G. (1991). Production of indole-3-acetic acid analogues by strains of *Bacillus cereus* in relation to their influence on seedling development. Soil Biol. Biochem. 23: 401-403.

Sessitsch, A., Reiter, B. and Berg, G. (2004). Endophytic bacterial communities of field-grown potato plants and their plant-growth-promoting and antagonistic abilities. Can. J. Microbiol. 50: 239-249.

Sessitsch, A., Coenye, T., Sturz, A.V., Vandamme, P., Barka, E.A., Salles, J.F., Van Elsas, J.D., Faure, D., Reiter, B., Glick, B.R., Wang-Pruski, G. and Nowak, J. (2005). *Burkholderia phytofirmans* sp. nov., a novel plant-associated bacterium with plant-beneficial properties. Int. J. Syst. Evol. Microbiol. 55: 1187-1192.

Sevilla, M., Burris, R.H., Gunapala, N. and Kennedy, C. (2001). Comparison of benefit to sugarcane plant growth and $^{15}N_2$ incorporation following inoculation of sterile plants with *Acetobacter diazotrophicus* wild-type and Nif$^-$ mutants strains. Mol. Plant Microbe Interact. 14: 358-366.

Shaharoona, B., Arshad, M. and Zahir, Z.A. (2006). Effect of plant growth promoting rhizobacteria containing ACC-deaminase on maize (*Zea mays* L.) growth under axenic conditions and on nodulation in mung bean (*Vigna radiata* L.). Lett. Appl. Microbiol. 42: 155-159.

Sisler, E.C., Serek, M., Dupille, E. and Goren, R. (1999). Inhibition of ethylene response by 1-methylcyclopropene and 3-methylcyclopropene. Plant Growth Regulat. 24: 105-111.

Skoog, F. and Miller, C.O. (1957). Chemical regulation of growth and organ formation in plant tissues cultured *in vitro*. Symp. Soc. Exp. Biol. 54: 118-130.

Smalle, J. and Van Der Straetem, D. (1997). Ethylene and vegetative development. Plant Physiol. 100: 593-605.

Smidt, M. and Kosuge, T. (1978). The role of indole-3-acetic acid accumulation by alpha methyl tryptophan-resistant mutants of *Pseudomonas savastanoi* in gall formation on oleanders. Physiol. Plant Pathol. 13: 203-214.

Stearns, J.C. and Glick, B.R. (2003). Transgenic plants with altered ethylene biosynthesis or perception. Biotechnol. Adv. 21: 193-210.

Sturtevant, D.B. and Taller, B.J. (1989). Cytokinin production by *Bradyrhizobium japonicum* Plant Physiol. 89: 1247.

Sun, X., Griffith, M., Pasternak, J.J. and Glick, B.R. (1995). Low temperature growth, freezing survival, and production of antifreeze protein by the plant growth promoting rhizobacterium *Pseudomonas putida* GR12-2. Can. J. Microbiol. 41: 776-784.

Surico, G., Comai, L. and Kosuge, T. (1984). Pathogenicity of strains of *Pseudomonas syringae* pv. *savastanoi* and their indoleacetic acid-deficient mutants on olive and oleander. Phytopathol. 74: 490-493.

Taller, B.J. and Wong, T.Y. (1989). Cytokinins in *Azotobacter vinelandii* culture medium. Appl. Environ. Microbiol. 55: 266-267.

Tamas, I.A. (1995). Hormonal regulation in apical dominance. In: Plant Hormones (ed.) P.J. Davies. Kluwer Academic Publishers, Dordrecht, The Netherlands, pp. 572-797.

Thimann, K.V. and Lane, R.H. (1938). After-effects of treatment of seed with auxin. Am. J. Bot. 25: 535-543.

Thomashow, L.S., Reeves, S. and Thomashow, M.F. (1984). Crown gall oncogenesis: evidence that a T-DNA gene from the *Agrobacterium* Ti plasmid pTiA6 encodes an enzyme that catalyzes synthesis of indoleacetic acid. Proc. Natl. Acad. Sci. USA 81: 5071-5075.

Tien, T.M., Gaskins, M.H. and Hubbell, D.H. (1979). Plant growth substances produced by *Azospirillum brasilense* and their effect on the growth of pearl millet. Appl. Environ. Microbiol. 37: 1016-1024.

Tsuchisaka, A. and Theologis, A. (2004). Unique and overlapping expression patterns among the *Arabidopsis* 1-amino-cyclopropane-1-carboxylate synthase gene family members. Plant Physiol. 136: 2982-3000.

Turnbull, G.A., Morgan, J.A., Whipps, J.M. and Saunders, J.R. (2001). The role of bacterial motility in the survival and spread of *Pseudomonas fluorescens* in soil and in the attachment and colonization of wheat roots. FEMS Microbiol. Ecol. 36: 21-31.

Uthede, R.S., Koch, C.A. and Menzies, J.G. (1999). Rhizobacterial growth and yield promotion of cucumber plants inoculated with *Pythium aphanidermatum*. Can. J. Plant Pathol. 21: 265-271.

Van Onckelen, H., Prinsen, E., Inzé, D., Rüdelsheim, R., Van Lijsebettens, M., Follin, A., Schell, J., Van Montagu, M. and De Greef, J. (1986). *Agrobacterium* T-DNA gene 1 codes for tryptophan 2-monooxygenase activity in tobacco crown gall cells. FEBS Lett 198: 357-360.

Van Peer, R. and Schippers, B. (1998). Plant growth responses to bacterization with selected *Pseudomonas* spp. strains and rhizosphere microbial development in hydroponic cultures. Can. J. Microbiol. 35: 456-463.

Van Rhijn, P. and Vanderleyden, J. (1995). The rhizobium-plant symbiosis. Microbiol. Rev. 59: 124-142.

Vande Broek, A., Lambrecht, M., Eggermont, K. and Vanderleyden, J. (1999). Auxins upregulate expression of the indole-3-pyruvate decarboxylase gene in *Azospirillum brasilense*. J. Bacteriol. 181: 1338-1342.

Vedder-Weiss, D., Jurkevitch, E., Burdman, S., Weiss, D. and Okon, Y. (1999). Root growth, respiration and beta-glucosidase activity in maize (*Zea mays*) and common bean (*Phaseolus vulgaris*) inoculated with *Azospirillum brasilense*. Symbiosis 26: 363-377.

Vesper, S.J. (1987). Production of pili (fimbriae) by *Pseudomonas fluorescens* and correlation with attachment to corn roots. Appl. Environ. Microbiol. 53: 1397-1405.

Walker, T.S., Bais, H.P., Grotewold, E. and Vivanco, J.M. (2003). Root exudation and rhizosphere biology. Plant Physiol. 132: 44-51.

Wang, Y., Brown, H.N., Crowley, D.E. and Szaniszlo, P.J. (1993). Evidence for direct utilization of a siderophore, ferrioxamine B, in axenically grown cucumber. Plant Cell Environ. 16: 579-585.

Wang, C., Knill, E., Glick, B.R. and Defago, G. (2000). Effect of transferring 1-amino-cyclopropane-1-carboxylic acid (ACC) deaminase genes into *Pseudomonas fluorescens* strain CHA0 and its *gacA* derivative CHA96 on their growth-promoting and disease-suppressive capacities. Can. J. Microbiol. 46: 898-907.

Wang, C., Wang, D. and Zhou, Q. (2004). Colonization and persistence of a plant growth-promoting bacterium *Pseudomonas fluorescens* strain CS85, on roots of cotton seedlings. Can. J. Microbiol. 50: 475-481.

Wang, Y., Ohara, Y., Nakayashiki, H., Tosa, Y. and Mayama, S. (2005). Microarray analysis of the gene expression profile induced by the endophytic plant growth-promoting rhizobacteria, *Pseudomonas fluorescens* FPT9601-T5 in *Arabidopsis*. Mol. Plant Microbe Interact. 18: 385-396.

Waters, C.M. and Bassler, B.L. (2005). Quorum sensing: cell-to-cell communication in bacteria. Annu. Rev. Cell Dev. Biol. 21: 319-346.

Xie, H., Pasternak, J.J. and Glick, B.R. (1996). Isolation and characterization of mutants of the plant growth-promoting rhizobacterium *Pseudomonas putida* GR12-2 that overproduce indoleacetic acid. Curr. Microbiol. 32: 67-71.

Yamada, T. (1993). The role of auxin in plant-disease development. Annu. Rev. Phytopathol 31: 253-273.

Yan, Z., Reddy, M.S. and Kloepper, J.W. (2003). Survival and colonization of rhizobacteria in a tomato transplant system. Can. J. Microbiol. 49: 383-389.

Yang, T., Law, D.M. and Davies, P.J. (1993). Magnitude and kinetics of stem elongation induced by exogenous indole-3-acetic acid in intact light-grown pea seedlings. Plant Physiol. 102: 717-724.

Yu, J., Penaloza-Vazquez, A., Chakrabarty, A.M. and Bender, C.L. (1999). Involvement of the exopolysaccharide alginate in the virulence and epiphytic fitness of *Pseudomonas syringae* pv. *syringae*. Mol. Microbiol. 33: 712-720.

5

Mycorrhizal Symbiosis—An Indispensable Component of the Plant Culture

Milan Gryndler

INTRODUCTION

Cultivated plants, a source of nutrition for humans, are dependent on supply of water and minerals obtained from soil, a natural growth medium for the great majority of plant species. Soil is extremely rich in many forms of living organisms. These organisms constitute really very rich cenosis, the complexity of which by far exceeds the complexity of the richest aboveground plant cover. Soil biota involve microorganisms of different life strategies such as saprotrophs feeding on dead organic matter, autotrophs that do not need organic nutrients, various parasitic organisms harmful for their host organisms, and symbiotic organisms that may collaborate with other components of soil biota, for example, plant roots, through mycorrhizal symbiosis.

Mycorrhizal symbiosis - a functional coexistence of a fungus with roots of a host plant - is very common in nature (Trappe, 1987; Peterson and Farquahar, 1994). Mycorrhizal symbiosis is a widespread natural phenomenon and a common nutritional mechanism for the vast majority of cultivated plants, hence mycorrhizal fungi merit earnest attention of growers. Mycorrhizal fungi colonizing root tissues profoundly alter the physiological states of the plant and the fungus, resulting in formation of new complex organs called mycorrhiza. In Volume I of this series (Microbial Biotechnology in Horticulture), Bagyaraj et al. (2006) have described the uses of mycorrhizal fungi in increasing productivity in horticulture. Some more information is provided in this chapter on functioning of mycorrhizal symbiosis in agriculture, forestry and horticulture.

SYMBIOSIS OF HORTICUTURAL PLANTS WITH MYCORRHIZAL FUNGI

Several dissimilar types of mycorrhizal symbiosis have been recognized (Peterson and Farquahar, 1994). While all these types cannot be discussed in detail, their essential characteristics are mentioned here because they have interesting and important practical applications.

The most common type of mycorrhizal symbiosis is arbuscular mycorrhizal (AM) symbiosis (**Fig. 5.1**, right). It is formed by vascular plants (either herbaceous or woody). Some colonization of lower plants has been noted as well. This type of mycorrhizal symbiosis is most typical for many agricultural plants, such as cereals (including rice), potato, sugarcane, maize and beans. Only a minority of plants, like brassicaceous plants (mustard and rape) and several other cultivated plants (sugar beet) are nonmycorrhizal or weakly mycorrhizal (Harley and Harley, 1987). Thus, AM symbiosis is probably the most important type of symbiosis from the point of view of gardeners and agriculturalists.

AM fungi penetrate the rhizodermis and cortical cells forming typical intracellular structures, highly dichotomically branched arbuscules. Arbuscules, having extremely large specific surfaces, ensure the most intimate communication between the fungus and its host. Arbuscules are not perennial structures. In several days they undergo lytic processes, probably driven by the host (Toth and Miller, 1984). Mycelium of AM fungi is primarily aseptate and, in the soil, often forms large, mostly globular, thick-walled resting spores. The life of soil mycelium of AM fungi is also counted in days (Staddon et al., 2003). It has recently been widely accepted that these fungi are obligate biotrophs, i.e. they need a living host for their carbon nutrition. The frequent genera of AM fungi are *Glomus, Gigaspora* and *Scutellospora*.

A very popular group of cultivated ornamental plants is the family *Orchidaceae* - orchids. This family forms a special type of mycorrhizae, typical for orchideoid mycorrhizal symbiosis. In this type of symbiosis, the fungus belonging to Basidiomycetes, in many cases to fungi of the imperfect genus *Rhizoctonia* (Pope and Carter, 2001), penetrates the rhizodermis and superficial layers of cortical cells. In deeper parts of the cortex it forms typical fungal structures - pelotons - which may be

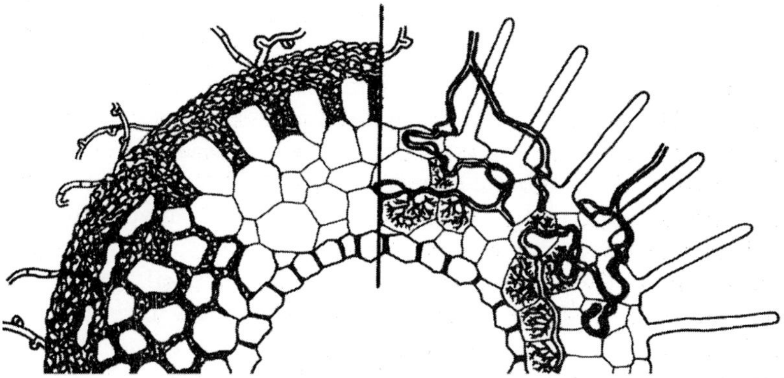

Fig. 5.1. Schematized transversal cutting of two most widespread types of mycorrhizae: ectomycorrhiza (left) and arbuscular mycorrhiza (right). Ectomycorrhiza is characterized in most cases by a hyphal mantle covering the surface of mycorrhizal root and Hartig net – a net of fungal hyphae in intracellular spaces of the root cortex. Unlike arbuscular mycorrhiza, ectomycorrhiza is morphologically distinct from nonmycorrhizal root. In arbuscular mycorrhiza, the fungus does not form a hyphal mantle and Hartig net, nor penetrates into the internal spaces of cortical cells. Instead, it forms highly branched structures called arbuscules, ensuring most intimate contact between both partners in symbiosis

described as compact tangles of hyphae inside the plant cells (Hadley and Williamson, 1972). These structures are typical for major sub-type of orchideoid mycorrhizal symbiosis, the so-called tolypophagous sub-type. Dichotomically branched intracellular structures similar to arbuscules are not formed. Intracellular pelotons in orchideoid mycorrhizae undergo lysis in a manner similar to arbuscules in arbuscular mycorrhizae. Once the peloton is destroyed, the fungus may colonize the host cells again (Hadley, 1982). This cycle may be repeated several times. Such colonization occurs not only in the root cortex, but also in tissues of tiny germinating seeds of orchids, called protocorms. These almost imperceptible seeds do not have sufficient reserves of energy and usually cannot survive the first stages of germination without the intervention of mycorrhizal fungi (Arditti and Ghani, 2000). In fact, protocorms receive carbon nutrition from the mycorrhizal fungi. Generative reproduction of orchids in nature thus depends on symbiosis with mycorrhizal fungi and this is also reflected in artificial techniques of seed germination and seedlings cultivation (Stewart and Zettler, 2002; Sharma et al., 2003). Similarly, adult plants in some species of *Orchidaceae* lack active photosynthetic apparatus and thus have to rely on carbon nutrition gained from mycorrhizal fungus. This life strategy, where the plant is supplied with energy rich organic compounds by the symbiotic fungus, is called mycotrophy (mycoheterotrophy) (Leake, 1994).

Another type of mycorrhizal symbiosis is **ericoid** mycorrhizal symbiosis. This type of symbiosis is known in ericaceous plants (order *Ericales*) and is formed by *Ascomycetes* or their anamorphs (Read, 1974; Chambers et al., 2000; Whittaker and Cairney, 2001; Vralstad et al., 2002). The reports suggesting formation of ericoid mycorrhizae by basidiomycetes are not conclusive. Some ericaceous plants are cultivated as ornamentals (*Rhododendron* spp.) and mycorrhizal symbiosis of the ericoid type might, therefore, have some practical importance. The symbiotic fungus (e.g. ascomycete *Hymenoscyphus ericae*) penetrates the cell wall of the rhizodermis, forming pelotons inside and also within some cortical layers of the cells. In this type of mycorrhizal symbiosis, the fungus seems to dominate over its host. No physiological lysis of intracellular fungal structures occurs here (Smith and Read, 1997) and these structures can survive the death of host root tissue and take part in its decomposition (Kerley and Read, 1998).

Some isolates of ericoid mycorrhizal fungi may attack the root tissues of non-host plants. In this case, the colonization is not restricted to the rhizodermis and cortex. The fungus then destroys plasmalemma and behaves as a necrotrophically growing parasite (Bonfante-Fasolo et al., 1984). Such isolates of ericoid mycorrhizal fungi may represent a potential menace for non-ericoid plants, cultivated in farms or gardens after the ericoid plants, because ericoid mycorrhizal fungi may survive in soil even for decades in absence of a suitable host (Bergero et al., 2003).

Many woody plants live in **ectomycorrhizal symbiosis** (**Fig. 5.1**, left) caused by many basidiomycetes and ascomycetes, as well as with some zygomycetes (Trappe, 1962). The total number of species of ectomycorrhizal fungi is estimated to be about 5,000 (Molina et al., 1992). In ectomycorrhizal symbiosis (unlike the above described types of mycorrhizal symbiosis), the mycorrhizal fungus does not penetrate the cell wall of the host tissue, but instead forms a compact mycelial net in the intercellular spaces. This structure is called the 'Hartig net'. A mycelial layer often occurs on the surface of ectomycorrhizae and is called hyphal mantle.

Ectomycorrhizal fungi often form large mycelial colonies of different shapes in the soil (Read, 1992). The sizes of such colonies depend on fungal species and may reach several tens of meters. These colonies represent dynamic structures changing in time and affecting the soil environment, soil biota and plant cover.

FUNCTIONING OF MYCORRHIZAL SYMBIOSIS

The most important feature of mycorrhizal symbiosis from the viewpoint of horticultural and agricultural practice is that it affects growth of the host plants. Recently, several physiological mechanisms were considered to be involved in the response of plant growth by mycorrhizal fungi.

Transport of Soil Nutrients and the Role of Soil Saprotrophs

Mycelia effectively interconnect soil environments with the internal structures of the host root (**Fig. 5.2**) and mediate transport of nutrients from soil to the plant organs (George, 2000). This phenomenon has been extensively studied mainly in the case of phosphorus. Phosphorus is often

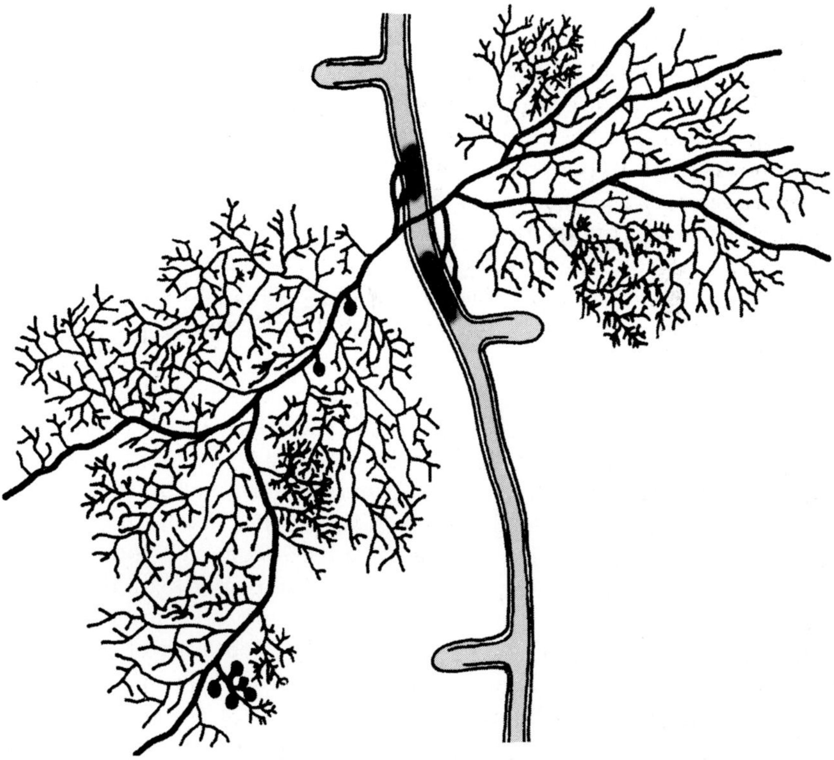

Fig. 5.2. Fine mycelia of mycorrhizal fungi reach a much larger soil volume than roots and may, therefore, more effectively absorb mineral nutrients than the root itself

present in the soil in low amounts as mobile phosphates, which are inaccessible to plants. Mycorrhizal fungi take up phosphate ions and make them easily available for plants. This effect mainly translates into an increase of soil volume where the plant can obtain its mineral nutrition (Paszkowski and Boller, 2002).

With regard to phosphorus uptake by mycorrhizal fungi, the cooperation with specific soil saprotrophic microorganisms capable of solubilizing insoluble forms of phosphate (mainly Ca-phosphate) has been reported. Simultaneous inoculation of AM fungi and phosphate solubilizing bacteria in phosphate deficient soils may be more effective in growth stimulation of mycorrhizal host plant than single inoculation by both inoculants (Barea et al., 1975), and may even lead to better utilization of phosphate fertilizers (Piccini and Azcon, 1987), which otherwise become rapidly unavailable to plants. The mechanisms of phosphate solubilization are not completely understood but probably consist in acidification of some soil microenvironments, which increases solubility of calcium phosphates. For example, phosphate solubilizing *Enterobacter agglomerans* increased acidity of the rhizosphere by accumulating organic acids (Kim et al., 1998). Similar effects have also been observed with the phosphate solubilizing saprotrophic microfungus *Penicillium bilaji* in greenhouse and field experiments (Kucey, 1987). This saprotrophic fungus alone was able to increase the phosphorus uptake by wheat plants, but mycorrhizal fungus further improved phosphate supply due to enhanced absorption and transport of newly solubilized phosphate. Phosphate solubilization by saprotrophic microfungi and synergism with mycorrhizal fungi has also been reported for *Aspergillus niger* and *Penicillium citrinum* (Omar, 1998).

Soil saprotrophic microorganisms possess an excessively wide range of metabolic capabilities, which implies that they may affect mycorrhizal fungi by production of many different biologically active compounds. Soil saprotrophic microflora is known to stimulate germination of spores of AM fungi and their mycelial growth (Giovannetti, 2000). In turn, AM fungi may stimulate development of some bacteria in close proximity to mycelium, which has been demonstrated for bacteria of the genus *Paenibacillus* (Mansfeld-Giese et al., 2002). These bacteria are capable of stimulating mycelial growth substantially (Hildebrandt et al., 2002). Interestingly, bacteria of the genus *Paenibacillus* have also been detected on ectomycorrhizae together with the genera *Burkholderia* and *Rhodococcus* (Poole et al., 2001), and are capable of stimulating the formation of ectomycorrhizae.

Host Plant Water Economy

Mycorrhizal symbiosis may also affect **water economy of the host plant**. Mycorrhizal mycelium may increase the mobility of water in the system soil-mycorrhizal plant, which may result in higher water uptake by the plant and a corresponding decrease in soil humidity under mycorrhizal plants (Ebel et al., 1996; Koide and Wu, 2003).

However, the opposite water flux is also possible, as was observed for the ectomycorrhizal *Quercus agrifolia* (Querejeta et al., 2003). In this case, the mycorrhizal root system of the oak supplied mycelium with humidity at night in overdried superficial soil layers. Using such a mechanism, deep rooting plants might supply dry soil with humidity taken up in deeper layers via distribution by mycorrhizal mycelia.

The direction of water flux is determined by a physical parameter called water potential. As no specialized water transporters have been discovered till now in nature, water always flows from spaces with higher water potential to spaces with lower values of this physical parameter, and living organisms do not determine the flux direction. Soil biota could thus affect only soil mobility and retention ability (Augé et al., 2001) and this is the field where mycorrhizal symbiosis might positively affect plant growth and survival in ecosystems where water content in the soil is suboptimal.

If the stimulation effect of enhanced mineral nutrition or water uptake mediated by the mycorrhizal fungus is greater than the negative effect of energy loss caused by consumption of a portion of photoassimilated energy, then the net effect on plant biomass accumulation (growth) may be positive (**Fig. 5.3**, Findlay and Kendle, 2001).

Increased Resistance against Diseases and Pests

In addition to enhancing mineral nutrition, mycorrhizal plants often possess increased resistance against diseases and pests, and mycorrhizal fungi, thus, might represent potential for development of a biological means of plant protection. For example, an antagonistic relation between the AM fungus *Glomus mosseae* and the fungus *Rhizoctonia solani,* pathogenic for *Pisum sativum,* has been observed (Morandi et al., 2002). In this case, colonization of pea roots by the pathogenic fungus was significantly decreased in the presence of the mycorrhizal fungus, probably because the concentration of protective flavonoid pisatin in roots was increased. Interactions of AM fungi with pathogens were also observed for other host plants as well as for other harmful fungi (Hwang et al., 1992; Vestberg et al., 1994; St Arnaud et al., 1996; Lingua et al., 2002),

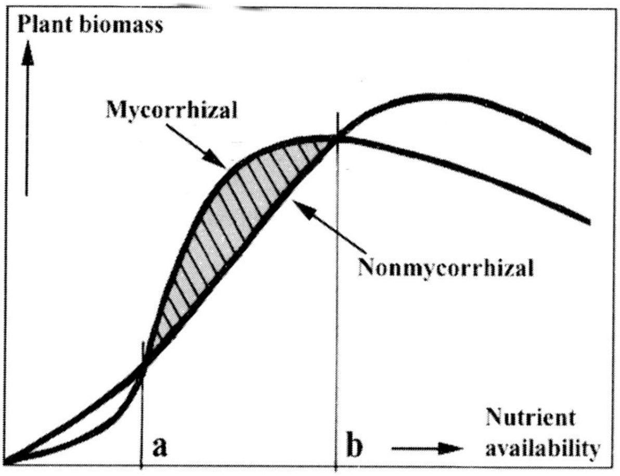

Fig. 5.3. When the effect of the availability a limiting factor (e.g. phosphorus) on plant growth is studied, the dose-response curves shown by mycorrhizal plants are dissimilar to those of control plants. Stimulation of plant growth by mycorrhizal fungi is often reached only within an interval of the phosphorus availability (a to b, shaded area). If its availability is lower than a, plant growth may be decreased because the fungus itself covers its needs for phosphorus preferentially. If it is higher than b, plant growth may be decreased (compared to growth of nonmycorrhizal plants at the same level of phosphorus availability) because plant benefits gained from symbiosis are lower than costs to be invested into the symbiosis (assimilates consumed by the mycorrhizal fungus)

but the mechanisms causing the affection of resistance of plants against fungal pathogens are generally not sufficiently elucidated.

Ectomycorrhizal fungi often produce antibiotic compounds which were first demonstrated through suppression of the pathogenic *Rhizoctonia solani* by ectomycorrhizal basidiomycete *Lactarius helvus* (Šašek and Musílek, 1968). Further, *Pinus* ectomycorrhizae formed by *Cenococcum graniforme* (root tips colonized by the symbiotic fungus) are effectively protected against *Rhizoctonia* sp. (Sen, 2001). Many other reports describing effects of myccorrhizal fungi beneficial to plants exposed to pathogenic fungi can be found in the mycological literature (Linderman, 2000).

Positive effects related to the damage caused by some plant pests have also been observed. For example, extensive efforts have been expended into the study of the interactions between mycorrhizal fungi and soil nematodes (Ingham, 1988). Some species of nematodes are responsible for serious damages of cultivated plants. AM fungi have been reported to decrease the population densities of harmful nematodes (Sitaramaiah and Sikora, 1982) and the synchronization of seasonal changes of population densities of arbuscular mycorrhizal fungi and nematodes (Rich and Schenck, 1981). Substantial improvements in growth of plants colonized by the nematode *Pratylenchus vulnus* were noted in the case of cherry (Pinochet et al., 1995) and several related fruit species (Camprubi et al., 1993; Calvet et al., 1995). The same beneficial effect of AM fungi was further observed with *Coffea* plants colonized by the nematode *Pratylenchus coffeae* (Vaast et al., 1998) and with *Musa* plants colonized by *Meloidogyne javanica* (Pinochet et al., 1996). The practical use of mycorrhizal inoculation in the struggle against harmful nematodes seems to be the most promising practical application of the mycorrhizal fungi as phytoprotective agent.

Stability of Soils

Mycorrhizal symbiosis may positively affect not only growth, health status and stress tolerance of host plants but also the stability of the soil itself (Miller and Jastrow, 2000). Ectomycorrhizal fungi produce a significant amount of biomass of fine mycelia which mechanically stabilizes soil structure (Caravaca et al., 2002). This results in increased resistance to soil erosion. Mycorrhizal fungi might also produce macromolecular compounds (polysaccharides and proteins), the chemical nature of which increases the stability of soil aggregates.

One such compound being described as a product of AM fungi is the glycoprotein glomalin (Wright and Upadhyaya, 1996). Glomalin accumulates in soil because of its relatively long lifetime (Rillig et al., 2001) and its concentration may reach 60 mg per gram soil such that it may represent a significant fraction of soil organic matter. Production of this compound is strongly affected by soil structure and probably represents a factor determining the stability of soil aggregates (Rillig and Steinberg, 2002) in addition to the mechanical action of mycelium.

However, some negative effects of mycorrhizal fungi on soil stability were also observed. Some ectomycorrhizal fungi produce antibiotic/allelopathic compounds, which may in some cases lead to a reduction of plant cover richness of the forest floor. Such an area with reduced plant cover, and having a radically modified soil microflora, is called the hydnosphere (Pacioni, 1991). Furthermore, some ectomycorrhizal fungi possess oxidase activities (Burke and Cairney, 2002) which enable them to utilize soil humic substances. Both effects may endanger soil stability. In some cases, the introduction of improperly chosen fungal strains into the rhizosphere of tree seedlings may cause losses of soil organic matter (Chapela et al., 2001).

POTENTIAL OF THE USE OF MYCORRHIZAL TECHNOLOGY IN AGRICULTURAL AND HORTICULTURAL SYSTEMS

Physiological features of mycorrhizal symbiosis offer potential for practical applications of mycorrhizal technology (inoculations and enhancement of mycorrhizal colonization of plants) in plant cultivation. This practical approach is most promising for low input or organic agricultural systems where the ability of mycorrhizal fungi to enhance plant mineral nutrition and water supply is essential.

Mycorrhizal Symbiosis: Conventional vs Organic Production Systems

Conventional, high input agricultural/horticultural production systems tend to be dominated by a simple management model, wherein a limitation in nutrient supply is directly followed by the application of that nutrient as a fertilizer (Atkinson et al., 2002). Such systems do not favor the use of mycorrhizal technology because they do not support the development of AM symbiosis (Ryan et al., 1994; Mader et al., 2000).

In organic farming, the nutrients come into the soil environment in organically bound forms, which may be mobilized and efficiently taken up by AM fungi. This utilization of organic forms of nitrogen (Hawkins et al., 2000) and phosphorus (Kahiluoto and Vestberg, 1998) has been observed.

Although the effects of organic matter on AM fungi are considered to be generally beneficial, in fact they may depend on the type of organic material used for soil amendment. An inhibition of plant mycorrhizal colonization was observed when various organic substances were added into the soil (Calvet et al., 1992), cellulose being an example of such inhibitory material (Avio and Giovannetti, 1988; Gryndler et al., 2002). However, the negative effects of cellulose on AM fungi were not persistent. After 11 months of cultivation, the cellulose-ammended substrate caused higher plant colonization by AM fungi and a more extensive development of mycorrhizal mycelium (Gryndler et al., 2002), compared to control. It is, therefore, probably reasonable not to apply large amounts of fresh organic matter directly into the soil, but rather to subject this material to a period of composting, which will prevent the possible negative effects of fast decomposition on AM fungi.

Further, amendment of substrates with peat may result in decreased effectiveness of mycorrhizal symbiosis or even the negative effect of plants to inoculation as shown, for example, for strawberries (Vestberg et al., 2005).

Mycorrhizal Fungi in Artificial Substrates

Inoculation with properly selected mycorrhizal fungi (Estaún et al., 2002) may be beneficial for plants mainly in substrates that do not contain their own populations of living mycorrhizal fungi. Substrates used in horticulture are often sterilized (steamed, fumigated) to destroy dangerous phytopathogenic organisms. However, mycorrhizal fungi most probably will not survive such a treatment of the substrate. Similarly, the substrates composed of various non-soil components (for example, sand, vermiculite and perlite) and different ceramic-based hydroponic substrates will lack mycorrhizal fungi. This is the situation where the application of mycorrhizal inoculum is reasonable. Under outdoor conditions, the mycorrhizal fungi may be absent from different mineral waste materials (e.g., clay, sand, subsoil and ash deposits and various industrial wastes) intended for cultivation of plants.

There is also an opportunity to apply mycorrhizal inoculum in situations where living (indigenous) mycorrhizal fungi are present in the substrate but are ineffective (do not support plant growth). In this case, the selection of mycorrhizal fungi is more problematic because the inoculant must be successful in competition with indigenous population. Various fungal isolates should be first tested for their compatibility with the same sterilized substrate in an microcosm experiment. The fungi should be able to grow in the sterilized substrate, produce extraradical mycelium and colonize host plants. Selected fungal isolates need to be then tested in microcosms in unsterile substrate for their desirable effects. The isolates passing this second selection are multiplied and tested in larger outdoor experiments. The results obtained under adequate field conditions are then indicative for final selection of effective isolates (**Fig. 5.4**). Selected isolates are multiplied and a technology of inoculum mass production is developed.

In some cases, it is not necessary to develop specialized (directed) mycorrhizal inocula (mainly if the problem to be resolved by inoculation consists of insufficient phosphorus concentration in the soil) and a commercially available inoculum may be used (Feldman and Grotkass, 2002). The commercial inocula containing generally effective fungal strains should have standardized concentrations of infective propagules and their efficacy (i.e. ability to stimulate plant growth under specified conditions) should be guaranteed (Von Alten et al., 2002).

Recently, the inoculation of plants mainly with AM fungi and ectomycorrhizal fungi has been translated into practice. Processes for handling these mycorrhizal fungi in culture are described in detail by many authors, but are comprehensively summarized by Brundrett et al.

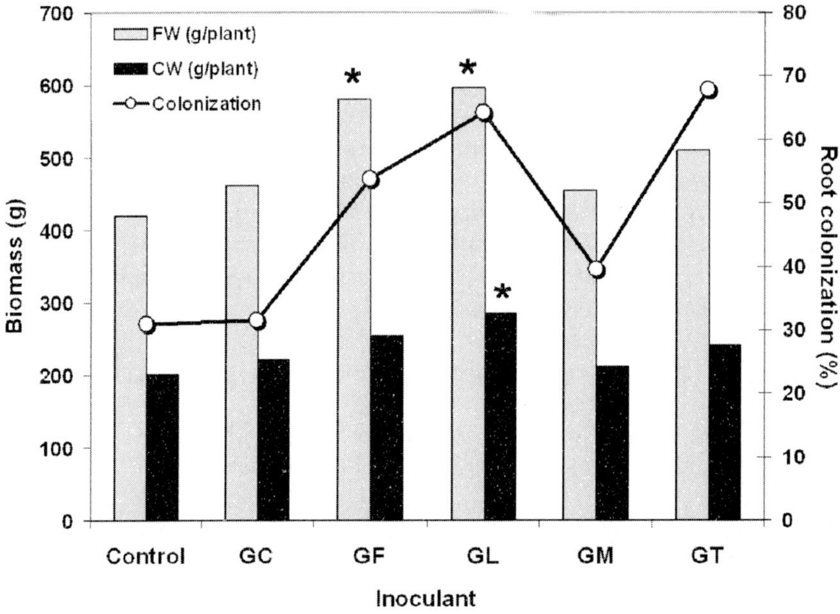

Fig. 5.4. Effect of inoculation of maize plants with arbuscular mycorrhizal fungi on total shoot biomass (FW) and cob biomass (CW) in nonsterile soil under field conditions (g per plant). Mean values of 24 replicates significantly different from the control are marked by asterisks. Inoculants: GC – *Glomus caledonium*, GF – *Glomus fasciculatum*, GL – *Glomus claroideum*, GM – *Glomus mosseae*, GT – *Glomus tenuis*. Mycorrhizal colonization (percentage of the root length colonized) correlated with plant biomass accumulated. Plants were fertilized with 100 kg N/ha, no phosphorus fertilizer was applied

(1996). Processes for production of inocula of AM fungi are different from those used for production of ectomycorrhizal fungi, because the latter can be often cultivated on artificial media, whereas AM fungi are obligate biotrophs which have to be cultivated on living host only. This means that the cultivation of these fungi is possible, for example, in pot cultures (**Fig. 5.5**), in hydroponics, or in more advanced cultivation systems such as those involving RiT-DNA transformed roots instead of intact host plants.

Promising Opportunities for Use of AM Fungal Inoculants

Inocula of AM fungi may be successfully used in multiplication of micropropagated plants (Taylor and Harrier, 2003) during the step of transferring the plantlets from a sterile environment *in vitro* to nonsterile substrates, where the mycorrhizosphere is formed *de novo*. At this step of cultivation of the plant, the introduction of AM fungi is easy and effective. This has been shown, for example, in cases of *Leucaena leucocephala* (Naqvi

Fig. 5.5. Pot cultures of arbuscular mycorrhizal fungi in a growth chamber with maize used as the host plant

and Mukerji, 1998), *Sesbania sesban* (Subhan et al., 1998), potato (Yao et al., 2002), citrus (Quatrini et al., 2003), and avocado (Vidal et al., 1992).

Positive effects on seed-reproduced plants have also been observed frequently, for example, in case of *Carica papaya* (Cruz et al., 2000) and watermelon (*Citrullus lanatus*, Kaya et al., 2003) where significant growth stimulation by inoculation has been observed on both well-watered and drought-stressed plants.

Stimulation of plant growth by inoculation with arbuscular mycorrhizal fungi has been observed many times and is connected, in most cases, with insufficient phosphorus accessibility (Lekkberg and Koide, 2005).

Promising Possibilities for Use of Ectomycorrhizal Fungal Inoculants

The most important practical result of the inoculation with ectomycorrhizal fungi is the improvement of thermotolerance (probably connected with resistance to drought stress) of ectomycorrhizal pine plantations. This was successful in large areas in the United States, Brazil, Mexico, several states of Africa, South Korea, Phillippines and France. Inoculated plants grew better and were able to more effectively exploit the fertilizer applied (Marx et al., 1985). *Pisolithus tinctorius* (*arhizus*) has been recognized as a very effective symbiotic fungus and industrial processes of production of this inoculum have been developed (Marx et al., 1982). Many further reports describing the beneficial effects of ectomycorrhizal fungi on host trees can be found in the literature (Castellano, 1996).

Ectomycorrhizal fungi may be also used to improve rooting of woody plant species and to increase the efficiency of their vegetative multiplication (Niemi et al., 2000), which may decrease the cost of multiplication and improve the physiological status of the plants. Further, artificial introduction of ectomycorrhizal fungi into the rhizosphere of woody plants may be reasonable in forest plantations (Castellano, 1996), mainly in soils disturbed by humans.

Factors Limiting Practical Applications of Mycorrhizal Inocula

The use of practical inoculations of plant cultures by mycorrhizal fungi is currently limited by several factors. Firstly, the inoculum of arbuscular mycorrhizal fungi is often too **expensive** for massive inoculations in the field. However, at least in some cases, the expensive commercial inoculum may be successfully replaced by promotion of populations of indigenous mycorrhizal fungi. Indigenous AM fungi can be much better adjusted to the target environments than strains or selected isolates supplied in commercial inocula. The multiplication of indigenous populations may be promoted by establishment of correctly watered and fertilized (**Fig. 5.6**) pot culture with nonsterile target soil as a substrate planted with a proper host plant (clover, maize, sudan grass, rye grass, etc.). After a reasonable cultivation period (usually three-five months), the plants are harvested and the soil is used as inoculum. Such inocula may show substantially higher plant growth stimulation than pure inoculum (Quatrini et al., 2003). Multiplication of indigenous mycorrhizal fungi thus represents an interesting alternative to the use of commercial inocula, but it must be free of pathogenic microorganisms, which can be sometimes multiplied together with mycorrhizal fungi (Fortuna et al., 1998). If plant species different from the target are used as a host for multiplication of inoculum, this problem may be minimized. AM fungi may be also isolated from the target soil, multiplied separately in pasteurized substrate and mixed together to prepare inocula for experimentation (Caravaca et al., 2003). In any event, if the inoculum of the indigenous fungi is produced, the

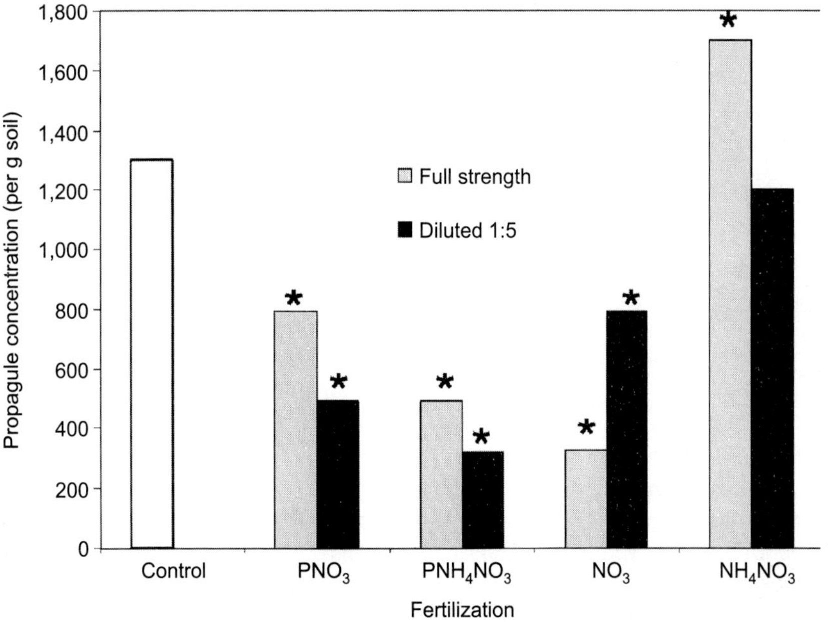

Fig. 5.6. Effect of host plant mineral nutrition on production of infective propagules (diaspores) of *Glomus claroideum* in soil based pot culture after three months of cultivation. Treatments marked NO_3 received 36 mg $MgSO_4 \cdot 7H_2O$, 15 mg $Ca(NO_3)_2.4H_2O$ and 10 mg KNO_3 per plant, weekly. Besides this, phosphate fertilized treatments (P) received 0.6 mg KH_2PO_4 and ammonium fertilized treatments (NH_4) received 1.2 mg NH_4NO_3 per plant, weekly. Dissolved nutrients were applied in 10 ml water. The above doses represent full strength fertilization (gray columns), decreased doses of fertilizer (the same volume of the 1:5 diluted full strength solution) have also been used. The results indicate that fungal propagule production is strongly affected by mineral nutrition

persons managing the culture should undergo specialized training to obtain the basic knowledge and skills necessary for acceptable results.

Ectomycorrhizal inoculum can be produced by cultivation of fungi in pure culture, but the resulting biomass is often sensitive to drying and cannot be stored for longer periods before usage. Basidiospore preparations obtained from field-collected fruitbodies are also used as active components of inoculum. In this case, the vitality of the produced inoculum is prolonged, but its quantity depends on the amount of field collected fungal fruitbodies.

The use of mycorrhizal fungi is also limited by application of fungicides and some other pesticides. Residual concentrations of fungicides may dramatically reduce mycorrhizal colonization of host roots (Trappe et al., 1984), even when these are applied at recommended levels. Fungicides (Carbendazim) may suppress absorption of phosphorus by extraradical mycelium (Kling and Jakobsen, 1997). Obviously, regular application of nonselective fungicides is not compatible with normal functioning of mycorrhizal symbiosis.

EDIBLE MYCORRHIZAL FUNGI IN FORESTRY AND HORTICULTURE

Many ectomycorrhizal fungi produce macroscopic fruitbodies which have significant value for human alimentation and may represent an important source of protein in some regions. A list of edible ectomycorrhizal fungi important at the global scale has been published by Yun et al. (2001a). Some mycorrhizal fungi are cultivated. A simple approach to cultivation involves seeding of commercially useless fruitbodies of different fungal species at the sites where they naturally occur (Pilz and Molina, 2002).

An important example of a mushroom which is cultivated and consumed by man is *Tricholoma matsutake* 'matsutake'. Planting of host pine plants inoculated with *Tricholoma matsutake* mainly in Japan and Korea is restricted only to areas with particular soil properties. This mushroom has been consumed by traditional Japanese society for more than 1,000 years. Forest management has facilitated efforts to improve the harvest of fruitbodies of this very valuable fungus (Ito and Ogawa, 1979) but the total annual harvest is gradually declining.

Another economically important fungus is the black truffle of Perigord (*Tuber melanosporum*), a species which is cultivated on inoculated trees (oaks, hazels and some other woody hosts) in calcareous soil. No particular soil properties are required, hence this culture can be used in most calcareous soils if climatic conditions are favorable (Hall and Yun, 2001; Kagan-Zur et al., 2001). This manner of cultivation is called true semiculture. The annual harvest of *Tuber melanosporum* is also declining, probably as a result of a combination of unfavorable social, economic and ecological reasons.

Under similar semiculture conditions, successful fructifications of *Lactarius deliciosus* and *Rhizopogon rubescens* under inoculated plantation of *Pinus radiata* (Yun et al., 2001b) and of pot cultured *Cantharellus cibarius* (Danell, 2001) were obtained.

The strategy of a semicultivation approach to production of fruitbodies of ectomycorrhizal fungi is probably also applicable to other fungal species not mentioned here. This may open a possibility for cultivation of ectomycorrhizal fungi as an additional product in forestry and horticulture. Because semiculture demands large land areas and involves manual operations, it is expensive and only economical for production of the fungal species producing the most expensive fruitbodies (*Tuber melanosporum*). However, the situation is different if production of fruitbodies could be planned as an auxilliary product to timber or fruit production cultures. In this case, investing some funds in inoculated plants and irrigation technology appears to make sense. Pine seedlings inoculated by *Suillus* spp. are ocassionally available in the European market, but potential for supply of semicultures of many other fungal species may be achieved in future.

CONTRIBUTION OF GENE TECHNOLOGY TO MYCORRHIZAL BIOTECHNOLOGY

The tools of molecular genetics are being exploited to modify important properties of many industrial organisms. In principle, genetic modifications of mycorrhizal fungi might also significantly improve the performance of inoculated plants in comparison with plants inoculated with mycorrhizal fungi of wild genotypes.

However, the genomic sequences of AM fungi are still not characterized (Bever and Wang, 2005) and currently available data on the regulation of gene expression and functioning of genes in these organisms is limited. Obviously, such information is essential for any serious practical application of this technology. Hence successful genetic manipulations of AM fungi resulting in desirable effects on plant physiology should not be expected in the near future.

So far, genetics of ectomycorrhizal fungi have been addressed at a more classical level, because some basidiomycetous and ascomycetous fungi have been used as model organisms in studies in general genetics. While this is not the case for any of the ectomycorrhizal fungi, there is greater potential to exploit applied genetic research of ectomycorrhizal fungi than there is in the case of AM fungi.

However, the potential use of genetically modified microorganisms (mycorrhizal fungi), even when practically efficient, is a subject of legal restrictions in many countries because of possible serious risks connected with their unexpected behavior in the environment. This raises doubts regarding potential translating the use of genetically modified mycorrhizal fungi into practice.

As indicated in Chapter 6 in this Volume, the method of monoxenic cultivation of AM fungi on RiT-DNA transformed roots has been developed as an alternative to pot cultured non sterile inocula. This represents an application of genetic transformation of the host plant. In this case, genetic transformation is achieved using infection with

Agrobacterium strains containing a plasmid whose DNA is incorporated into the plant genome. If the resulting transformed root tissue finally does not contain *Agrobacterium* cells and bacterial DNA, which is incorporated into the plant genome, is of natural origin (i.e. genetically unmodified bacterial strains are used for transformation), then the resulting root tissue does not represent a threat to the environment and may be used for monoxenic production of inoculum of AM fungi. This represents an invaluable application of genetic transformation in mycorrhizal biotechnology.

Recently, techniques of molecular genetics have been used to trace ectomycorrhizal (Murata et al., 2005) or AM (Jansa et al., 2002) fungi of interest in soils. This very important application of molecular methods provides the means for understanding the fate of inoculants in target soils. In this case, no environmental risk can be supposed.

CONCLUSION

Application of mycorrhizal technology in agriculture, horticulture and forestry is a promising tool which could decrease inputs necessary for management of plant cultures, improve yield of some products and decrease risks connected with cultivation of plants in environments with suboptimal water accessibility. The use of inoculation with mycorrhizal fungi is still restricted to some specific areas of plant production because of insufficient knowledge of the biology (ecology) of the symbiosis and of mycorrhizal fungi in particular. However, significant practical results have already been achieved.

Another important obstacle is the relatively high price of mycorrhizal inocula as a result of challenges in the cultivation of the symbiotic fungi. Methods of cultivation of mycorrhizal fungi are continuously improving so that mycorrhizal inocula might be widely accessible in future.

The use of mycorrhizal technology is highly desirable because it might be helpful in establishing sustainable and eco-friendly agricultural systems, which represent the only possible solution of future problems related to human nutrition and the stability of ecosystems.

REFERENCES

Arditti, J. and Ghani, A.K.A. (2000). Tansley Review No. 110: Numerical and physical properties of orchid seeds and their biological implications. New Phytol. 145: 367-421.

Atkinson, D., Baddeley, J.A., Goicoechea, A., Green, J., Sánchez-Díaz, M. and Watson, C.A. (2002). Arbuscular mycorrhizal fungi in low-input agriculture. In: Mycorrhizal Technology in Agriculture. From Genes to Bioproducts (eds.) S. Gianinazzi, H. Schüepp, J.M. Barea and K. Haselwandter, Birkhäuser Verlag, Berlin, Germany, pp. 211-222.

Augé, R.M., Stodola, A.J.W., Tims, J.E. and Saxton, A.M. (2001). Moisture retention properties of a mycorrhizal soil. Plant Soil 230: 87-97.

Avio, L. and Giovannetti, M. (1988). Vesicular-arbuscular mycorrhizal infection of lucerne roots in a cellulose amended soil. Plant Soil 112: 99-104.

Bagyaraj, D.J., Tharun, C. and Balakrishna, A.N. (2006). Utilization of arbuscular mycorrhizal fungi in horticulture. In: Microbial Biotechnology in Horticulture, Volume I (eds.) R.C. Ray and O.P. Ward, Science Publishers, Enfield, NH, USA, pp. 21- 48.

Barea, J.M., Azcon, R. and Hayman, D.S. (1975). Possible synergistic interactions between Endogone and phosphate solubilizing bacteria in low phosphate soils. In: Endomycorrhizas (eds.) F.E. Sanders, B. Mosse and P.B. Tinker, Academic Press, London, UK, pp. 409-418.

Bergero, R., Girlanda, M., Bello, F., Luppi, A.M. and Perotto, S. (2003). Soil persistence and biodiversity of ericoid mycorrhizal fungi in the absence of the host plant in a Mediterranean ecosystem. Mycorrhiza 13: 69-75.

Bever J.D. and Wang, M. (2005). Hyphal fusion and multigenomic structure. Nature 433: E3.

Bonfante-Fasolo, P., Gianinazzi-Pearson, V. and Martinengo, L. (1984). Ultrastructural aspects of endomycorrhiza in the Ericaceae. IV. Comparison of infection by *Pezizella ericae* in host and non-host plants. New Phytol. 98: 329-333.

Brundrett, M., Bougher, A., Dell, B., Grove, T. and Malajczuk, N. (1996). Working with mycorrhizas in forestry and agriculture. Australian Centre for International Agricultural Research, Monograph No. 32, Canberra, Australia , pp. 374.

Burke, R.M. and Cairney, J.W.G. (2002). Laccases and other polyphenol oxidases in ecto- and ericoid mycorrhizal fungi. Mycorrhiza 12: 105-116.

Calvet, C., Estaun, V. and Camprubi, A. (1992). Germination, early mycelial growth and infectivity of a vesicular-arbuscular mycorrhizal fungus in organic substrates. Symbiosis 14: 149-159.

Calvet, C., Pinochet, J., Camprubi, A. and Fernandez, C. (1995). Increased tolerance to the root lesion nematode *Pratylenchus vulnus* in mycorrhizal micropropagated BA-29 quince rootstock. Mycorrhiza 5: 253-258.

Camprubi, A., Pinochet, J., Calvet, C. and Estaun, V. (1993). Effects of the root-lesion nematode *Pratylenchus vulnus* and the vesicular-arbuscular mycorrhizal fungus *Glomus mosseae* on the growth of tree plum rootstock. Plant Soil 153: 223-229.

Caravaca, F., Garcia, C., Hernández, C. and Roldán, A. (2002). Aggregate stability changes after organic amendment and mycorrhizal inoculation in the afforestation of a semiarid site with *Pinus halepensis*. Appl. Soil Ecol. 19: 199-208.

Caravaca, F., Barea, J.M., Palenzuela, J., Figueroa, D., Alguacil, M.M. and Roldán, A. (2003). Establishment of shrub species in a degraded semiarid site after inoculation with native or allochthonous arbuscular mycorrhizal fungi. Appl. Soil Ecol. 22: 103-111.

Castellano, M.A. (1996). Outplanting performance of mycorrhizal inoculated seedlings. In: Concepts in Mycorrhizal Research, (ed.) K.G. Mukerji, Kluwer Academic Publishers B.V., Dordrecht, The Netherlands, pp. 223-301.

Chambers, S.M., Liu, G. and Cairney, J.W.G. (2000). ITS-rDNA sequence comparison of ericoid mycorrhizal endophytes from *Woolsia pungens* (Cav.) F. Muell. *(Epacridaceae)*. Mycol. Res. 104: 169-175.

Chapela, I.H., Osher, L.J., Horton, T.R. and Henn, M.R. (2001). Ectomycorrhizal fungi introduced with exotic pine plantations induce soil carbon depletion. Soil Biol. Biochem. 33: 1733-1740.

Cruz, A.F., Ishii, T. and Kadoya, K. (2000). Effects of arbuscular mycorrhizal fungi on tree growth, leaf water potential and levels of 1-aminocyclopropane-1-carboxylic acid and ethylene in the roots of papaya under water-stress conditions. Mycorrhiza 10: 121-123.

Danell, E. (2001). Current research on chantarelle cultivation in Sweden. Proceedings of the Second International Conference on Edible Mycorrhizal Mushrooms, Christchurch, New Zealand, 3-6 July 2001, CD-ROM.

Ebel, R.C., Welbaum, G.E., Gunatilaka, M., Nelson, T. and Auge, R.M. (1996). Arbuscular mycorrhizal symbiosis and nonhydraulic signalling of soil drying in *Vigna unguiculata* (L.) Walp. Mycorrhiza 6: 119-127.

Estaún, V., Camprubí, A. and Joner, E. (2002). Selecting arbuscular mycorrhizal fungi for field application. In: Mycorrhizal Technology in Agriculture. From Genes to Bioproducts (eds.) S. Gianinazzi, H. Schüepp, J.M. Barea and K. Haselwandter, Birkhäuser Verlag, Berlin, Germany, pp. 249-259.

Feldmann, F. and Grotkass, C. (2002). Directed inoculum production - shall we be able to design populations of arbuscular mycorrhizal fungi to achieve predictable symbiotic effectiveness? In: Mycorrhizal Technology in Agriculture. From Genes to Bioproducts (eds.) S. Gianinazzi, H. Schüepp, J.M. Barea, and K. Haselwandter, Birkhäuser Verlag, Berlin, Germany, pp. 261-279.

Findlay, C.M. and Kendle, A.D. (2001). Towards a mycorrhizal application decision model for landscape management. Landsc. Urb. Planning 56: 149-160.

Fortuna, P., Morini, S. and Giovannetti, M. (1998). Effects of arbuscular mycorrhizal fungi on in vivo root initiation and development of micropropagated plum shoots. J. Hortic. Sci. Biotechnol. 73: 19-28.

George, E. (2000). Nutrient uptake. Contribution of arbuscular mycorrhizal fungi to plant mineral nitrition. In: Arbuscular Mycorrhizas: Physiology and Function (eds.) Y. Kapulnik and D.D. Douds, Jr., Kluwer Academic Publishers, Dordrecht, The Netherlands, pp. 307-343.

Giovannetti, M. (2000). Spore germination and pre-symbiotic mycelial growth. In: Arbuscular Mycorrhizas: Physiology and Function (eds.) Y. Kapulnik and D.D. Douds, Jr., Kluwer Academic Publishers, Dordrecht, The Netherlands, pp. 47- 68.

Gryndler, M., Vosátka, M., Hršelová, H., Chvátalová, I. and Jansa, J. (2002). Interaction between arbuscular mycorrhizal fungi and cellulose in growth substrate. Appl. Soil Ecol. 19: 279-288.

Hadley, G. and Williamson, B. (1972). Features of mycorrhizal infection in some Malayan orchids. New Phytol. 71: 1111-1118.

Hadley, G. (1982). Orchid mycorrhiza. In: Orchid Biology - Reviews and Perspectives II (ed.) J. Arditti, Cornell University Press, Ithaca, NY, USA, pp. 83-118.

Hall, I. and Yun, W. (2001). Truffles and other edible mycorrhizal mushrooms - some new crops for the Southern Hemisphere. Proceedings of the Second International Conference on Edible Mycorrhizal Mushrooms, 3-6 July 2001, Christchurch, New Zealand, CD-ROM.

Harley, J.L. and Harley, E.L. (1987). A check-list of mycorrhiza in the British flora. New Phytol. 105(S): 1-102.

Hawkins, H.J., Johansen, A. and George, E. (2000). Uptake and transport of organic and inorganic nitrogen by arbuscular mycorrhizal fungi. Plant Soil 226: 275-285.

Hildebrandt, U., Janetta K. and Bothe, H. (2002). Towards growth of arbuscular mycorrhizal fungi independent of a plant host. Appl. Environ. Microbiol. 68: 1919-1924.

Hwang, S.F., Chang, K.F. and Chakravarty P. (1992). Effects of vesicular arbuscular mycorrhizal fungi on the development of *Verticillium* and *Fusarium* wilts of alfalfa. Plant Disease 76: 239-243.

Ingham, R.E. (1988). Interaction between nematodes and vesicular-arbuscular mycorrhizae. Agric. Ecosyst. Environ. 24: 169-182.

Ito, T. and Ogawa, M. (1979). Cultivating method of the mycorrhizal fungus, *Tricholoma matsutake* (Ito et Imai) Sing. II. Increasing number of Shiro (fungal colony) of *T. matsutake* by thinning the understory vegetation. J. Jap. Forest Soc. 61: 163-173.

Jansa J., Mozafar, A., Anken, T., Ruh, R., Sanders, I.R. and Frossard, E. (2002). Diversity and structure of AMF communities as affected by tillage in a temperate soil. Mycorrhiza 12: 225-234.

Kagan-Zur, V., Freeman, S., Luzzati, Y., Roth-Bejerano, N. and Shabi, E. (2001). Survival of introduced *Tuber melanosporum* at two sites in Israel as measured by occurrence of mycorrhizas. Plant Soil 229: 159-166.

Kahiluoto, H. and Vestberg, M. (1998). The effect of arbuscular mycorrhiza on biomass production and phosphorus uptake from sparingly soluble sources by leek (*Allium porrum* L.) in Finnish field soils. Biol. Agric. Hortic. 16: 65-85.

Kaya, C., Higgs, D., Kirnak, H. and Tas, I. (2003). Mycorrhizal colonisation improves fruit yield and water use efficiency in watermelon (*Citrullus lanatus* Thunb.) grown under well-watered and water-stressed conditions. Plant Soil 253: 287-292.

Kerley, S.J. and Read, D.J. (1998). The biology of mycorrhiza in the Ericaceae. XX. Plant and mycorrhizal necromass as nitrogenous substrates for ericoid mycorrhizal fungus *Hymenoscyphus ericae* and its host. New Phytol. 139: 353-360.

Kim, K.Y., Jordan, D. and McDonald, G.A. (1998). Effect of phosphate solubilizing bacteria and vesicular arbuscular mycorrhizae on tomato growth and soil microbial activity. Biol. Fertil. Soils 26: 79-87.

Kling, M. and Jakobsen, I. (1997). Direct application of carbendazim and propiconazole at field rates to the external mycelium of three arbuscular mycorrhizal fungal species: Effect on ^{32}P-transport and succinate dehydrogenase activity. Mycorrhiza 7: 33-37.

Koide, R.T. and Wu, T. (2003). Ectomycorrhizas and retarded decomposition in a *Pinus resinosa* plantation. New Phytol. 158: 401-407.

Kucey, R.M.N. (1987). Increased phosphorus uptake by wheat and field beans inoculated with a phosphorus-solubilizing *Penicillium bilaji* strain and with vesicular arbuscular mycorrhizal fungi. Appl. Environ. Microbiol. 53: 2699-2703.

Leake, J.R. (1994). Tansley review No. 69: The biology of myco-heterotrophic ('saprophytic') plants. New Phytol. 127: 171-216.

Lekkberg, Y. and Koide, R.T. (2005). Is plant performance limited by abundance of arbuscular mycorrhizal fungi? A meta-analysis of studies published between 1988 and 2003. New Phytol. 168: 189-204.

Linderman, R.G. (2000). Effects of mycorrhizas on plant tolerance to diseases. In: Arbuscular Mycorrhizas: Physiology and Function (eds.) Y. Kapulnik and D.D. Douds, Jr., Kluwer Academic Publishers, Dordrecht, The Netherlands, pp. 345-365.

Lingua, G., D´Agostino, G., Massa, N., Antosiano, M. and Berta, G. (2002). Mycorrhiza-induced differential response to a yellow disease in tomato. Mycorrhiza 12: 191-198.

Mader, P., Edenhofer, S., Boller, T., Wiemken, A. and Niggli, U. (2000). Arbuscular mycorrhizae in a long-term field trial comparing low-input (organic, biological) and high-input (conventional) farming systems in a crop rotation. Biol. Fertil. Soils 31: 150-156.

Mansfeld-Giese, K., Larsen, J. and Bodker, L. (2002). Bacterial populations associated with mycelium of the arbuscular mycorrhizal fungus *Glomus intraradices*. FEMS Microbiol. Ecol. 41: 133-140.

Marx, D.H., Ruehle, J.H., Kenney, D.S., Cordell, C.E., Riffle, J.W., Molina, R.J., Pawuk, W.H., Navratil, S., Tinus, R.W. and Goodwin, O.C. (1982). Commercial vegetative inoculum of *Pisolithus tinctorius* and inoculation techniques for development of ectomycorrhizae on container-grown tree seedlings. Forest Sci. 28: 373-400.

Marx, D.H., Hedin, A. and Toe, S.F.P. (1985). Field performance of *Pinus caribaea* var. *hondurensis* seedlings with specific ectomycorrhizae and fertilizer after three years on a savanna site in Liberia. Forest Ecol. Management 13: 1-25.

Miller, R.M. and Jastrow, J.D. (2000). Mycorrhizal fungi influence soil structure. In: Arbuscular Mycorrhizas: Physiology and Function, (eds.) Y. Kapulnik and D.D. Douds, Jr., Kluwer Academic Publishers, Dordrecht, The Netherlands, pp. 3-18.

Molina, R., Massicotte, H. and Trappe, J. (1992). Specificity phenomena in mycorrhizal symbiosis: Community-ecological consequences and practical implications. In: Mycorrhizal Functioning: An Integrative Plant-Fungal Process (ed.) M.F. Allen, Routledge, Chapman & Hall, New York, USA, pp. 357-423.

Morandi, D., Gollotte, A. and Camporotta, P. (2002). Influence of an arbuscular mycorrhizal fungus on the interaction of a binucleate *Rhizoctonia* species with Myc^+ and Myc^- pea roots. Mycorrhiza 12: 97-102.

Murata, H., Ohta, A., Yamada, A., Narimatsu, M. and Futamura, N. (2005). Genetic mosaics in the massive persisting rhizosphere colony "shiro" of the ectomycorrhizal basidiomycete *Tricholoma matsutake*. Mycorrhiza 15: 505-512.

Naqvi, N.S. and Mukerji, K.G. (1998). Mycorrhization of micropropagated *Leucaena leucocephala* (Lam.) de Witt. Symbiosis 24: 103-114.

Niemi, K., Salonen, M., Ernstsen, A., Heinonen-Tanski, H. and Häggman, H. (2000). Application of ectomycorrhizal fungi in rooting of Scots pine fascicular shoots. Can. J. Forest Res. 30: 1221-1230.

Omar, S.A. (1998). The role of rock-phosphate-solubilizing fungi and vesicular-arbuscular-mycorrhiza (VAM) in growth of wheat plants fertilized with rock phosphate. World J. Microbiol. Biotechnol. 14: 211-218.

Pacioni, G. (1991). Effects of tuber metabolites on the rhizospheric environment. Mycol. Res. 95: 1355-1358.

Paszkowski, U. and Boller, T. (2002). The growth defect of *ltr1*, a maize mutant lacking lateral roots, can be complemented by symbiotic fungi or high phosphate nutrition. Planta 214: 584-590.

Peterson, R.L. and Farquahar, M.L. (1994). Mycorrhizas - Integrated development between roots and fungi. Mycologia 86: 311–326.

Piccini, D. and Azcon, R. (1987). Effect of phosphate-solubilizing bacteria and vesicular-arbuscular mycorrhizal fungi on the utilization of Bayovar rock phosphate by alfalfa plants using a sand-vermiculite medium. Plants Soil 101: 45-50.

Pilz, D. and Molina, R. (2002). Commercial harvests of edible mushrooms from the forests of the Pacific Northwest United States: Issues, management, and monitoring for sustainability. Forest Ecol. Management 155: 3-16.

Pinochet, J., Calvet, C., Camprubi, A. and Fernandez, C. (1995). Interaction between the root-lesion nematode *Pratylenchus vulnus* and the mycorrhizal association of *Glomus intraradices* and Santa Lucia 64 Cherry rootstock. Plant Soil 170: 323-329.

Pinochet, J., Calvet, C., Camprubi, A. and Fernandez, C. (1996). Interactions between migratory endoparasitic nematodes and arbuscular mycorrhizal fungi in perennial crops: A review. Plant Soil 185: 183-190.

Poole, E.J., Bending, G.D., Whipps, J.M. and Read, D.J. (2001). Bacteria associated with *Pinus sylvestris–Lactarius rufus* ectomycorrhizas and their effects on mycorrhiza formation in vitro. New Phytol. 151: 743-751.

Pope, E.J. and Carter, D.A. (2001). Phylogenetic placement and host specificity of mycorrhizal isolates belonging to AG-6 and AG-12 in the *Rhizoctonia solani* species complex. Mycologia 93: 712-719.

Quatrini, P., Gentile, M., Carmi, F., DePasquale, F. and Puglia, A.M. (2003). Effect of native arbuscular mycorrhizal fungi and *Glomus mosseae* on acclimatization and development of micropropagated *Citrus limon* (L.) Burm. J. Hortic. Sci. Biotechnol. 78: 39-45.

Querejeta, J.I., Egerton-Warburton, L.M. and Allen, M.F. (2003). Direct nocturnal water transfer from oaks to their mycorrhizal symbionts during severe soil drying. Oecologia 134: 55-64.

Read, D.J. (1974). *Pezizella ericae* sp. nov., the perfect state of a typical mycorrhizal endophyte of Ericaceae. Trans. Brit. Mycol. Soc. 63: 381-419.

Read, D.J. (1992). The mycorrhizal fungal community with special reference to nutrient mobilization. In: The Fungal Community. Its Organization and Role in the Ecosystem. 2nd edn. (eds.) G.C. Caroll and D.T. Wicklow, Marcel Dekker, New York, USA, pp. 631-652.

Rich, J.R. and Schenck, N.C. (1981). Seasonal variations in populations of plant-parasitic nematodes and vesicular-arbuscular mycorrhizae in Florida field corn. Plant Dis. 65: 804-807.

Rillig, M.C., Wright, S.F., Nichols, K.A., Schmidt, W.F. and Torn, M.S. (2001). Large contribution of arbuscular mycorrhizal fungi to soil carbon pools in tropical forest soils. Plant Soil 233: 167-177.

Rillig, M.C. and Steinberg, P.D. (2002). Glomalin production by an arbuscular mycorrhizal fungus: A mechanism of habitat modification. Soil Biol. Biochem. 34: 1371-1374.

Ryan, M.H., Chilvers, G.A. and Dumaresq, D.C. (1994). Colonisation of wheat by VA-mycorrhizal fungi was found to be higher on a farm managed in an organic manner than on a conventional neighbour. Plant Soil 160: 33-40.

Šašek, V. and Musílek, V. (1968). Antibiotic activity of mycorrhizal basidiomycetes and their relation to the host-plant parasites. Čes. Mykol. 22: 50-55.

Sen, R. (2001). Multitrophic interactions between a *Rhizoctonia* sp. and mycorrhizal fungi affect Scots pine seedling performance in nursery soil. New Phytol. 152: 543-553.

Sharma, J., Zettler, L.W., VanSambeek, J.W., Ellersieck, M.R. and Starbuck, C.J. (2003). Symbiotic seed germination and mycorrhizae of federally threatened *Platanthera praeclara* (Orchidaceae). Amer. Midl. Naturalist 149: 104-120.

Sitaramaiah, K. and Sikora, R.A. (1982). Effect of the mycorrhizal fungus *Glomus fasciculatum* on the host-parasite relationship of *Rotylenchus reniformis* in tomato. Nematologica 28: 412-419.

Smith, S.E. and Read, D.J. (1997). Mycorrhizal Symbiosis. 2nd. ed., Academic Press, London, UK, pp. 605.

Staddon, P.L., Ramsey, C.B., Ostle, N., Ineson, P. and Fitter, A.H. (2003). Rapid turnover of hyphae of mycorrhizal fungi determined by AMS microanalysis of ^{14}C. Science 300: 1138-1140.

St. Arnaud, M., Hamel, C., Vimard, B., Caron, M. and Fortin, J.A. (1996). Inhibition of *Fusarium oxysporum* f. sp. *dianthi* in the non-VAM species *Dianthus caryophyllus* by co-culture with *Tagetes patula* companion plants colonized by *Glomus intraradices*. Can. J. Bot. 75: 998-1005.

Stewart, S. L. and Zettler, L. W. (2002). Symbiotic germination of three semi-aquatic rein orchids (*Habenaria repens, H. quinquiseta, H. macroceratitis*) from Florida. Aquac. Bot. 72: 25-35.

Subhan, S., Sharmila, P. and Pardha Saradhi, P. (1998). *Glomus fasciculatum* alleviates transplantation shock of micropropagated *Sesbania sesban*. Plant Cell Rep. 17: 268-272.

Taylor, J. and Harrier, L.A. (2003). Beneficial influences of arbuscular mycorrhizal (AM) fungi on the micropropagation of woody and fruit trees. In: Micropropagation of Woody Trees and Fruits (eds.) S.M. Jain and K. Ishii, Kluwer Academic Publishers, Dordrecht, The Netherlands, pp. 129-150.

Toth, R. and Miller, R.M. (1984). Dynamics of arbuscule development and degeneration in a *Zea mays* mycorrhiza. Am. J. Bot. 71: 449-460.

Trappe, J.M. (1962). Fungus associates of ectotrophic mycorrhizae. Bot. Rev. 23: 528-606.

Trappe, J.M. (1987). Phylogenetic and ecologic aspects of mycotrophy in the angiosperms from an evolutionary standpoint. In: Ecophysiology of VA Mycorrhizal Plants (ed.) G.R. Safir, CRC Press, Boca Raton, Florida, USA, pp. 5-25.

Trappe, J.M., Molina, R. and Castellano, M. (1984). Reactions of mycorrhizal fungi and mycorrhiza formation to pesticides. Annu. Rev. Phytopathol. 22: 331-359.

Vaast, P., Caswell-Chen, E.P. and Zasoski, R.J. (1998). Influences of a root-lesion nematode, *Pratylenchus coffeae*, and two arbuscular mycorrhizal fungi, *Acaulospora mellea* and *Glomus clarum* on coffee (*Coffea arabica* L.). Biol. Fertil. Soils 26: 130-135.

Vestberg, M., Palmujoki, H., Parikka, P. and Uosukainen, M. (1994). Effect of arbuscular mycorrhizas on crown rot (*Phytophthora cactorum*) in micropropagated strawberry plants. Agric. Sci. Finland. 3: 289-295.

Vestberg, M., Saari, K., Kukkonen, S. and Hurme, T. (2005). Mycotrophy of crops in rotation and soil amendment with peat influence the abundance and effectiveness of indigenous arbuscular mycorrhizal fungi in field soil. Mycorrhiza 15: 447-458.

Vidal, M.T., Azcón-Aguilar, C. and Barea, J.M. (1992). Mycorrhizal inoculation enhances growth and development of micropropagated plants of avocado. Hort. Sci. 27: 785-787.

Von Alten, H., Blal, B., Dodd, J.C., Feldmann, F. and Vosátka, M. (2002). Quality control of arbuscular mycorrhizal inoculum in Europe. In: Mycorrhizal Technology in Agriculture. From Genes to Bioproducts (eds.) S. Gianinazzi, H. Schüepp, J.M. Barea and K. Haselwandter, Birkhäuser Verlag, Basel, Boston, Berlin, Germany, pp. 281-296.

Vralstad, T., Schumacher, T. and Taylor, F.S. (2002). Mycorrhizal synthesis between fungal strains of the *Hymenoscyphus ericae* aggregate and potential ectomycorrhizal and ericoid hosts. New Phytol. 153: 143-152.

Whittaker, S.P. and Cairney, J.W.G. (2001). Influence of amino acids on biomass production by ericoid mycorrhizal endophytes from *Woolsia pungens* (Epacridaceae). Mycol. Res. 105: 105-111.

Wright, S.F. and Upadhyaya, A. (1996). Extraction of an abundant and unusual protein from soil and comparison with hyphal protein of arbuscular mycorrhizal fungi. Soil Sci. 161: 575-586.

Yao, M.K., Tweddell, R.J. and Désilets, H. (2002). Effect of two vesicular-arbuscular mycorrhizal fungi on the growth of micropropagated potato plantlets and on the extent of disease caused by *Rhizoctonia solani*. Mycorrhiza 12: 235-242.

Yun, W., Buchanan, P. and Hall, I. (2001a). A list of edible ectomycorrhizal mushrooms. Proceedings of the Second International Conference on Edible Mycorrhizal Mushrooms, Christchurch, New Zealand, 3-6 July 2001, CD-ROM.

Yun, W., Hall, I.R., Dixon. C., Hance-Halloy, M., Strong, G. and Brass, P. (2001b). The cultivation of *Lactarius deliciosus* (saffron milk cap) and *Rhizopogon rubescens* (shoro) in New Zealand. Proceedings of the Second International Conference on Edible Mycorrhizal Mushrooms, Christchurch, New Zealand, 3-6 July 2001, CD-ROM.

6

Application of Arbuscular Mycorrhizal Fungi in the Production of Fruits and Ornamental Crops

Mahaveer P. Sharma* and
Alok Adholeya

INTRODUCTION

Environmental and economic concerns associated with increased use of chemical fertilizers and pesticides have stimulated interest in bio-based agriculture as an alternative to conventional agriculture. Bio-based agriculture involves the use of farm-generated inputs or applies biological systems such as management of arbuscular mycorrhizal (AM) fungi for sustainable plant production by better exploitation of soil resources. Reganold et al. (1990) have stated "soil is a complex, fragile medium that must be protected and nurtured to ensure long-term productivity and stability".

The living components of the soil, made up of plant roots and soil organisms, play a vital role in the development and maintenance of soil structure, nutrient cycling and plant health (Sylvia, 2000). Microbial communities in the soil are composed of bacteria, actinomycetes, fungi, algae, protozoa, nematodes and micro-arthropods (Sylvia et al., 1998). The density of these organisms is usually much greater in the rhizosphere than in the bulk soil (de Ridder-Duine et al., 2005). A dominant component of the rhizosphere microbial community consists of the mycorrhizal fungi

*Corresponding Author

that form beneficial symbioses with the fine roots of plants (Smith and Read, 1997). Mycorrhizal fungi, the most common fungal association formed nearly in all cultivated plants, be they agricultural, horticultural or forestry plant species, are gaining importance as colonizers not only of roots and rhizosphere, but also of the bulk soil (Barea et al., 1993; Bethlenfalvay and Barea, 1994; Lovato et al., 1995). The different types of mycorrhizal fungi present in nature have been discussed in the previous chapter (Gryndler, Chapter 5 in this Volume).

Mycorrhizal fungi have proved to be potential candidates for supplementing chemical fertilizers and for protection against environmental and cultural stress (Barea and Jeffries, 1995). The most important mycorrhizal response is the increased efficiency of mineral uptake, especially of poorly mobile ions, and many reports deal with the nutritional benefits that plants derive from mycorrhizal associations (Gianinazzi et al., 1990; Barea, 1991; Gaur et al., 1998; Sharma and Adholeya, 2000, 2004; Rea and Tullio, 2005).

Most agricultural soils are usually deficient in phosphorus (P), an element that is vital but is generally the most limited as compared to other major nutrients (Mikanová and Kubát, 2006). Moreover, when applied to soils, P is fixed quickly and immobilized (Mosse, 1973), hence plants are unable to utilize it. Mycorrhizal fungi facilitate its mobilization/ solubilization and increase the uptake of P in plants (Harley and Smith, 1983; Bolan, 1991; Li et al., 1991). Inoculation of plants with mycorrhizal fungi during the seedling stage and then transplantation in well-manured fields can certainly substitute the chemical fertilizer inputs, particularly P, to a large extent (Bansal and Mukerji, 1994; Bagyaraj et al., 2006). In this way, the production achieved may be sustained without affecting long-term soil productivity and fertility.

Ornamental and fruit plants are generally grown first in nursery or potting mixtures to produce seedlings. Micropropagation of fruit and ornamental plants is also increasingly being adopted. Augmentation of potential mycorrhizal fungi in the system of seedling production is being encouraged because it eliminates the indigenous microorganisms including mycorrhizal fungi. Having realized the potential benefits of AM fungi in horticulture, a large number of trials were conducted involving fruits and ornamentals in response to AM fungi **(Table 6.1)**. However, before applying mycorrhizal fungi, there are many aspects such as screening of potential mycorrhizal isolates, environmental conditions and effective dosage of mycorrhizal inoculum application, which are to be taken in to consideration to guarantee the benefits of AM fungi (Jeffries et al., 2003). Bagyaraj et al. (2006) in Volume 1 of this Series: Microbial Biotechnology in Horticulture, discussed the general utilization of microbial technology in horticultural crops. This current chapter

Table 6.1. Some examples of application of AM fungi to horticultural plants [*Glomus* (*Gl.*); *Gigaspora* (G.)]

Plant Specie	*AM Fungi Used*	*Reference*
Annona cherimoya	*Gl. deserticola*	Azcon-Aguilar et al., 1996
Apple	*Gl. intraradices*	Plenchette et al., 1981
	Gl. fasciculatum, Gl. mosseae, Gl. intraradices	Branzanti et al., 1992
	Gl. epigaeum	Granger et al., 1983
Apple, Peach, and Plum root stocks	*Glomus* sp.	Sbrana et al., 1994
Asparagus	*Gl. versiforme, Gl. vesiculiferum*	Parent et al., 1993
Avocado	*Gl. fasciculatum*	Vidal et al. 1992
	Gl. deserticola, Gl. mosseae	Azcon-Aguilar et al., 1992
	Gl. mosseae	Gribaudo et al., 1996
	Several AM fungi	da Silveira et al., 2002
Balsam, Callistephus, Petunia	Mixed native *Glomus* sp.	Gaur et al., 2000a
Banana	*Gl. fasciculatum*	Jaizme-Vega, 1992
	Gl. fasciculatum, Gl. intraradices, Gl. versiforme, Gl. macrocarpum	Declerck et al., 1994
	Acaulospora scrobiculata, Gl. clarum, Gl. etunicatum	Yano-Melo et al., 1999
	G. proliferum, Gl. versiforme, Gl. intraradices	Jaizme-Vega et al., 2003
Blackberry, Passion fruit, Apple, Pineapple	*Glomus* sp. AM1-8	Lin et al., 1987
Cashew	*Gl. aggregatum, Gl. fasciculatum, Gl. mosseae*	Ananthakrishnan et al., 2004
Cassava	*Gl. deserticola, Gl. clarum, Gl. fasciculatum*	Azcon-Aguilar et al., 1997
Citrus	*Gl. fasciculatus*	Menge et al., 1981
Citrus lemon	*Gl. mosseae* (BEG 116)	Quatrini et al., 2003
Coleus, Petunia, Salvia, Viola	*Gl. inraradices*	Koide et al., 1999
Date palm	*Gl. fasciculatum, Gl. intraradices*	Bouhired et al., 1992
Different genotypes of Marigold (*Tagetes* sp.)	Several AM fungi	Linderman and Davis, 2004
Dracaena, Syngonium	Mixed native *Glomus* sp.	Gaur and Adholeya, 1999
Ficus benjamina	*Gl. mosseae* and co-inoculation with *Trichoderma* and *Bacillus* sp.	Srinath et al., 2003

(*Table 6.1. Contd.*)

(*Table 6.1. Contd.*)

Gerbera	*Glomus* sp.	Wen, 1991
Germander, Lavender, *Tagetes*	*Gl. intraradices*	Linderman and Davis, 2003
Grapevine	*Glomus* sp. *Gl. aggregatum*	Schreiner, 2003; Cheng and Baumgartner, 2004 Aguin et al., 2004
Guava	*Gl. diaphanum, Gl. albidum, Gl. claroides*	Estrada-Luna et al., 2000
Hortensia	*Gl. intraradices*	Varma and Schuepp, 1994b
Kiwi, Grapevine	*Gl. fasciculatum, Gl. caledonium, Gl. monosporum, Gl. constrictum, Gl. versiforme*	Schubert et al., 1990
Leek	AM fungi	Kahiluoto and Vestberg, 1998
Lilium	Mixed native *Glomus* sp., *Gl. inraradices*	Varshney et al., 2002
Litchi	AM fungi	Janos et al., 2001
Mango root stock	Several AM fungi	Reddy and Bagyaraj, 1994
Oil palm	*Glomus* sp.	Blal et al., 1990
Ornamental trees	AM fungi	Klingeman et al., 2002
Mulberry, Papaya	*Gl. fasciculatum* *Gl. mosseae, Gl. fasciculatum* Several AM fungi	Mamatha et al., 2002 Mohandas and Das, 1992 Reddy et al., 1996
Pear seedlings	*Gl. intraradices, Gl. deserticola*	Gardiner and Christensen, 1991
Pineapple	*Gl. clarum, Scutellospora pellucida, Glomus* sp.	Guillemin et al., 1992
Plum root stock	*Gl. mosseae, Gl. caledonium, Gl. coronatum*	Fortuna et al., 1992
Prunus avium	*Gl. intraradices, Gl. fasciculatum, Gl. caledonium*	Cordier et al., 1996
Raspberry	*Gl. intraradices* *Gl. mosseae*	Varma and Schuepp, 1994a Campos-Mota et al., 2004
Sour orange	*Gl. clarum*	Ortas et al., 2002
Strawberry	Native mixed AM fungi *Glomus intraradices* Finn 98, *Glomus* sp. Finn 128, *G. geosporum* *Gl. inraradices, Glomus* sp. V21/88, *Gl. macrocarpum* *Gl. fistulosum*	Sharma and Adholeya, 2004 Williams et al., 1992 Vestberg, 1992a, b Cassells et al., 1996

(*Table 6.1. Contd.*)

Table 6.1. Contd.

	Gl. intraradices	Elmeskaoui et al., 1995
	Gl.intraradices, Gl. versiforme, Gl. mosseae	Khanizadeh et al., 1995
	Gl. intraradices	Varma and Schuepp, 1994a
	Several AM fungi	Taylor and Harrier, 2001
Tetraclinis articulata	*Gl. fasciculatum*	Morte et al., 1996
Walnut	*Gl. mosseae, Gl. intraradices*	Dolcetsanjuan et al., 1996
Watermelon	*Gl. clarum*	Kaya et al., 2003

emphasizes the applications of mycorrhiza in micropropagated fruit and ornamental plants, and the various strategies for large-scale inoculum production.

MANAGEMENT OF AM INOCULATION

Screening and Selection of Fungal Isolates

The AM fungi should be evaluated for their ability to improve plant growth in its natural soil by comparing it with disinfected soil, and if possible, with disinfected soil consisting one or more AM fungal isolates. Gianinazzi-Pearson et al. (1990) proposed that a site intended for plant production should initially be tested for the presence of AM fungi through infectivity tests, like the most probable number (MPN) technique (Porter, 1979), and infectivity potential (Liu and Luo, 1994; Gaur et al. 1998). It is generally accepted that AM associations lack specificity and that most AM fungi can form symbiosis with the majority of plants, at least to some degree (Vestberg and Estaun, 1994). However, the mycorrhizal responsiveness of the plant may vary with the fungal species, its potential and soil characteristics. This might be interpreted as a kind of functional host specificity (Clarke and Mosse, 1981; Plenchette et al., 1982; Rea and Tullio, 2005). A number of methodological approaches and parameters have been proposed (Lovato et al., 1995) to interpret host specificity. The effectiveness of AM in promoting plant growth and in protecting against plant pathogens/or abiotic stress such as salinity, are the most common considerations (Azcon-Aguilar and Barea, 1997; Rea and Tullio, 2005).

"Baiting," the AM fungi with a host plant, is usually the starting point of the selection process. Spores, infected roots, or even bulk soils, are used to infect the "bait plants". This last approach is important, since many fungal species may be efficient colonizers and growth promoters, but not good spore producers (Walker, 1992). This technique has been used to select AM fungi, which were highly efficient in promoting growth and reducing variability of oil palm clones (Blal and Gianinazzi-Pearson, 1989).

Inoculum Production of AM Fungi

Conventional Methods (in vivo *Multiplication)*

Since AM fungi are obligate symbionts, they are always produced on roots. There are various techniques currently used to culture AM fungi on hosts such as on-farm production, nursery production (Gaur and Adholeya, 2000, Gaur et al., 2000a, b), nutrient film technique (Mosse and Thompson, 1984), and aeroponics (Jarstfer and Sylvia, 1995). There are many factors which govern AM development besides the host plant (Sreenivasa and Bagyaraj, 1988), such as temperature (Furlan and Fortin, 1973), light (Ferguson and Menge, 1982), pot size, soil fertility (Menge et al., 1978), and the particle size of the growth substrate (Gaur and Adholeya, 2000).

The most frequently used technique for increasing propagule number has been the propagation of AM fungi on a suitable host in disinfested soil using pot cultures. Hepper (1984) reviewed procedures for disinfestation and for germinating spores. Williams (1990) mentioned detailed methods for reducing contamination of colonized root pieces. The most effective methods use chlorine compounds, surfactants and a combination of antibacterial agents. The required practice at the Mycorrhiza laboratory at TERI, New Delhi (India) is to routinely decontaminate spores by incubation in 2% chloramine T, 200 mg l^{-1} of streptomycin sulfate for 15 minutes followed by four to five rinses in distilled water.

Other factors for creating a conducive environment for culturing of AM fungi are a balance of light intensity, adequate moisture, and moderate temperature without detrimental addition of fertilizers or pesticides (Jarstfer and Sylvia, 1992; Al-Karaki et al., 1998; Bereau et al., 2000; Whitbeck, 2001). Any moisture conditions which inhibit primary root growth will reduce the development of mycorrhizal colonization. Similarly, soil temperature is important for the growth of fungi and indirectly affects the host chosen (Braunberger et al., 1997). Detailed methods for pot culturing and extensive discussions about these methods are provided by Jarstfer and Sylvia (1992). Cultures reaching high propagule density (10 spores per gram) after a number of multiplication cycles can be stored using suitable methods after air-drying (Kuszala et al., 2001). Furthermore, AM fungi have been cultured with plant hosts in different substrates such as sand, peat, expanded clay, perlite, vermiculite, soilrite (Mallesha et al., 1992), rock wool (Heinzemann and Weritz, 1990) and glass beads (Redecker et al., 1995). Apart from the above environmental and physical factors, biological factors also play important roles in rapid culturing of AM fungi in pots. Recently, Bhowmik and Singh (2004) demonstrated the role of plant growth promoting rhizobacteria in enhancing the mycorrhizal colonization of *Gl. mosseae* in pots on *Chloris gayana.*

Apart from nursery and pot cultures, AM production is also produced aeroponically (Sylvia and Hubbell, 1986). The aeroponic system was adopted for mycorrhiza production by the utilization of seedlings with roots pre-colonized by an AM fungus and the use of modified Hoagland's nutrition with a very low P level (Hoagland and Arnon, 1950). In the aeroponic system, colonization and sporulation were superior to those reported in soil-based pot culture. Sylvia and Jarstfer (1992) further determined the viability and density of aeroponically produced inocula after shearing. *Entrophospora kentinensis* was successfully propagated with bahia grass and sweet potato in an aeroponic system by Wu et al. (1995). Mohammad et al. (2000) reported the production of *Glomus intraradices* in an aeroponic system, where they compared the conventional atomizing disc with the ultrasonic nebulizer technology as misting sources.

Hydroponics or nutrient film technique (NFT) was adapted for the AM fungal inoculum production by Mosse and Thompson (1984). Culture host plants are placed on an inclined tray over which flows a layer of nutrient solution. As in the aeroponic culture, seedlings should be pre-colonized in another medium. Recently, Lee and George (2005) proposed a modified nutrient film technique for large-scale production of AM fungal biomass with the help of improved aeration by intermittent nutrient supply, optimum P supply, and use of glass beads as support materials.

In vitro/Axenic Cultivation of AM Fungi (Root Organ Culture)

The root organ culture (ROC) is a most attractive mass multiplication method for providing a pure, viable, rapid and contamination free inoculum using less space and has an advantage over the pot culture multiplication/conventional system (Tiwari et al., 2003). The mono-specific strains available can be used directly as starting material for large-scale inoculum production, a sole petri dish culture being enough to generate several thousand spores and meters of hyphae within four months (Dalpe and Monreal, 2004).

In vitro culture of AM fungi was achieved for the first time in early 1960s (Mosse, 1962). Since then, various pioneering steps were taken with the objective of axenic culturing of AM fungi. Becard and Fortin (1988),

Declerck et al. (1996), Plenchette et al. (1996), and Fortin et al. (2002) presented a detailed evaluation of the ROC technique and reported the basic improvements necessary for AM fungus colonization of roots.

Recently, Douds (2002) reported monoxenic culture of *Gl. intraradices* with Ri T-DNA transformed roots in two-compartment petri dishes as a very useful technique both for physiological studies and the production of clean fungal tissues. Fortin et al. (2002) developed successful propagation of some AM fungal strains on ROC, which allowed the cultivation of monoxenic strains that could be used either directly as inoculum or as starting inoculum for large-scale production.

Host/Root Cultures

In vitro propagation on ROC consists of excised roots that proliferate under axenic conditions on a synthetic nutrient medium supplemented with vitamins, minerals, and carbohydrates. The excised roots are Ri T-DNA-transformed, which have been used to obtain colonized root cultures. It is for the first time that the use of Ri T-DNA transformed roots of *Daucus carota* as host by *Agrobacterium rhizogenes* has successfully resulted in an increase of spore production of *Gl. mosseae* (Mugnier and Mosse, 1987; Krandashov et al., 2000).

A. rhizogenes inserts in transformed root copies of T-DNA (transfer DNA), which are found in a large plasmid of *A. rhizogenes* (Chilton et al., 1982). The transformed roots have the ability to grow profusely and vigorously (negative geotropism), and to provide opportunity for AM fungi hyphae to make contact (Becard and Fortin, 1988).

Becard and Fortin (1988) evaluated the ROCs for AM colonization in roots and identified rate limiting factors in the *G. margarita* colonized transformed roots of carrot. Continuous cultures of vigorous ROCs (Ri T-DNA-transformed) have been obtained through transformation of roots by the soil bacterium *A. rhizogenes* (Tepfer, 1989) that provided the new way to obtain mass production of roots in a very short span of time. Several subcultures (three to four) are necessary in this medium enriched with antibiotics such as carbenicillin or ampicillin, to obtain free-living roots without bacteria (Diop, 2003) and for establishing dual culture.

Fungus Inoculum, Culture Media and Growth

Usually, extraradical spores and infected root fragments are used for developing monoxenic cultures. Attempts have also been made to establish sporocarps of *Gl. mosseae in vitro* cultures of AM fungi (Budi et al., 1999). In most cases, purified (Furlan et al., 1980) and surface sterilized spores (Becard and Piche, 1992) isolated from the field, or from traps have proved successful for establishing dual cultures under *in vitro* conditions. Spore sterilization can also be achieved with a solution containing chloramine T (oxidizing agent) and Tween 20 (surfactant) (Fortin et al., 2002), and rinsed in a streptomycin-gentamycin antibiotic solution (Becard and Piche, 1992). While the success of spore germination is solely dependent on sterilization, the presence of root exudates and 2% CO_2 can stimulate germination and postgermination hyphal growth (Poulin et al., 1993).

The infected roots, which come from trap culture roots, can be used as a section of vesicles for *in vitro* establishment. The root sections can be disinfected in an ultrasonic processor under aseptic conditions and incubated in MSR (Modified Strullu-Romand) medium (Declerck et al., 1998). Hyphal regrowth from root pieces is usually observed within two to fifteen days (Fortin et al., 2002). Using segments of vesicles have more advantages over spores as far as contamination is concerned, but recovery of propagules is low compared to spores. Therefore, the vesicles are rarely used for routine inoculation for *in vitro* establishment (Fortin et al., 2002). The most widely used media for establishing *in vitro* AM root cultures are M (Minimal) (Becard and Fortin, 1988; Fortin et al., 2002) and MSR (Declerck et al. 1998) media (**Table 6.2**).

There are a number of AM species which have been successfully grown on ROC: *G. margarita, Gl. fasciculatum, Gl. macrocarpum, Gl. intraradices, G. gigantia, Gl. versiformis, Gl. caledonium, Gl. clarum, Gl. fistulosum* and *Gl. etunicatu* (Chabot et al., 1992; Diop et al., 1992; Douds and Becard, 1993; Declerck et al., 1998; Gryndler et al., 1998; Pawlowska et al., 1999; Karandashov et al., 2000). Regular subculturing is required to maintain higher infectivity. Continuous cultures can be obtained by transferring mycorrhizal roots to fresh medium either with (St.-Arnaud et al., 1996) or without (Declerck et al., 1996) spores. It is always preferable to use actively growing roots (Declerck et al., 1996).

The monoxenic system is more cost effective than pot cultures especially in developing countries. Mass production of spores is a prerequisite and mathematical models may be useful as descriptive and predictive tools for sporulation dynamics (Declerck et al., 2001). The ROC mode of AM production provides a unique platform to understand the interaction with beneficial microbes besides investigating the molecular and biochemical basis of AM symbiosis. In India TERI, New Delhi has developed a mass production technology, ROC, for AM fungi under industrial conditions which has been commercialized (Adholeya, 2003).

Table 6.2. Composition of minimal modium (M) and modified Strullu-Romand (MSR) media*

	M Medium (μM)	*MSR (μM)*
B	24	30
Biotin	—	0.004
Ca	1,200	1,520
Cl	870	870
Cu	0.96	0.96
Cyanocabalamine	—	0.29
Fe	20	20
Gellan agent (g L^{-1})	5	3
Glycine (mg L^{-1})	3	—
I	4.5	—
K	1,735	1,650
Mg	3,000	3,000
Mn	30	11
Mo	0.01	0.22
Myo-inositol (mg L^{-1})	50	—
N (NO_3)	3,200	3,800
N (PO_4^-)	—	180
Na	20	20
Nicotinic acid	4	8.10
P	30	30
Pantholenate Ca	—	1.88
pH (before autoclave)	5.5	5.5
Pyridoxine	0.49	4.38
S	3,000	3,013
Sucrose (g L^{-1})	10	10
Thiamine	0.3	2.96
Zn	9	1

**Source:* Fortin et al., 2002

Techniques of AM Inoculation

The first step in AM fungi inoculum management is to evaluate the density of mycorrhizal propagules in the soil or substrate, using infectivity tests with suitable host plants (Gaur et al., 1998). For inoculation, three main factors are important for the application of mycorrhizal plants: (1) time, (2) dosage and form of mycorrhizal inoculum, substrate to be used, and (3) choice of mycorrhizal inoculum (Barea et al., 1993; Lovato et al., 1996). To get quick and effective AM symbiosis and to initiate the primary infection, the inoculation should be done during sowing in mother beds. At the time of transplanting, the seedlings would be already treated with AM fungi and an additional second dose of AM applied during transplanting will further guarantee the success of inoculation. For example, the best time to inoculate micropropagated plants is just at the time of transplanting, as has been demonstrated for oil palm, pineapple and grapevine (Blal and Gianinazzi-Pearson, 1989). However, Vidal et al. (1992) and Azcon-Aguilar et al. (1992) observed that inoculation of avocado plants gave the best results when the platelets have passed a four-week period of acclimatization in a soilless potting mix before being transferred to a soil-sand mix and inoculated with AM fungi. Recently, Sharma and Adholeya (2004) implemented a micropropagated strawberry field trial, where they inoculated AM fungi during transferring juveniles from rooting media to polyethylene bags consisting of unfertilized sandy loam soil and kept until hardened. The second dose of inoculum was applied during transplanting in the field. Similarly, for foliage plants of *Draceana* and *Syngonium,* AM inoculation was done during the weaning and hardening stages (Gaur and Adholeya, 1999). The inoculation of micropropagated plants in groups together in trays containing AM fungi mixed into the weaning substrate has been encouraged to reduce the amount of inoculum, but got higher survival (Lovato et al., 1996).

Various forms of inocula are used, the most common being a bulk inoculum containing a mixture of spores, colonized roots, hyphae and substrate from the pot containers, usually grown in sterilized soil or soilless media (Jarstfer and Sylvia, 1992). Thus, for each plant species, in addition to the selection of AM fungi, there is a need to determine the best substrate, and optimal forms and time of application of the AM fungi, in order to obtain maximum benefit.

Horticultural Practices Influencing Inoculation

Substrates

The substrates used as growing media for horticultural plants are important not only for the development of the symbiosis but also for plant growth. Successful AM colonization has been reported in substrates containing sand, gravel, peat, expanded clay, perlite, bark, sawdust, vermiculite or a combination of these (Gianinazzi et al., 1990). In the case of field manipulation of AM fungi, the site soil or substratum should be tested for its receptivity to inoculation (Gianinazzi et al., 1990). The fungi should be tested under conditions similar to those in the field. Vestberg and Estaun (1994) emphasized that receptivity to the AM fungi of substrates used in potting mixes in commercial nurseries has apparently not assumed much importance earlier. Calvet et al. (1992) found that certain types of peat and composted substrates had a negative effect on the establishment of the AM symbiosis, although the germination and early mycelial growth were not affected, indicating a biological cause of inhibition. The differential effect of native and pure strains of AM fungi was reported on micropropagated citrus lemon by Quatrini et al. (2003). Linderman and Davis (2003) showed a better AM symbiosis and growth of several ornamental plants using coconut dust (coir)-amended peat substrate, rather than in peat alone. Vidal et al. (1992) found that the symbiosis could be established in peat-sand mixes, although soil sand mixes were more conducive to AM development of micropropagated avocado plants. Recent study conducted by Schreiner (2003) to determine grapevine rootstock variability in forming functional mycorrhizae showed that small differences in the ability to form mycorrhizae exist among rootstocks, but other factors like crop load, soil fertility and moisture, have large impacts on root colonization by AM fungi.

Fertilization and Pesticides

During AM management, fertilization should be taken into account to get maximum benefit in terms of growth enhancement and maintenance of soil health (Lovato et al., 1995). The reduction of P fertilization due to AM fungi has been reported in highly mycorrhiza dependent micropropagated plants like grapevine (Ravolanirina et al., 1989) or apple (Branzanti et al., 1992). The P fertilization of 40 mg l^{-1} caused the same growth response as the use of AM fungi with no P in soil based substrate. Similarly, Sharma and Adholeya (2004), working with micropropagated strawberry, showed that native mixed AM fungi inoculated at 71 kg P ha^{-1} produced comparable fruit yield with non-mycorrhizal plants grown at 106 kg P ha^{-1}. Thus, the mycorrhizal application could save about 35 kg P ha^{-1} application. Williams et al. (1992) showed that growth of mycorrhizal strawberry plants required only 25% of the minimal recommended fertilizer rate to reach the same level as non-mycorrhizal plants receiving 100% recommended fertilizer use. The enhanced P uptake in plants was due to the increase in surface absorptive area explored by hyphae in the soil, which increased the access to water, which was retained in pores and inaccessible to roots. Phosphorus uptake may also be enhanced by the modification of the root system of AM colonized roots. Schellenbaum et al. (1991) observed that roots of grapevine and plane trees colonized by AM fungi were more branched than those of non-mycorrhizal plants. AM fungi were also found to enhance the uptake of nutrients other than P such as nitrogen, potassium, copper and zinc (Frey and Schuepp, 1993). Ikram et al. (1994) observed an increase in copper concentrations in mycorrhizal *Hevea brasiliensis* rootstock in micro-plots, which did not receive P. The AM symbiosis could result in a decrease in Cu and Zn in plants growing under higher P fertilization (O'Keefe and Sylvia, 1991). Therefore, appropriate fertilization regimes should be carefully established with regard to quantities, nutrient composition and application schedules (Lovato et al., 1995).

Other aspects to consider in the management of horticultural plants are the use of pesticides and soil disinfection for controlling soilborne pathogens. Pesticides may affect AM fungi in several ways, as reviewed by Trappe et al. (1984). Besides the harmful effects on the environment, some of the pesticides, especially fungicides, may affect the development of AM fungi. For example, pentachloronitrobenzene was found to reduce AM colonization and captan seems to stimulate hyphal growth through soil (Schuepp and Bodmer, 1991). It has been demonstrated that certain pesticide treatments are compatible with AM fungi. Captan appears to stimulate the development of AM fungi mycelium in horticultural substrates (Guillemin and Gianinazzi, 1992). In some cases, it has been shown that fosetyl-Al appears to have no significant effect on AM formation (Aziz et al., 1990). Metalaxyl has been also recommended for AM inoculum production (Seymour, 1994). Venkateswarlu et al. (1994) reported that carbofuran (2,3-dihydro-2,2-dimethyl-7-benzofuranyl methylcarbamate) at concentrations of 0.5 to 2 kg ha^{-1} did not show any detrimental effect on mycorrhizal development by *Glomus clarum* in groundnut; only at 5 kg ha^{-1} did it significantly inhibit AM colonization. The use of carbofuran for large-scale production of "clean" AM inocula has been suggested (Sreenivasa and Bagyaraj, 1989) since carbofuran (20 mg l^{-1}) application was found to enhance AM colonization. Hence, the compatibility of pesticides with AM fungi would lead to a decrease in pesticide use and consequently to a reduction in the impact of pesticides on the environment (Lovato et al., 1995).

AM FUNGI IN MICROPROPAGATION

Currently, micropropagation is the most widely and successfully used commercial technology for mass production of horticultural plants: ornamentals, fruits, vegetables, plantation crops and spices (Hooker et al., 1994; Vestberg and Estaun, 1994). In nature, mycorrhizal fungi are an integral part of the plant, assuring that satisfactory growth and development occurs in a microbially-rich and nutrient-poor environments (Lovato et al., 1996). However, micropropagation technology obviously eliminates all microorganisms from plant tissues, including mycorrhizal fungi. The absence of mycorrhiza, therefore, requires the use of nutrient-rich and microbially poor environments to guarantee growth during hardening and outplanting. This is presently assured through application of fertilizers and pesticides to the artificial substrates. The importance of mycorrhizal fungi in enhancing plant growth by improving nutrient uptake, increasing the rhizosphere soil volume and alleviating biotic and abiotic stresses has been realized. It is perhops possible that AM fungi would improve the growth and establishment of micropropagated plants and can save inputs for hardening and fertilizers (Gianinazzi et al., 1990). Thus, research is being conducted worldwide to integrate both biotechnological approaches, i.e., inoculation and micropropagation of plant species of interest (Varma and Schuepp, 1995, 1996; Lovato et al., 1996). Mycorrhiza inoculation in micropropagated plants is generally done after roots are formed which limits inoculation to three suitable stages: (1) *in vitro* - rooting phase; (2) *ex vitro* – immediately after the rooting phase at the beginning of the acclimatization period; and (3) *ex vitro* - after the acclimatization phase, the post acclimatization period under greenhouse conditions (Vestberg and Estaun, 1994).

The rooting phase is normally conducted in agar-based media. AM symbiosis has been synthesized on agar culture using seedlings and root organ cultures (Mugnier and Mosse, 1987; Becard and Fortin, 1988; Chabot et al., 1992). The appropriate medium for the growth of both the micropropagated plants and the fungus has to be found for each plant-fungus combination. Pons et al. (1983) and Ravolanirina et al. (1989) achieved a functional mycorrhizal symbiosis in *Prunus avium* and *Vitis venifera* in rooting media. However, mycorrhizal symbiosis in the rooting phase is not always possible because most plants develop only primary roots at the rooting stage (Barea et al., 1992) and AM fungi only infect secondary roots (Brundrett et al., 1985). Elmeskaoui et al. (1995) have developed a method based on a system for culturing of the AM fungus *Gl. intraradices* with Ri T-DNA transformed carrot roots or non-transformed tomato roots. After root induction, micropropagated plantlets were grown on cellulose plugs, in contact with the mycorrhizal root organ culture. Subsequently, they were placed in growth chambers under enriched (5000 mg l^{-1}) CO_2 and fed with minimal medium. After 20 days of tripartite culture, all the plantlets were mycorrhizal. In contrast to *in vitro* inoculation, inoculation at the stage in which the axenically grown plantlets were transferred into the potting media was found to be more feasible and convincing (Schubert et al., 1990).

Mycorrhizal inoculation for *ex vitro* plants can be performed at any of the following stages:

(1) after the rooting stage *in vitro*
(2) at the beginning of the acclimatization phase, or

(3) after the acclimatization during the hardening stage under greenhouse conditions (Azcon-Aguilar and Barea, 1997). Various studies have been conducted demonstrating AM efficacy at the *in vitro* and *ex vitro* steps mentioned above. The micropropagated crop plants include:

Apple, peach, and plum root stocks (Sbrana et al., 1994)

Avocado (Vidal et al., 1992)

Banana (Jaizme-Vega et al., 1991; Jaizme-Vega, 1992, 2003; Declerck et al., 2002)

Citrus lemon (Quatrini et al., 2003)

Draceana and *Syngonium* (Gaur and Adholeya, 1999)

Ficus (Srinath et al., 2003)

Grapevines (Ravolanirina et al., 1989; Schubert et al., 1990)

Guava (Estrada-Luna et al., 2000)

Hortensia (Varma and Schuepp, 1994b)

Lilium (Varshney et al., 2002).

Oil palm (Blal et al., 1990)

Pineapple (Varma and Schuepp, 1994a; Jaime-Vega and Azcon, 1995)

Raspberries (Varma and Schuepp, 1994a)

Strawberry (Niemi and Vestberg, 1992; Williams et al., 1992; Taylor and Harrier, 2001; Sharma and Adholeya, 2004)

In most of the studies, the time of AM inoculation has been worked out at the beginning of the hardening phase, during the weaning stage or during acclimatization stage. A better response to inoculation was observed on varieties of plants when inoculation was done at the beginning of the hardening phase (Azcon-Aguilar et al., 1992; Vidal et al., 1992; Williams et al., 1992; Sharma and Adholeya, 2004). On the other hand, AM inoculation of foliage plants, *Dracaena* and *Syngonium*, during the weaning stage has been demonstrated (Gaur and Adholeya, 1999). Therefore, to ensure effective AM symbiosis, it is essential to optimize AM inoculation with respect to the stage of micropropagated plant to be inoculated *vis-a-vis* AM species across the gradient of fertilizer/substrate/soil conditions. The criteria for selecting efficient AM fungi for micropropagated plants are the same as for conventionally produced plants/seedlings, which has been already described earlier in the text (Section: Screening and Selection of Fungal Isolates).

With regard to AM efficacy, recently Marin et al. (2003) demonstrated the comparative efficacy of two AM fungal species to micropropagated persimmon (*Diospyros kaki*) 'Rojo Brillante'. They showed that *Gl. mosseae* did not enhance, but instead depressed growth and decreased root colonization percentage, whereas shoot and root growth enhancements were observed for plants colonized by *Gl. intraradices*. Similarly, Taylor and Harrier (2001) also compared nine species of AM fungi on development and mineral nutrition of micropropagated *Fragaria ananassa* cv. *Elvira* (strawberry). The nine species of AM fungi were *Gl. clarum, Gl. etunicatum, Gl. intraradices, Gigaspora rosea, G. gigantea, G. margarita, Scutellospora calospora, Sc. heterogama* and *Sc. persica*. They found that no positive shoot growth responses to AM fungal inoculation were observed. Root growth enhancements were observed for plants colonized by *Gl. intraradices* and *G. margarita*. Shoot and root growth depressions were observed for *G. rosea* and all species of *Scutellospora*. Furthermore, individual isolates of AM fungi had unique effects on the mineral status of the strawberry plants. In general, Mn and Mg concentrations were significantly increased within the shoot tissue of plants colonized by all isolates of AM fungi. Moreover, plants colonized by *Gl. clarum* and *G. rosea,* had significantly increased concentrations of 7 and 9 out of the 12 mineral nutrients within the shoot tissue, respectively (Taylor and Harrier, 2001).

The potential of AM fungi for micro propagated plants has been studied by many workers, who concluded the possibility of 25-50% saving of fertilizer inputs and hardening cycle (Williams et al., 1992; Hooker et al., 1994; Sharma and Adholeya, 2004). The substrate used for hardening and weaning of micropropagated plants should ensure both the development of the symbiosis and plant performance. The commonly used substrates are peat, perlite and vermiculite (Vestberg and Estaun, 1994). Sterilized sandy soil with low to medium fertility of NPK nutrients can be used as a substrate for hardening the *ex vitro* strawberry micropropagated plants (Sharma and Adholeya, 2004). Vidal et al. (1992) showed that symbiosis could be developed in peat-sand mixes of avocado plants. Schubert et al. (1990) also found that micropropagated grapevine grown in peat-based medium showed better AM establishment.

Quatrini et al. (2003) carried out a study and prepared a comparison of AM fungi *Gl. mosseae* (BEG 116) and native mixed AM fungi on the survival and growth of micropropagated lemon plants (*Citrus limon* [L.] Burm. 'Zagara Bianca') during the weaning phase to a final age of 16 months. The native AM fungal mixture was obtained from the rhizosphere soil of a citrus grove (Sicily, Italy) and was found to be more infective than *Gl. mosseae* alone. It significantly increased plant height, root and shoot weight, leaf area, P content and shoot/root ratio at the end of the weaning phase, i.e., 17 weeks after *in vivo* transferring and inoculation. A severe growth depression was observed in plants inoculated with *Gl. mosseae*. During the early weaning phase, plant survival was reduced only by the native AM fungal mix inoculum, while at the end of the experiment (69 weeks) plant survival was reduced by all the treatments except for the native AM fungi mix inoculum. A rhizosphere effect, other than mycorrhizal, was detected in the rhizosphere control inoculated with the sievate of the native AM fungi inoculum, in particular on root development.

OPTIMIZATION OF AM EFFICACY UNDER DIFFERENT CONDITIONS

AM efficacy is influenced by a wide range of environmental and cultural factors (Azcon-Aguilar and Barea, 1997). Generally, AM fungi respond well under low to medium soil fertility; however, AM associations have been reported even at higher levels of P and N in soils (Hayman, 1983). Therefore, selection of substrates governs the efficacy of AM fungi. The increased amount of soluble P fertilizers has been found to reduce the external mycelium, number of entry points and overall development of AM fungi (Barea, 1991).

Shrestha et al. (1996) studied four AM fungi, i.e., *Gl. ambisporum, Gl. fasciculatum, Gl. mosseae* and *Gigaspora ramisporophora* at low concentrations of P on Satsuma mandarin (*Citrus unshiu* Marc. cv. Okitsu wase) trees which were grafted on trifoliate orange (*Poncirus trifoliata* Raf) rootstock. The trees that were inoculated with the AM fungi grew larger and had better fruit quality as compared with non-AM control trees. The fruits of the former were larger, had higher sugar content in the juice, and better peel colour in both 1992 and 1993 than in the later years. After water stress treatment of 10 days, the water stress tolerance of Satsuma mandarin trees was improved by the inoculation of an AM fungus (*G. ramisporophora*). The AM efficacy differences due to environmental factors have been reported by Lovato et al. (1992) by using commercial fungal inocula applied to three pineapple varieties, which enhanced growth differently in acid and alkali soils. They showed that the natural AM strains which originated from acid or alkali soils were more efficient than isolates which originated from neutral soils. Similarly, the forms of AM inocula (commercial and formulations) used should be optimized. Gaur et al. (1998) compared various types and formulations of AM inocula and showed that soil-based inocula and soil beads produced the highest response in *Polianthes* and *Capsicum. Gl. intraradices* resulted in the highest yield in both *Polianthes* and *Capsicum*. Among the commercial inocula tested, only Mycorise enhanced the growth and fruit yield. Sheared root inoculum of *Gl. intraradices,* despite achieving high colonization in roots, exhibited yield enhancements that were lower than those with soil-based formulations. The mixed indigenous culture produced the highest number of spores and propagules, and commercial inocula the lowest.

From the agronomic and cultural point of view, the P fertilizer dosage, which has a positive effect on AM symbiosis, should be optimized. Usually, at lower P dosage, mycorrhizal plants produce relatively greater biomass than those of plants grown at higher P dosage. The optimization of P helps in ascertaining P requirement for the plant as well as for the mycorrhiza, in terms of getting beneficial symbiosis. There is a level of P where plants do not derive additional benefit from mycorrhiza over non-mycorrhizal plants (Barea et al., 1993). Ortas et al. (2002) demonstrated mycorrhizal dependency on citrus plants in relation to P and Zn nutrition. They showed that sour orange was strongly mycorrhizal dependent. Nevertheless, with increasing P and Zn supply, mycorrhizal dependency gradually decreased. The decrease in mycorrhizal dependency was more pronounced for P requirement rather than Zn requirement.

Apart from fertilizers and AM species, the AM response also varied due to horticultural practices being followed during nursery preparation and transplanting in field, because AM fungi are ubiquitous and indigenous fungi naturally infect the plants in the field. Under such conditions the natural AM may persist and provide the benefits (Azcon-Aguilar and Barea, 1997). Plenchette et al. (1981) reported that apple seedlings inoculated before being transplanted into the field showed increased growth after three months. Vestberg (1992a, b) also reported similar results on strawberry.

Field soils of four middle Tennessee and two eastern Tennessee nurseries were surveyed for their mycorrhizal inoculum potential, P and K concentrations, and soil pH. AM fungi, which colonized seedlings of a *Sorghum bicolor* trap-crop, were recovered from all soils. Tissue samples were taken from young roots of three economically important tree species grown in nursery field soils: red maple (*Acer rubrum* L. 'October Glory'), flowering dogwood (*Cornus florida* L. 'Cherokee Princess'), and Kwanzan cherry (*Prunus serrulata* Lindl.'Kwanzan'). AM fungi, regardless of soil type, soil pH, or P or K concentration, had colonized young roots of all three species. Unless they are interested in establishing exotic mycorrhizae, ornamental nursery producers in Tennessee do not need to supplement field soils with these beneficial fungi (Klingeman et al., 2002). Schreiner (2003) studied mycorrhizal colonization of various types of grapevine rootstocks under field conditions to determine rootstock variability in forming functional mycorrhizae. AM colonization was generally found above 60% of fine root length for all rootstocks, although significant differences due to rootstock and time of sampling were evident.

All these results showed that the interactions between plants and AM fungi are complex and their expression for response is strongly dependent on environmental factors (Lovato et al., 1995). Therefore, there is a prerequisite to understand the environmental and cultural factors in the management practices with the possible and effective management of AM fungi.

EFFICACY OF AM FUNGI IN COMMERCIAL PRODUCTION OF FRUITS AND ORNAMENTALS

Natural systems plants are normally colonized by AM fungi and thus have less requirements for external application of AM fungi as compared to systems like micropropagation, where during the *in vitro* stage these fungi are removed along with other microbes (Hooker et al., 1994). Furthermore, substrates used in the *post vitro* stages of the micropropagation process are normally treated to remove pathogens, which also results in removal of AM fungi. Thus, plants produced using micropropagation methods will not have the benefits of AM symbiosis and will only benefit by re-introduction of the AM fungi. Considering the high cost of production in ornamental and fruit crops, it seems that the application of AM fungi may help provide the many benefits without much of an increase in the cost of production. Thus, the production of ornamental and fruit crops is a possible area in which application of AM fungi is feasible as biofertilizers, bioprotectors and biostimulators, and will most likely become a viable technology (Lovato et al., 1995). Bio-inoculants are available in different formulations e.g., powders, tablets/ pellets, gel beads and balls (Tiwari et al., 2003). A list of commercial formulations of AM inoculants available for exploitation in horticultural plants is presented in **Table 6.3**.

Beneficial effects of mycorrhiza on fruit and ornamental crops include promotion of seedling growth, reduction of phosphate requirement, achievement of increased survival rate and improved growth of micropropagated plantlets (Ponton et al., 1990; Vestberg, 1992a, b). For micro propagated strawberry plants, AM inoculation was found to reduce external P requirements and save P inputs (Williams et al., 1992; Hooker et al., 1994; Sharma and Adholeya, 2004). Blal et al. (1990) showed that mycorrhizal oil palm micro plants obtain P from the same pool in the soil as non-mycorrhizal plants, but found that the coefficient of fertilizer utilization is manyfolds higher. AM inoculation efficacy was found to vary according to the stage in the micropropagation process: *in vitro*, weaning and post weaning (hardening). The results showed that inoculation during the weaning, phase imparted best colonization of

Table 6.3. List of commercial sources of AM inoculants*

Company	*City/Country*
Accelerator Horticultural Products	Ohio, USA
AgBio Inc.	Westminister, Colorado, USA
Becker –Underwood	Ames, Iowa, USA
BioGrowTM	North America
Biological Crop Protection Ltd.	Kent, UK
Bio-organics	Medellin, Columbia
Bio-Organics Supply	Camarillo, California, USA
Biorize	Dijon, France
BioScientific, Inc.	Avondale, Arizona, USA
Cadila Pharmaceuticals Ltd.	Ahmedabad, India
Central Glass Co.	Tokyo, Japan
EcoLife Corporation	Moorpark, California, USA
First Fruits	Sarasota, Florida, USA
Global Horticare	Lelystad, The Netherlands
Idemitsu Kosan Co.	Sodegaura, Chile
J.H. Biotech, Inc.	Ventura, California, USA
KCP Sugar and Industries Corporation Ltd.	Andhra Pradesh, India
Majestic Agronomics Pvt. Ltd.	Himachal Pradesh, India
MicroBio Ltd.	Royston, Hertz, UK
Mikro-Tek Inc.	Ontario, Canada
Mycorrhizal Applications	Grants Pass, Oregon, USA
MycorTM VAM Mini Plug^{TM}	North America
N-Viron Sdn Bhd	Malaysia city
Plant Health Care, Inc.	Pennsylvania, USA
PlantWorks Ltd.	Sittingbourne, UK
Premier Horticulture	Red Hill, Pennsylvania, USA
Premier Tech	Quebec, Canada
Reforestation Technologies	Salinas, California, USA
Roots Inc.	Independence, Montana, USA
T & J Enterprises	Spokane, Washington, USA
TIPCO, Inc.	Knoxvile, Tennessee, USA
Tree of Life Nursery	San Juan Capistrano, California, USA
Tree Pro West	Lafayette, Indiana, USA
Triton Umweltschutz GmbH	Bitterfeld, Germany

**Source:* Tiwari et al., 2003

micropropagated plants (Ravolanirina et al., 1989; Gaur and Adholeya, 1999).

Varshney et al. (2002) demonstrated the effect of three different species of AM fungi at four different available levels of P on the growth, flowering, P uptake and root colonization in micropropagated bulblets of *Lilium* sp. (Asiatic hybrid cultivar 'Gran Paradiso'). Growth of the inoculated bulblets was significantly better than that of the uninoculated ones, in terms of various growth parameters including shoot length, number of leaves, leaf area, bulblet size and weight, and in P uptake. In other studies, gerbera plant colonized by *Gl. etunicatum* flowered 16 days earlier than the non-mycorrhizal control (Wen, 1991) and zinnia inoculated with *G. mosseae* flowered 24 days before the control (Cheng, 1989). Gerbera plants inoculated with *Gl. mosseae* produced more flowers, which lasted three days longer than non-mycorrhizal plants in the vase (Wen, 1991). The increased vase life of flowers from mycorrhizal plants may be due to the greater development of water conducing tissues in mycorrhizal plants than in non-mycorrhizal plants (Wen, 1991; Chang, 1992, 1994).

AboulNasr (1996) demonstrated AM fungi, *Gl. etunicatum* response applied in nursery during the sowing of seeds of *Tagetes erecta* and *Zinnia elegans.* It was found that AM inoculation had positive effects (faster flowering and an increased number of flowers) on both *Tagetes* and *Zinnia* as compared with the non-inoculated ones.

CONCLUSION

AM symbiosis should be considered as an essential component for promoting plant health and productivity. AM technology is feasible and viable mainly for crops which involve a transplanting stage, as in horticultural systems where plants are produced in nursery beds, containers/plugs or by micropropagation (Hooker et al., 1994). The symbiotic combinations of both AM fungi and plants should be evaluated in order to select the right combination for beneficial AM symbiosis. Identifying the right substrate/soil, plant and efficient AM fungi will help in reducing the inputs, especially the dependence on inorganic fertilizers. The high production costs involved in generating micropropagated plants, especially during *post vitro* phase, can be minimized with the application of AM fungi. For example, some woody plants which are difficult to root (oak, apple, plum, pear, avocado, etc.) exhibit improvements in survival rates when inoculated with AM fungi. Another problem that could be resolved by the AM fungi is the dormancy that some micropropagated plants (*Prunus* and *Malus*) present once they have been acclimatized. Thus, the production of ornamental and fruit crops is probably the area in which the application of AM fungi may be best exploited as biofertilizers, bioregulators and bioprotectors, and will most likely become a viable technology.

Once the importance of AM technology for plant growth and development is understood by the grower, commercial nurseries will have to include use of AM fungi in a standard protocol in their plant production system in response to the demand. A perspective for the future should be the development of integrated biotechnologies in which not only mycorrhizal fungi, but also other organisms of the rhizosphere capable of promoting plant growth or protection - such as plant growth promoting bacteria and symbiotic bacteria, plant antagonists, should be incorporated into the system of plant production including micropropagated systems. It is necessary to develop management practices for plant establishment and for better functioning of mycorrhizal fungi. Parameters such as growth substrate/soil, dosage of fertilizers, stages for AM inoculation like scheduling of weaning and transplanting, should be optimized to maximize the AM benefit for the productivity of fruits and ornamental plants, as discussed in this chapter.

REFERENCES

AboulNasr, A. (1996). Effects of vesicular-arbuscular mycorrhiza on *Tagetes erecta* and *Zinnia elegans*. Mycorrhiza 6: 61-64.

Adholeya, A. (2003). Commercial production of AMF through industrial mode and its large-scale application. 4th Int. Conf. Mycorrhizae. Montreal, Quebec Canada, 10-15 August 2003, No. 240.

Aguin, O., Mansilla, J.P., Vilarino, A. and Sainz, M.J (2004). Effects of mycorrhizal inoculation on root morphology and nursery production of three grapevine rootstocks. Am. J. Enol. Viticul. 55: 108-111.

Ananthakrishnan, G., Ravikumar, R., Girija, S. and Ganapathi, A. (2004). Selection of efficient arbuscular mycorrhizal fungi in the rhizosphere of cashew and their application in the cashew nursery. Scientia Hort. 100: 369-375.

Al-Karaki, G.N., Al-Raddad, A. and Clark, R.B. (1998). Water stress and mycorrhizal isolate effects on growth and nutrient acquisition of wheat. J. Plant Nutr. 21: 891-902.

Azcon-Aguilar, C., Barcelo, A., Vidal, M.T. and del la Vina, G. (1992). Further studies on the influence of mycorrhizae on growth and development of micropropagated avocado plants. Agronomie 12: 837-840.

Azcon-Aguilar, C., Padilla, I.G., Encina, C.L., Azcon, R. and Barea, J.M. (1996). Arbuscular mycorrhizal inoculation enhances plant growth and changes root system morphology in micropropagated *Annona cherimola* Mill. Agronomie 16: 647-652.

Azcon-Aguilar C. and Barea, J.M. (1997). Applying mycorrhiza biotechnology to horticulture: Significance and potentials. Scientia Hort. 68: 1-24.

Azcon-Aguilar, C., Cantos, M., Troncoso, A. and Barea, J.M. (1997). Beneficial effect of arbuscular mycorrhiza on acclimatization of micropropagated cassava plantlets. Scientia Hort. 7291): 63-71.

Aziz, T., Yuen, J.E. and Habte, M. (1990). Response of pineapple to mycorrhizal inoculation and fosetyl-Al treatment. Comm. Soil Sci. Plant Anal. 21: 2309-2317.

Bagyaraj, D.J., Tharun, C. and Balakrishna, A.N. (2006). Utilization of arbuscular mycorrhizal fungi in horticulture. In: Microbial Biotechnology in Horticulture, Volume 1 (eds.) R.C. Ray and O.P. Ward, Science Publishers, Enfield, NH, USA, pp. 21- 48.

Bansal, M. and Mukerji, K.G. (1994). Efficacy of root litter as a biofertilizer. Biol. Fertil. Soils 18: 228-230.

Barea, J.M. (1991). Vesicular arbuscular mycorrhizae as modifiers of soil fertility. Adv. Soil Sci. 15: 1-40.

Barea, J.M., Azcon, R., and Azcon-Aguilar, C. (1992). VAM fungi in nitrogen-fixing systems. In: Methods in Microbiology (eds.) J.R. Norris, D.J. Read and A.K. Varma, Academic Press, London, UK, pp. 167-189.

Barea, J.M; Azcon, R. and Azcon-Aguilar, C. (1993). Mycorrhiza and crops. In: Advances in Plant Pathology, Volume 9. Mycorrhiza: A Synthesis (ed.) I. Tommerup, Academic Press, London, UK, pp. 167-189.

Barea, J.M. and Jeffries, P. (1995). Arbuscular mycorrhizas in sustainable soil plant systems. In: Mycorrhiza: Structure, Function, Molecular Biology and Biotechnology (eds.) A. Varma and B. Hock, Springer-Verlag, Berlin, Germany, pp. 521-560.

Becard, G. and Fortin, J.A. (1988). Early events of vesicular arbuscular mycorrhiza formation on Ri T-DNA transformed roots. New Phytol. 108: 211-218.

Becard, G. and Piche, Y. (1992). Establishment of AM in root organ cultures: Review and proposed methodology. In: Techniques for the Study of Mycorrhiza (eds.) J. Norris, D. Read and A. Verma, Academic Press, NY, USA, pp. 89-108.

Bereau, M., Barigah, T.S., Louisanna, E. and Garbaye, J. (2000). Effects of endomycorrhizal development and light regimes on the growth of *Dicorynia guianensis* Amshoff seedlings. Anals Forest Sci. 57: 725-733.

Bethlenfalvay, G.J. and Barea, J. M. (1994). Mycorrhizae in sustainable agriculture. I. Effects on seed yield and soil aggregation. Am. J. Alternat. Agric. 9: 157-161.

Bhowmik, S.N. and Singh, C.S. (2004). Mass multiplication of AM inoculum: Effect of plant growth-promoting rhizobacteria and yeast in rapid culturing of *Glomus mosseae*. Curr. Sci. 86: 705-709.

Blal, B. and Gianinazzi-Pearson, V. (1989). Interest of mycorrhiza for production of micropropogated oil palm clones. Agric. Ecosyst. Environ. 29: 39-43.

Blal, B., Morel, C., Gianinazzi-Pearson, V., Fardeau, J.C. and Gianinazzi, S. (1990). Influence of VA mycorrhizae on phosphate fertilizer efficiency in two tropical acid soils planted with micropropagated oil palm (*Elaeis guineensis* Jacq.). Biol. Fertil. Soils 9: 43-48.

Bolan, N.S. (1991). A critical review on the role of mycorrhizal fungi in the uptake of phosphorus by plants. Plant Soil 134: 189-207.

Bouhired, L., Gianinazzi, S. and Gianinazzi-Pearson, V. (1992). Influence of endomycorrhizal inoculation on the growth of *Phoenix dactyfera*. In: Micropropagation, Root Regeneration and Mycorrhizas. Joint meeting between COST 87 and COST 8.10, Dijon, France, 53.

Branzanti, B., Gianinazzi-Pearson, V. and Gianinazzi, S. (1992). Influence of phosphate fertilization on the growth and nutrient status of micropropagated apple infected with endomycorrhizal fungi during the weaning stage. Agronomie 12: 841-846.

Braunberger, P.G., Abbott, L.K. and Robson, A.D. (1997). Early vesicular-arbuscular mycorrhizal colonization in soil collected from an annual clover-based pasture in a Mediterranean environment: soil temperature and the timing of autumn rains. Aust. J. Agric. Res. 48: 103-110.

Brundrett, M.C., Piche, Y. and Peterson, R.L. (1985). A developmental study of the early stages in VAM formation. Can. J. Bot. 63: 184-194.

Budi, S.W., Blal, B. and Gianinazzi, S. (1999). Surface-sterilization of *Glomus mosseae* sporocarps for studying endomycorrhization in vitro. Mycorrhiza 9: 65-68.

Calvet, C., Barea, J.M. and Pera, J. (1992). In vitro interactions between the vesicular arbuscular mycorrhizal fungus *Glomus mosseae* and some saprophytic fungi isolated from organic substrates. Soil Biol. Biochem. 24: 775-780.

Campos-Mota, L., Baca-Castillo, G.A., Jaen-Contreras, D., Muratalla-Lua, A. and Acosta-Hernandez, R. (2004). Fertirrigation and mycorrhiza in red raspberry cultured on tepetate. Agrociencia 38: 75-83.

Cassells, A.C., Mark, G.L. and Periappuram, C. (1996). Establishment of arbuscular mycorrhizal fungi in autotrophic strawberry cultures in vitro. Comparison with inoculation of microplants in vivo. Agronomie 16: 625-632.

Chabot, S., Becard, G. and Piche, Y. (1992). Life cycle of *Glomus intraradix* in root organ culture. Mycologia 84: 315-321.

Chang, D.C.N. (1992). Studies and prospect of horticultural arbuscular mycorrhizae. Taiwan. Sci. Agric. 40: 45-52.

Chang, D.C.N. (1994). What is the potential for management of vesicular-arbuscular mycorrhizae in horticulture? In: Management of Mycorrhizas in Agriculture, Horticulture and Forestry (eds.) A.D. Robson, L.K. Abbott and N. Malajczuk, Kluwer Academic Publishers, Dordrecht, The Netherlands, pp. 187- 190.

Cheng, S.F. (1989). Single pore culture of AM fungi and their effects on three flower crops. Ph.D. thesis, Department of Horticulture, National Taiwan University, Taiwan, Republic of China.

Cheng, X.M. and Baumgartner, K. (2004). Arbuscular mycorrhizal fungi-mediated nitrogen transfer from vineyard cover crops to grapevines. Biol. Fertil. Soils 40: 406-412.

Chilton, M.D., Tefer, D.A., Petit, A., David, C., Casse-Delbart, F. and Tempe, J. (1982). *Agrobacterium rhizogenes* inserts T-DNA into the genomes of the host plant root cells. Nature 295: 432-434.

Clarke, C. and Mosse, B. (1981). Plant growth responses to VA mycorrhiza. XII. Field inoculation responses at two soil P levels. New Phytol. 87: 695-703.

Cordier, C., Trouvelot, A., Gianinazzi, S. and Gianinazzipearson, V. (1996). Arbuscular mycorrhiza technology applied to micropropagated *Prunus avium* and to protection against *Phytophthora cinnamomi*. Agronomie 16: 679-688.

Dalpé, Y. and Monreal, M. (2004). Arbuscular mycorrhiza inoculum to support sustainable cropping systems. Online. Crop Management doi: 10.1094/CM-2004-0301-09-RV.

da Silveira, S.V., de Souza, P.V.D., Bender, R.J. and Koller, O.C. (2002). Effect of arbuscular mycorrhizae on cv. Carmem avocado plants. Comm. Soil Sci. Pl.. Anal. 33: 1323-1333.

de Ridder-Duine, A.S., Kowalchuk, G.A., Gunnewiek, P.J.A.K., Smant, W., van Veen, J.A. and de Boer, W. (2005). Rhizosphere bacterial community composition in natural stands of *Carex arenaria* (sand sedge) is determined by bulk soil community composition. Soil Biol. Biochem. 37: 349- 357.

Declerck, S., Delvaux, B. and Plenchette, C. (1994). Growth response of micropropagated banana plants to AM inoculation. Fruits 49: 103-109.

Declerck, S., Strullu, D.G. and Plenchette, C. (1996). In vitro mass-production of the arbuscular mycorrhizal fungus, *Glomus versiforme*, associated with Ri T-DNA transformed carrot roots. Mycol. Res. 100: 1237-1242.

Declerck, S., Strullu, D.G. and Plenchette, C. (1998). Monoxenic culture of the intraradical forms of *Glomus* sp. isolated from a tropical ecosystem: A proposed methodology for germplasm collection. Mycologia 99: 579-585.

Declerck, S., D'Or, D., Cranenbrouck, S. and Leboulenge, E. (2001). Modeling and sporulation dynamics of AM fungi in monoxenic culture. Mycorrhiza 11: 1178-1187.

Declerck, S., Risede, J.M. and Delvaux, B. (2002). Greenhouse response of micropropagated bananas inoculated with in vitro monoxenically produced arbuscular mycorrhizal fungi. Scientia Hort. 93: 301-309.

Diop, T.A., Becard, G. and Piche, Y. (1992). Long term in vitro culture of an endomycorrhizal fungus, *Gigaspora margarita* on Ri TDNA transformed root of carrot. Symbiosis 12: 249-259.

Diop, T.A. (2003). In vitro cultures of AM fungi: advances and future prospects: Mini review. African J. Biotechnol. 2: 692-697.

Dolcetsanjuan, R., Claveria, E., Camprubi, A., Estaun, V. and Calvet, C. (1996). .Micropropagation of walnut trees (*Juglans regia* L.) and response to arbuscular mycorrhizal inoculation. Agronomie 16: 639-645.

Douds, D.D. Jr and Becard, G. (1993). Competitive interactions between *Gigaspora margarita* and *G. gigantia* in vitro: Proceedings of 9th NACOM, 8-12 August 1993 Guelph, Ontario, Canada.

Douds, D.D. Jr. (2002). Increased spore production by *Glomus intraradices* in the split-plate monoxenic culture system by repeated harvest, gel replacement, and re-supply of glucose to the mycorrhiza. Mycorrhiza 12: 163-167.

Elmeskaoui, A., Damont, J.P., Poulin, M.J., Piche, Y. and Desjardins, Y. (1995). A tripartite culture system for endomycorrhizal inoculation of micropropagated strawberry plantlets in vitro. Mycorrhiza 5: 313-319.

Estrada-Luna, A.A., Davies, F.T. and Egilla, J.N. (2000). Mycorrhizal fungi enhancement of growth and gas exchange of micropropagated guava plantlets (*Psidium guajava* L.) during ex vitro acclimatization and plant establishment. Mycorrhiza 10: 1-8.

Ferguson, J.J. and Menge, J.A. (1982). The influence of light intensity and artificially extended photoperiod upon infection and sporulation of *Glomus fasciculatus* on Sudan grass and on root exudation of Sudan grass. New Phytol. 92: 183-191.

Fortin, J.A., Becard, G., Declerck, S., Dalpe, Y., St Arnaud, M., Coughlan, A.P. and Piche, Y. (2002). Arbuscular mycorrhiza on root-organ cultures. Can. J. Bot. 80: 1-20.

Fortuna, P., Citernesi, S., Morini, S., Giovannetti, M. and Loreti, F. (1992). Infectivity and effectiveness of different species of arbuscular mycorrhizal fungi in micropropagated plants of Mr S 2/5 plum rootstock. Agronomie 12: 825-829.

Frey, B. and Schuepp, H. (1993). A role of vesicular-arbuscular (VA) mycorrhizal fungi in facilitating interplant nitrogen transfer. Soil Biol. Biochem. 25: 651-658.

Furlan, V. and Fortin, J.A. (1973). Formation of endomycorrhizae by Endogone calospora on *Allium cepa* under three temperature regimes. Nat. Can. 100: 467-447.

Furlan, V., Bartschi, H. and Fortin, J.A. (1980). Media for density gradient extraction of endomycorrhizal spores. Trans. Brit. Mycol. Soc. 75: 336-338.

Gardiner, D.T. and Christensen, N.W. (1991). Pear seedlings responses to phosphorus, fumigation and mycorrhizal inoculation. J. Hort. Sci. 66: 775-780.

Gaur, A., Adholeya, A. and Mukerji, K.G. (1998). A comparison of AM fungi inoculants using *Capsicum* and *Polianthes* in marginal soil amended with organic matter. Mycorrhiza 7: 307-312.

Gaur, A. and Adholeya A (1999). Mycorrhizal effects on the acclimatization, survival, growth and chlorophyll of micropropagated *Syngonium* and *Dracaena* inoculated at weaning and hardening stages. Mycorrhiza 9: 215-219.

Gaur, A. and Adholeya, A. (2000). Effects of the particle size of soil-less substrates upon AM fungus inoculum production. Mycorrhiza 10: 43-48.

Gaur, A., Adholeya, A. and Mukerji, K.G. (2000a). On-farm production of VAM inoculum and vegetable crops in marginal soil amended with organic matter. Trop. Agric. (Trinidad) 1: 21-26.

Gaur, Anupama, Gaur, A. and Adholeya, A. (2000b). Growth and flowering in *Petunia hybrida, Callistephus chinensis* and *Impatiens balsamina* inoculated with mixed AM inocula or chemical fertilizers in a soil of low P fertility. Scientia Hort. 84: 151-162.

Gianinazzi, S., Trouvelot, A. and Gianinazzi-Pearson, V. (1990). Role and use of mycorrhizas in horticultural crop production. XXIII International Horticulture Congress, 27 August-1 September 1990, Florence, France, pp.25-30.

Gianinazzi-Pearson, V., Gianinazzi, S. and Trouvelot, A. (1990). Potentialities and procedures for the use of endomycorrhizas with special emphasis on high value crops. In:

Biotechnology of Fungi for Improving Plant Growth (eds.) J.M. Whipps and R.D. Lumsden, Cambridge University Press, Cambridge, UK, pp. 41-54.

Granger, R.L., Plenchette, C. and Fortin, J.V. (1983). Effect of VA endomycorrhizal fungus (*Glomus epigaeum*) on the growth and leaf mineral content of two apple clones propagated in vitro. Can. J. Plant Sci. 63: 551-555.

Gribaudo, I., Zanetti, R., Morte, M.A., Previati, A. and Schubert, A. (1996). Development of mycorrhizal infection in *in vitro* and *in vivo*-formed roots of woody fruit plants. Agronomie 16: 621-624.

Guillemin, J.P. and Gianinazzi, S. (1992). Fungicides interactions with AM fungi in *Ananas comosus* grown in a tropical environment. In: Mycorrhizas in Ecosystems (eds.) D.J. Read, D.H. Lewis, A.H. Fitter and I.J. Alexanders, CAB International, Wallingford, UK, pp. 381.

Guillemin, J.P., Gianinazzi, S. and Trouvelot, A. (1992). Screening of arbuscular endomycorrhizal fungi for establishment of micropropagated pineapple plants. Agronomie 12: 831-836.

Gryndler, M., Hrselova, H., Chvatalova, I. and Vosatka, M. (1998). In vitro proliferation of *Glomus fistulosum* intraradical hyphae from mycorrhizal root segments of maize. Mycol. Res. 102: 1067-1073.

Harley, J.L. and Smith, S.E. (1983). Mycorrhizal Symbiosis. Academic Press, New York, USA, pp. 483.

Hayman, D. S. (1983). The physiology of vesicular-arbuscular endomycorrhizal symbiosis. Can. J. Bot. 61: 944–963.

Heinzemann, J. and Weritz, J. (1990). Rockwool: A new carrier system for mass multiplication of vesicular-arbuscular mycorrhizal fungi. Angewandte Bot. 64: 271-274.

Hepper, C.M. (1984). Isolation and culture of VA mycorrhizal (VAM) fungi. In: VA mycorrhiza (eds.) C.L. Powell and D.J. Bagyaraj, CRC Press Inc., Boca Raton, Florida, USA, pp. 95-112.

Hoagland, D.R and Arnon, D.I. (1950). The water-culture method for growth of plants without soil. Univ. of California, Calif. Agric. Exp. Stn. Circ. No. 347.

Hooker, J.E., Gianinazzi, S., Vestberg, M., Barea, J.M. and Atkinson, D. (1994). The application of arbuscular mycorrhizal fungi to micropropagation systems: an opportunity to reduce chemical inputs. Agric. Sci. Fin. 3: 227-232.

Ikram, A., Jensen, E.S. and Jakobsen, I. (1994). No significant transfers of N and P from *Pueraria phaseoloides* to *Hevea brasiliensis* via hyphal links of arbuscular mycorrhiza. Soil Biol. Biochem. 26: 1541-1547.

Jaizme-Vega, M.C., Galan Sauco, V. and Cabrera Cabrera, J. (1991). Preliminary results of VAM effects on banana under field conditions. Fruits 46: 19-22.

Jaizme-Vega, M.C. (1992). AM inoculation of micropropagated banana plantlets (*Musa acuminata* Colla AAA). In: Micropropagation, Root Regeneration and Mycorrhizas. Joint meeting between COST 87 and COST 8.10, Dijon, France, pp. 45.

Jaizme-Vega, M.C. and Azcon, R. (1995). Responses of some tropical and subtropical cultures to endomycorrhizal fungi. Mycorrhiza 5: 213-217.

Jaizme-Vega, M.C., Rodriguez-Romero, A.S., Hermoso, C.M. and Declerck, S. (2003). Growth of micropropagated bananas colonized by root-organ culture produced arbuscular mycorrhizal fungi entrapped in Ca-alginate beads. Plant Soil 254: 329-335.

Janos, D.P., Schroeder, M.S., Schaffer, B. and Crane, J.H. (2001). Inoculation with arbuscular mycorrhizal fungi enhances growth of *Litchi chinensis* Sonn. trees after propagation by air-layering. Plant Soil 233: 85-94.

Jarstfer, A.G. and Sylvia, D.M. (1992). Inoculum production and inoculation strategies for vesicular-arbuscular mycorrhizal fungi. In: Soil Microbial Ecology: Application in

Agriculture and Environmental Management (ed.) B. Meting, Marcel Dekker, New York, USA, pp. 349-377.

Jarstfer, A.G. and Sylvia, D.M. (1995). Aeroponic culture of VAM fungi. In: Mycorrhiza Structure, Function, Molecular Biology and Biotechnology (eds.) A. Varma and B. Hock, Springer Verlag, Berlin, Germany, pp. 521-559.

Jeffries, P., Gianinazzi, S., Perotto, K. and Barea, J. M. (2003). The contribution of arbuscular mycorrhizal fungi in sustainable maintenance of plant health and soil fertility. Biol. Fertil. Soils 37: 1-16.

Kahiluoto, H. and Vestberg, M. (1998). The effect of arbuscular mycorrhiza on biomass production and phosphorus uptake from sparingly soluble sources by leek (*Allium porrum* L.) in Finnish field soils. Biol. Agric. Hort. 16: 65-85.

Kaya, C., Higgs, D., Kirnak, H. and Tas, I. (2003). Mycorrhizal colonization improves fruit yield and water use efficiency in watermelon (*Citrullus lanatus* Thunb.) grown under well-watered and water-stressed conditions Plant Soil 253: 287-292.

Khanizadeh, S., Hamel, C., Kianmehr, H., Buszard, D. and Smith, D.L. (1995). Effect of three vesicular-arbuscular mycorrhizae species and phosphorus on reproductive and vegetative growth of three strawberry cultivars. J. Plant Nutr. 18: 1073-1079.

Klingeman, W.E., Auge, R.M. and Flanagan, P.C. (2002). Arbuscular mycorrhizal assessment of ornamental trees grown in Tennessee field soils. Hort. Sci. 37: 778-782.

Koide, R.T., Landherr, L.L., Besmer, Y.L., Detweiler, J.M. and Holcomb, E.J. (1999). Strategies for mycorrhizal inoculation of six annual bedding plant species Hort. Sci. 34: 1217-1220.

Krandashov, V.E., Kuzourina, I.N., Hawkins, H.J. and George, E. (2000). Growth and sporulation of the AM fungus *Glomus caledonium* in dual culture with transformed carrot roots. Mycorrhiza 10: 23-28.

Kuszala, C., Gianinazzi, S. and Gianinazzi-Pearson, V. (2001). Storage conditions for the long-term survival of AM fungal propagules in wet sieved soil fractions. Symbiosis 30: 287-299.

Lee, Y.J. and George, E. (2005). Development of a nutrient film technique culture system for arbuscular mycorrhizal plants. Hort. Sci. 40: 378-380.

Li, X.L., Marschner, H. and George, E. (1991). Acquisition of phosphorus and copper by VA mycorrhizal hyphae and root to shoot transport in white clover. Plant Soil 136: 49-57.

Lin, M.T., Lucena, F.B., Mattos, M.A.M., Paiva, M., Assis, M. and Caldas, L.S. (1987). Green house production of mycorrhizal plants of nine transplanted crops. In: Mycorrhizae for the Next Decade Practical Application and Research Priorities (eds.) D.M. Sylvia, L.L. Hung and J.H. Graham, 7th NACOM, USA, pp. 281.

Liu, R.J. and Luo, X.S. (1994). A new method to quantify the inoculum potential of arbuscular mycorrhizal fungi. New Phytologist 128: 89-92.

Linderman, R.G. and Davis, E.A. (2003). Arbuscular mycorrhiza and growth responses of several ornamental plants grown in soilless peat-based medium amended with coconut dust (coir). Hort. Technol. 13: 482-487.

Linderman, R.G. and Davis, E.A. (2004). Varied response of marigold (*Tagetes* spp.) genotypes to inoculation with different arbuscular mycorrhizal fungi. Scientia Hort. 99: 67-78.

Lovato, P.E., Guillemin, J.P. and Gianinazzi, S. (1992). Application of commercial arbuscular endomycorrhizal fungal inoculants to the establishment of micropropagated grapevine rootstock and pineapple plants. Agronomie 12(10): 873-880.

Lovato, P. E., Schuepp, H. and Giganinazzi, S. (1995). Application of arbuscular mycorrhizal fungi in orchard and ornamental plants. In: Mycorrhiza Structure, Function, Molecular Biology and Biotechnology (eds.) A. Varma and B. Hock, Springer-Verlag, Berlin, Germany, pp. 521-559.

Lovato, P.E., Gianinnazi-Pearson, V., Trouvelot, A. and Gianinnazi, S. (1996). The state of art of mycorrhizas and micropropagation. Adv. Hort. Sci. 10: 46-52.

Mallesha, B.C., Bagyaraj, D.J. and Pai, G. (1992). Perlite-soilrite mix as a carrier for mycorrhiza and rhizobia to inoculate *Leucaena leucocephala*. Leucaena Res. Rep. 13: 32-33.

Mamatha, G., Bagyaraj, D.J. and Jaganath, S. (2002). Inoculation of field-established mulberry and papaya with arbuscular mycorrhizal fungi and a mycorrhiza helper bacterium. Mycorrhiza 12: 313-316.

Marin, M., Mari, A., Ibarra, M. and Garcia-Ferriz, L. (2003). Arbuscular mycorrhizal inoculation of micropropagated persimmon plantlets. J. Hort. Sci. Biol. 78: 734-738.

Menge, J.A., Steirle, D., Bagyaraj, D.J., Johnson, E.L.V. and Leonard, R.T. (1978). Phosphorus concentrations in plants responsible for inhibition of mycorrhizal infection. New Phytol. 80: 575-578.

Menge, J.A., Jarrel, W.M., Labanauskas, C.K., Ojala, J.C., Huszar, C., Johnson, E.V. and Sibert, D. (1981). Predicting mycorrhizal dependency of troyer citrange on *Glomus fasciculatum* in California citrus soils and nursery mixes. Soil Sci. Soc. Am. J. 46: 762-768.

Mikanová, O. and Kubát, J. (2006). Phosphorus solubilizing microorganisms and their role in plant growth promotion. In: Microbial Biotechnology in Agriculture and Aquaculture, Volume II, (ed.) R.C. Ray, Science Publishers, Enfield, NH, USA, pp. 111-145.

Mohammad, A., Khan, A.G. and Kuek, C. (2000). Improved aeroponic culture of inocula of arbuscular mycorrhizal fungi. Mycorrhiza 9: 337-339.

Mohandas, S. and Das, S. (1992). Effect of VAM inoculation on plant growth, nutrient level and root phosphate activity in Papaya (*Carica papaya* cv. Coorg honey dew). Fertilizer Res. 31: 263-267.

Morte, M.A., Diaz, G. and Honrubia, M. (1996). Effect of arbuscular mycorrhizal inoculation on micropropagated *Tetraclinis articulata* growth and survival Agronomie 16: 633-637.

Mosse, B. (1962). The establishment of AM fungi under aseptic conditions. J. Gen. Microbiol. 27: 509-520.

Mosse, B. (1973). Advances in study of vesicular-arbuscular mycorrhiza. Annu. Rev. Phytopathol. 11: 171-196.

Mosse, B. and Thompson, J.P. (1984). Vesicular-arbuscular endomycorrhizal inoculum production. I. Exploratory experiments with beans (*Phaseolus vulgaris*) in nutrient flow culture. Can. J. Bot. 62: 1523-1530.

Mugnier, J. and Mosse, B. (1987). AM infection in transformed root-inducing T-DNA roots grown axenically. Phytopathol. 77: 1045-1050.

Niemi, M. and Vestberg, M. (1992). Inoculation of commercially grown strawberry with VA-mycorrhizal fungi. Plant Soil 144: 33-142.

O'Keefe, D.M. and Sylvia, D.M. (1991). Mechanisms of the vesicular-arbuscular mycorrhizal plant-growth response. Handbook of Applied Mycology 1: 35-53.

Ortas, I., Ortakci, D., Kaya, Z., Cinar, A. and Onelge, N. (2002). Mycorrhizal dependency of sour orange in relation to phosphorus and zinc nutrition. J. Pl. Nutr. 25: 1263-1279.

Parent, S., Desjardins, Y., Caron, M. and Lamarre, M. (1993). Growth of asparagus transplants inoculated with AM fungi for rapid assessment of infection. Trans. Brit. Mycol. Soc. 55: 158-160.

Pawlowska, T.E., Douds, D.D. Jr. and Charvat, I. (1999). In vitro propagation and life cycle of AM fungus, *Glomus etunicatum*. Mycol. Res. 103: 1549-1556.

Plenchette, C., Furlan, V. and Fortin, J.A. (1981). Growth stimulation of apple trees in unsterilized soil under field conditions with VA mycorrhiza inoculation. Can. J. Bot. 59: 2003-2008.

Plenchette, C., Furlan, V. and Fortin, J. A. (1982). Effect of different endomycorrhizal fungi on five host plants grown in calcined montmorillonite clay. J. Am. Soc. Hort. Sci. 107: 535-538.

Plenchette, C., Declerck, C., Diop, T.A. and Strullu, D.G. (1996). Infectivity of monoaxenic subcultures of the arbuscular mycorrhizal fungus *Glomus versiforme* associated with Ri-T-DNA-transformed carrot root. Appl. Microbiol. Biotechnol. 46: 545-548.

Pons, F., Gianinazzi-Pearson, V., Gianinazzi, S. and Navatel, J.C. (1983). Studies of VA mycorrhizae in vitro: Mycorrhizal synthesis of axenically propagated wild cherry (*Prunus avium* L.) plants. Plant Soil 71: 217-221.

Ponton, F., Piche, Y., Parent, S. and Caron, M. (1990). The use of vesicular-arbuscular mycorrhizae in Boston fern production. 2. Evaluation of four inocula. Hort. Sci. 25: 416-419.

Poulin, M.J., Bel-Rhlid, R., Piche, Y. and Chenevert, R. (1993). Flavonoids released by carrot (*Daucus carota*) seedlings stimulate hyphal development of VAM in the presence of optimal CO_2 enrichment. J. Chem. Ecol. 19: 2317-2327.

Porter, W.M. (1979). The "most probable number" method of enumerating infective propagules of VAM fungi in soil. Aust. J. Soil Res. 17: 515-519.

Quatrini, P., Gentile, M., Carimi, F., De Pasquale, F. and Puglia, A.M. (2003). Effect of native arbuscular mycorrhizal fungi and *Glomus mosseae* on acclimatization and development of micropropagated *Citrus lemon* (L.) Burm. J. Hort. Sci. Biotechol. 78: 39-45.

Ravolanirina, F., Gianinazzi, S., Trouvelot, A. and Carren, M. (1989). Production of endomycorrhizal explants of micropropagated grapevine rootstocks. Agric. Ecosyt. Environ. 29: 323-332.

Rea, E. and Tullio, M. (2005). The agro-industrial productions valorization: The possible role of the arbuscular mycorrhizal (AM) fungi. In: Microbial Biotechnology in Agriculture and Aquaculture, Volume I, (ed.) R.C. Ray, Science Publishers, Enfield, NH, USA, pp. 101-124.

Reddy, B. and Bagyaraj, D.J. (1994). Selection of efficient VAM fungi for inoculating the Mango root stock cultivar Nekkare. Scientia Hort. 59: 69-73.

Reddy, B. Bagyaraj, D.J. and Mallesha, B.C. (1996). Selection of efficient VA mycorrhizal fungi for papaya. Biol. Agric. Hort. 13: 1-6.

Redecker, D., Thierfelder, H. and Werner, D. (1995). A new cultivation system for arbuscular-mycorrhizal fungi on glass beads. Angewandte Bot. 69: 189-191.

Reganold, J.P., Papendick, R.I. and Parr, J.F. (1990). Sustainable agriculture. Scientific American 262: 112-120.

Sbrana, C., Giovanetti, M. and Vitagliano, C. (1994). The effect of mycorrhizal infection on survival and growth renewal of micropropagated fruit rootstocks. Mycorrhiza 5: 153-156.

Schellenbaum, L., Berta, G., Ravolanirina, F., Tisserant, B., Gianinazzi, S. and Fitter, A.H. (1991). Influence of endomycorrhizal infection on root morphology in a micropropagated woody plant species (*Vitis vinifera* L.). Ann. Bot. 68: 135-142.

Schuepp, H. and Bodmer, M. (1991). Complex responses of VA-mycorrhizae to xenobiotic substances. Toxicol. Environ. Chem. 30: 193-199.

Schreiner, R.P. (2003). Mycorrhizal colonization of grapevine rootstocks under field conditions. Am. J. En. Viticul. 54: 143-149.

Schubert, A., Mazzitelli, M., Ariusso, O. and Eynard, I. (1990). Effects of vesicular-arbuscular mycorrhizal fungi on micropropagated grapevines - influence of endophyte strain, P fertilization and growth medium. Vitis 29: 5-13.

Seymour, N.P. (1994). Phytotoxicity of fosetyl Al and phosphonic acid to maize during production of vesicular-arbuscular mycorrhizal inoculum. Plant Dis. 78: 441-446.

Sharma, M.P. and Adholeya, A. (2000). Enhanced growth and productivity following inoculation with indigenous AM fungi in four varieties of onion in an alfisol. Biol. Agric. Hort. 18: 1-14.

Sharma, M.P. and Adholeya, A. (2004). Influence of arbuscular mycorrhizal fungi and phosphorus fertilization on the post-vitro growth and yield of micropropagated strawberry in an alfisol. Can. J. Bot. 82: 322-328.

Shrestha, Y.H., Ishii, T., Matsumoto, I. and Kadoya, K. (1996). Effects of vesicular-arbuscular mycorrhizal fungi on satsuma mandarin tree growth, water stress tolerance, and on fruit development and quality. J. Soc. Hort. Sci. 64: 801-807.

Smith, S.E. and Read, D.J. (1997). Mycorrhizal symbiosis. Academic Press, San Diego, CA, USA.

Sreenivasa, M.N. and Bagyaraj, D.J. (1988). *Chloris gayana* (Rhodes grass), a better host for the mass production of *Glomus fasciculatum* inoculum. Plant Soil 106: 289-290.

Sreenivasa, M.N. and Bagyaraj, D. (1989). Use of pesticides for mass production of vesicular-arbuscular mycorrhizal inoculum. Plant Soil 119: 127-132.

Srinath, J., Bagyaraj, D.J. and Satyanarayana, B.N. (2003). Enhanced growth and nutrition of micropropagated *Ficus benjamina* to *Glomus mosseae* co-inoculated with *Trichoderma harzianum* and *Bacillus coagulans*. World. J. Microbiol. Biotech. 19: 69-72.

St. Arnaud, M., Hamel, C., Vimard, B., Caron, M. and Fortin, J.A. (1996). Enhanced hyphal growth and spore production of the AM fungus, *Glomus intraradices* in an in vivo system in absence of host roots. Mycol. Res. 100: 328-332.

Sylvia, D.M. and D.H. Hubbell. (1986). Growth and sporulation of vesicular-arbuscular mycorrhizal fungi in acroponic and membrane systems. Symbiosis. 1: 259-267.

Sylvia, D.M., Hartel, P., Fuhrmann, J. and Zuberer, D. (1998). Principles and Applications of Soil Microbiology. Prentice Hall, Upper Saddle River, NJ. USA, pp. 550.

Sylvia D.M and Jarstfer, A.G. (1992). Sheared-root inocula of vesicular-arbuscular mycorrhizal fungi. Appl. Environ. Microbiol. 58(1): 229-232.

Sylvia, D.M. (2000). Short root densities, ectomycorrhizal morphotypes, and associated phosphatase activity in a slash pine plantation. J. Sust. Forestry 11: 83-93.

Taylor, J. and Harrier, L.A. (2001). A comparison of development and mineral nutrition of micropropagated *Fragaria ananassa* cv. *Elvira* (Strawberry) when colonized by nine species of arbuscular mycorrhizal fungi. Appl. Soil Ecol. 18: 205-215.

Tepfer, D. (1989). Ri T-DNA from *Agrobacterium rhizogenes*: A source of genes having applications in rhizosphere biology and plant development, ecology and evolution. In: Plant-Microbe Interactions, Volume. 3 (eds.) T. Kosuge and E.W. Nester, McGraw-Hill Publishing, New York, USA, pp. 294-342.

Tiwari, P., Adholeya, A. and Prakash, A. (2003). Commercialization of arbuscular mycorrhizal biofertilizer. In: Fungal Biotechnology in Agricultural, Food, and Environmental Applications (eds.) D.K.Arora, Marcel Dekker, I., New York, USA, 195-201.

Trappe, J.M., Molina, R. and Castellano, M. (1984). Reactions of mycorrhizal fungi and mycorrhizal formation to pesticides. Annu. Rev. Phytopathol 22: 331-359.

Varma, A. and Schuepp, H. (1994a). Infectivity and effectiveness of *Glomus intraradices* on micropropagated plants. Mycorrhiza 5: 29-37.

Varma, A. and Schuepp, H. (1994b). Positive influence of AM fungus on in vitro raised *Hortensia* plant lets. Angewandte Bot. 15: 108-115.

Varma, A. and Schuepp, H. (1995). Mycorrhization of the commercially important micropropagated plants. Crit. Rev. Biotech. 15: 313-328.

Varma, A. and Schuepp, H. (1996). Influence of mycorrhization on the growth of micropropagated plants. Concepts in Mycorrhizal Res. 19 (Part 2): 113-132.

Varshney, A., Sharma, M.P., Adholeya, A., Dhawan, V. and Srivastava, P.S. (2002). Enhanced growth of micropropagated bulblets of *Lilium* sp. inoculated with arbuscular mycorrhizal fungi at different P fertility levels in an alfisol. J. Hort.. Sci. Biotech. 77: 258-263.

Venkateswarlu, K., Al-Garni, S.M. and Daft, M.J. (1994). The impact of carbofuran soil application on growth and mycorrhizal colonization by *Glomus clarum* of groundnut. Mycorrhiza 5: 125-128.

Vestberg, M. (1992a). The effect of vesicula-arbuscular mycorrhizal inoculation on the growth and root colonization of ten strawberry cultivars. Agric. Sci. Fin. 1: 527- 535.

Vestberg, M. (1992b). Arbuscular mycorrhizal inoculation of micropropagated strawberry and field observations in Finland. Agronomie 12: 865-867.

Vestberg, M. and Estaun, V. (1994). Micropropagated plants, an opportunity to positively manage mycorrhizal activities. In: Impact of Mycorrhizas on Sustainable Agriculture and Natural Ecosystems (eds.) S. Gianinazzi and H. Schuepp, Birkhauser Publisher Verlag Basel/Germany, pp. 217-226.

Vidal, M.T. Azcon-Aguilar, C., Barea, J.M. and Pliego-Alfaro, F. (1992). Myorrhizal inoculation enhances growth and development of micropropagated plants of avocado. Hort. Sci. 27: 785-787.

Walker, T. (1992). Systematics and taxonomy of the AM fungi (Glomales) — a possible way forward. Agronomie 12: 877-897.

Wen, C.L. (1991). Effect of temperature and *Glomus* sp. on growth and cut flower quality of micropropagated *Gerbera jamesonii*. M.S. thesis, Department of Horticulture, National Taiwan University, Taiwan, Republic of China, pp. 158.

Whitbeck, J.L. (2001). Effects of light environment on vesicular-arbuscular mycorrhiza development in *Inga leiocalycina,* a tropical wet forest tree. Biotropica 33: 303-311.

Williams, P.G. (1990). Disinfecting vesicular arbuscular mycorrhizas. Mycol. Res. 94: 995-997.

Williams, S. C. K. Vestberg, M., Uosukainen, M., Dodd, J. C. and Jeffries, P. (1992). Effects of fertilizer and arbuscular mycorrhizal fungi on the post-vitro growth of micropropagated strawberry. Agronomie 12: 851-857.

Wu, C.G., Liu, Y.S. and Hung, L.L. (1995). Spore development of *Entrophospora kentinensis* in an aeroponic system. Mycologia 87: 582-587.

Yano-Melo, A.M., Saggin O.J. Jr, Lima-Filho, J. M., Melo, N. F. and Maia, L .C. (1999). Effect of arbuscular mycorrhizal fungi on the acclimatization of micropropagated banana plantlets. Mycorrhiza 9: 119-123.

7

Baculoviruses: Molecular Biology and Advances in Their Development as Biological Pest Control Agents in Agriculture and Horticulture

Seema Mishra

INTRODUCTION

Biopesticides are natural products with a capacity to control pests detrimental to agricultural and horticultural crops and produce. Their inherent property of being less toxic than conventional man-made chemical pesticides has made their application in the control of infestation of agricultural and horticultural crops all the more useful. Their capacity to decompose quickly and their precise specificity confined only towards the target insect and their efficacy in low concentrations, represent additional advantages over conventional chemical pesticides. The US Environmental Protection Agency (USEPA) has classified biopesticides into three main categories: (1) microbial pesticides, (2) plant-incorporated-protectants and (3) biochemical pesticides (http://www.epa.gov/pesticides/biopesticides/). Microbial pesticides are one of the three classes of biopesticides which constitute microorganisms (fungi, protozoa, viruses and bacteria) active against pests. Baculoviruses belong to this class of biopesticides. Biotechnologists have made good use of baculoviruses as expression vectors, biopesticides, and mammalian cell gene transfer vectors. Recently, two novel roles of baculoviruses have been discovered: to function as an adjuvant and to facilitate the activation of dendritic cells in immunizations (Hervas-Stubbs et al., 2007; Schutz et al., 2006). Baculoviruses have been shown to have strong adjuvant properties in mice with both the humoral and adaptive (cytotoxic T cells) immune responses against coadministered antigens. Besides, these baculoviruses are also inducer of dendritic cells' *in vivo* maturation and inflammatory cytokines' production. This study demonstrates a strong effect of baculoviruses on the mammalian immune system (Hervas-Stubbs et al., 2007). In a different study, it has been shown that baculovirus *Ac*NPV is able to activate human monocyte-derived dendritic cells. This is one of the criteria in vaccination strategies targeting dendritic cells, since activation of dendritic cells is crucial for the initiation of an adaptive immune response (Schutz et al., 2006). Their effectiveness as powerful expression vectors stems from the fact that they have strong promoters driving high-level expression of foreign genes at a very late phase in the infection process in a eukaryotic environment. About 100 mg of protein per 10^9 cells can be obtained. Genes from a wide variety of sources ranging from viruses and bacteria to higher eukaryotes can be expressed using Baculovirus Expression Vector System (BEVS) technology. Being noninfectious for vertebrates, including human beings, their direct handling is safe. Even nontarget insects are protected from baculoviral infection, thus rendering baculovirus an effective biopesticide in integrated pest management (IPM) programs where it is mandatory to conserve nontarget beneficial insects.

BACULOVIRUS BIOLOGY

Baculoviruses ('baculum' meaning stick) are rod-shaped, extremely small (measuring 40–50 nm in diameter and 200–400 nm in length), double-stranded DNA viruses, which infect insects and other arthropods. In as early as 1527, during studies on 'jaundice disease' of silkworms, interest spurted in the direction of using baculoviruses as pest control agents (Benz, 1986). These viruses are found to infect insects mostly of the order Lepidoptera, Hymenoptera and Coleoptera, and have also been found to infect members of the order Diptera, Thysanura, Trichoptera and from the class Crustacea. They have been reported from more than 400 insect species (Volkman et al., 1995). They exist in nature in the form of virions that are occluded within proteinaceous crystals, and this occlusion body is also called as polyhedron (plural polyhedra). The polyhedra have been registered by EPA as the active ingredient (Ref. **Table 7.1** for a list of baculovirus occlusion bodies registered as active ingredients) because these are precisely the agents that infect insect larvae. This proteinaceous coat helps prevent the virus from being degraded by sunlight exposure.

Table 7.1. Nuclear polyhedrosis virus (NPV) and Granulosis virus (GV) occlusion bodies registered as pesticide active ingredients with EPA (as of October 2002)

Pest	*Crop/Use Sites*	*Active Ingredient (Opp Code)*	*#Products*	*Registrant*	*Remarks*
Beet armyworm	Vegetables (various); Ornamentals; Other crops such as cotton, corn, peanuts	Nuclear polyhedrosis virus of beet armyworm (*Spodoptera exigua* NPV) (129078)	1	Certis USA, LLC Columbia, MD	Registered 3/95
Celery and cabbage looper, cotton bollworm, etc.	Field and greenhouse raw agricultural commodities and ornamentals	Nuclear polyhedrosis virus of celery looper (*Anagrapha falcifera* NPV) (127885)CAS # 50933-33-0	1	Certis USA, LLC Columbia, MD	Registered 8/02
Codling moth	Apples; Pears; Walnuts	Granulosis virus of codling moth (*Cydia pomonella* GV) (129090)	3	Biotepp. Inc.Sterling, VA Certis USA, LLC Columbia, MD Sumitomo Corp. of America, Baltimore, MD	Registered 8/95
Corn earworm/ Tobacco budworm (same pest)	Vegetables (various); Ornamentals (various); Other crops	Nuclear polyhedrosis virus of corn earworm (*Helicoverpa zea* NPV) (107300)	2	Certis USA, LLC Columbia, MD Syngenta Crop Protection, Inc. Greensboro, NC	Registered 2/95
Douglas fir tussock moth	Forest trees (various); Ornamental or non-commercial trees at other specified sites. Use limited to wide-area government-sponsored programs	Nuclear polyhedrosis virus of Douglas fir tussock moth (*Orgyia pseudotsugata* NPV) (107302)	1	USDA Forest Service, Morgantown, WV	Registered 8/76 Eligible for re-registration 9/96

(*Table 7.1. Contd.*)

(Table 7.1. Contd.)

Gypsy moth	Forest trees (various); Ornamental or non-commercial trees at other specified sites. Use limited to wide-area government-sponsored programs	Nuclear polyhedrosis virus of gypsy moth (*Lymantria dispar* NPV) (107303)	1	USDA Forest Service, Morgantown, WV	Registered 4/78 Eligible for re-registration 9/96
Indian meal moth	Nuts and dried fruits before packaging. Areas for processing, packaging, and storing nuts and dried fruits	Granulosis virus of Indian meal moth (PDF) (*Plodia interpunctella* GV) (108896)	1	AgriVir, LLC, Washington, DC	Registered 12/01

Data source: EPA website: http://www.epa.gov/pesticides/biopesticides/ingredients/factsheets/factsheet_107300.html

Baculoviruses, in the form of polyhedra, are ubiquitous in their distribution in the environment. The natural mode of infestation of insect pests by polyhedra present on plant foliage is through feeding by the insect larvae, adult insects being non-susceptible. The polyhedra thus ingested get solubilized in the alkaline condition of the insect's midgut and the virions are released. Replication of these virions within the nuclei of epithelial cells lining the midgut produces more virions, which then either get released in a budded form by 10-12 h pi (hours post infection), or late in the infection process when polyhedrin proteins begin to accumulate, the virions get occluded within the polyhedra and released in an occluded form. Upon death of the infected larvae, tissue liquefaction and cell rupture liberates large amounts of polyhedra in the soil from which they are again ingested by live insects (**Fig. 7.1**).

BACULOVIRUS MOLECULAR BIOLOGY

The baculoviridae are a large and diverse family, comprising nuclear polyhedrosis viruses (NPVs) and granulosis viruses (GVs), characterized by conserved molecular systems. Nucleopolyhedroviruses and granulosis viruses are characterized on the basis of occlusion of virions within polyhedra or granules, respectively. Occlusion bodies of NPVs contain many virions embedded as nucleocapsids, which may exist as single or multiple units per envelope. Occlusion bodies of granulosis viruses contain one or occasionally two virions. Chief among a number of shared common features are mechanisms defining cell entry, genome replication and late and very late gene transcription (Okano et al., 2006). The most studied among the baculoviruses is *Autographa californica* multinucleocapsid nuclear polyhedrosis virus (*Ac*MNPV or *Ac*NPV), which has become the gold standard with which other baculoviruses are compared. The baculoviral genome is circular, covalently closed, having double-stranded DNA of 88–200 kb (kilobases) and has been sequenced (Ayres et al., 1994). After the infection has been established, a cascade of gene expression, starting from early baculoviral gene expression events to DNA replication to late and very late gene expression, results in the production of progeny virions. All these processes occur sequentially and are dependent on each other. Transcription of genes whose products are essential for viral DNA replication occurs during the early phase, which starts immediately after virus inoculation and continues up to 5 to 6 h pi, and these genes are transcribed by the RNA polymerase encoded by the host. Baculoviral DNA replication begins five to nine hours after infection and is required for late gene transcription. Between 5 and 18 h pi, transcription of late phase genes results in products such as structural

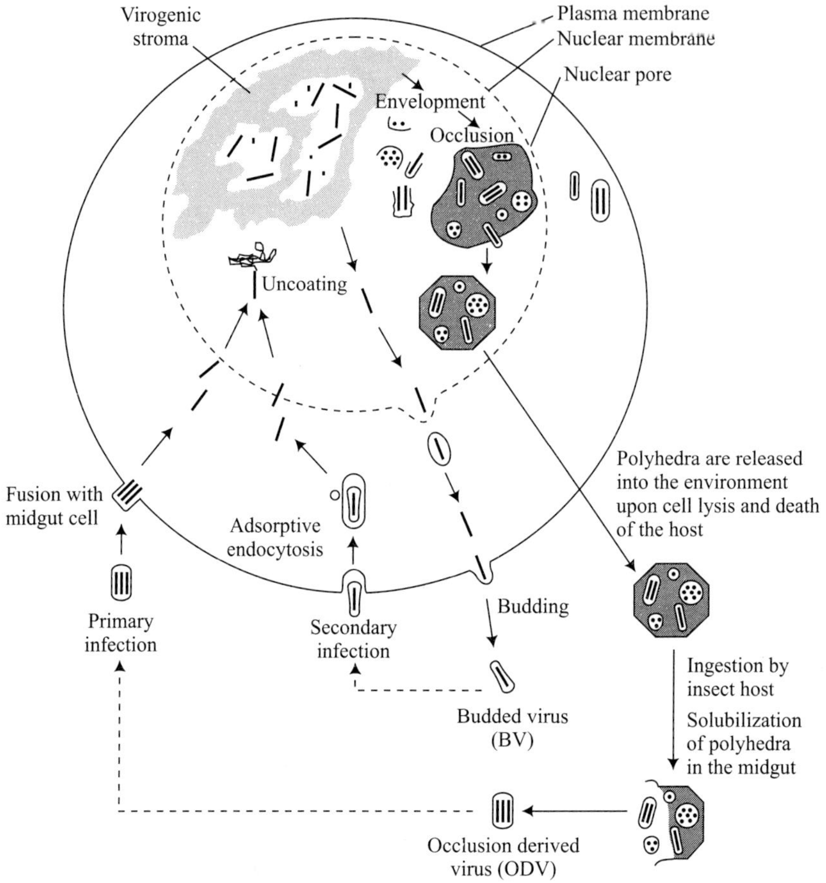

Fig. 7.1. Baculovirus life cycle

Source: **http://www.cheque.uq.edu.au/research/bioengineering/research/Baculovirus/Baculo_picturebook.html#b4**

proteins of the nucleocapsid. At around 20 h pi, transcription of the very late phase occlusion-specific genes (polyhedrin gene, *p10* gene) starts, which are involved in the viral occlusion process. The promoters of these genes are strong, as can be readily seen from the fact that polyhedrin protein occupies 50–75% of the total protein in an infected cell. Very late phase genes are also involved in the processes of nuclear and cellular lysis. A virus-encoded or virus-modified host RNA polymerase transcribes the late phase and very late phase genes. Host RNA polymerase, used to transcribe early genes, is sensitive to α-amanitin, an RNA polymerase II inhibitor (Hoopes and Rohrmann, 1991); whereas the virus-encoded or virus-modified host RNA polymerase transcribing late and very late baculovirus genes is α-amanitin- and tagetitoxin-resistant (Grula et al., 1981; Huh and Weaver, 1990; Glocker et al., 1993). The promoters of early baculoviral genes are similar to typical gene promoters that are responsive to RNA polymerase II. Most late and very late baculoviral gene promoters are characterized by the presence of an essential element, the ...TAAG... motif in promoter DNA sequences (Lu and Carstens, 1992; Eldridge et al., 1992a). In promoters of the hyperexpressed polyhedrin and *p10* genes, this motif is a part of a very well conserved sequence, ATAAGT/AATT (Weyer and Possee, 1989), which has been found to be responsible for the very high level of expression observed with these promoters. A conspicuous and novel feature of baculoviral genome is the presence of multiple copies of repeated sequences encompassing imperfect palindromes (a palindrome is a sequence of DNA bases that are the same in both forward and reverse direction on the two DNA strands, e.g., on one strand, the sequence is 5′...AGCT...3′ and on the second strand, it is 3′...TCGA...5′) flanked by direct repeats. These sequences are known as homologous regions (*hr*) and are found interspersed throughout the genome (Cochran and Faulkner, 1983). The *hr*s have been postulated to serve a dual role: (1) they function as enhancers of early, delayed early and very late viral gene transcription and (2) they are found to function as *cis*-acting replication origins in transient DNA replication assays. *Ac*NPV *hr*s, for example, are found interspersed at nine different regions of the genome (Possee and Rohrmann, 1997). A recent study utilizing recombinant viruses carrying deletion of different *hr* genes showed that each *hr* was dispensable for its activity and no specific *hr* region was absolutely essential for virus replication in cell culture (Carstens and Wu, 2007). Their activity as enhancers for either homologous or heterologous promoters is found to increase significantly upon insertion of multiple *hrs* copies in the genome (Venkaiah et al., 2004). Studies have shown that a homologous region named *hr*1 can also enhance transcription from non-baculoviral promoters like cytomegalovirus promoter and hsp70 promoter in mammalian cells (Viswanathan et al., 2003). *Hr*1 can also enhance transcription from the viral polyhedrin promoter in *Spodoptera frugiperda* insect cells and this role is independent of its function as a possible origin of replication (*ori*) (Habib et al., 1996). The binding of a host nuclear protein, *hr*1-binding protein (*hr*1-BP), is crucial for the enhancer activity (Habib and Hasnain, 1996). In addition, several non-*hr* regions, as well as early promoter regions in the baculoviral genome, are found to function as putative *ori*s (Lu et al., 1997; Habib and Hasnain, 2000). Both the host- and virus-encoded protein factors are involved in the transcriptional regulation of baculoviral early, late and very late gene promoters (Burma et al., 1994; Etkins et al., 1994; Jain and Hasnain, 1996; Hasnain et al., 1997; Ghosh et al., 1998). In addition to the involvement of host RNA polymerase and associated host transcription factors, early baculoviral gene expression is regulated by four baculoviral transregulators IE0, IE1, IE2 and PE38, all of which are immediate early genes. Late and very late genes are regulated by 18 *Ac*NPV late expression factor (LEF) genes. Very late gene transcription is also regulated by VLF-1 (very late factor-1) and one or more host factors including: 30-kDa (kilo Dalton) protein binding to p10 promoter; 30-kDa PPBP (polyhedrin promoter binding protein) binding to polyhedrin promoter; and a 200-kDa host factor interacting with a sequence located 21 to 34 bp upstream of TAAG motif in the polyhedrin gene promoter (Burma et al., 1994; Etkins et al., 1994; Jain and Hasnain, 1996). Involvement of novel members of the Sp family of transcription factors in transcription from the *polh* promoter has also been documented (Ramachandran et al., 2001). A novel viral factor, viral fibroblast growth factor (vfgf) homologous to cellular fgf, has been found to be encoded by all the lepidopteran baculovirus genome sequenced to date. That vfgf significantly increases the insect larvae mortality in insect bioassays has been shown by some recent studies (Detvisitsakun et al., 2007; Katsuma et al., 2006) and these studies reveal the potential of vfgf in being used for the improvement of baculovirus biopesticide. Baculoviral DNA replication is also dependent on both the baculoviral and host factors playing a part in the event. Five essential *Ac*NPV genes (*p143, ie-1, lef-1, lef-2* and *lef-3*) and five stimulatory *Ac*NPV genes (*dnapol, p35, ie-2, lef-7* and *pe38*) have been implicated in the replication process (Lu et al., 1997), while a host nuclear factor, a 38-kDa protein, *hr*1-BP, has been implicated in the enhancer activity of *hr*1 (Habib et al., 1996).

BACULOVIRUS AS A BIOLOGICAL CONTROL AGENT

The majority of baculoviruses, used and developed commercially as biopesticide, belong to the NPV category. Mostly baculoviruses isolated from Lepidoptera (butterflies and moths), Hymenoptera (sawflies only) and Coleoptera (beetles) have been researched for their suitability as a biological control agent since these present the best options for pest control. Of all baculoviruses, *Ac*NPV has the widest known host specificity, infecting more than 30 species of Lepidopteran insects (Granados and Federici, 1986), and so has the greatest potential for development as an insecticide. However, there is a report that *Anticarsia gemmatalis* multiple nuclear polyhedrosis virus isolate 2D is at present the most extensively used baculoviral biopesticide and its genome has been completely sequenced (Oliveira et al., 2006).

Developing countries are quite suitable for the large-scale development and use of these viral insecticides since production processes of these bio-insecticides are labor-intensive but not complex, and processing costs are low. Their narrow spectrum of activity and harmless nature towards beneficial insects, mammals, plants, birds and fishes make them particularly suitable for use in IPM practices. Currently, *in vivo* processes are used for bioinsecticide production using natural baculoviruses, usually implemented after the insecticidal activity of the baculovirus isolate has been established by several methods *in vitro*. Certain environmental conditions such as ultraviolet light, heat and physico-chemical properties of leaves from some plants, can cause inactivation of non-occluded baculoviruses, while wind and rain can wash away the polyhedra. Hence, an effective baculoviral formulation for insecticide application includes adjuvants, wetting agents, spreaders, stickers and UV masking agents. The viral formulation is mostly in the form of concentrated wettable powder. Two reports show that formulations made with natural ingredients such as pregelatinized corn flour and lignin (Tamez-Guerra et al., 2000; Behle et al., 2003) were found to help retain insecticidal activity of baculoviruses even after sunlight exposure. Release of infected insects in the field, baculoviral sprays and baculoviral dust represent some of the application methods applied to crops (for further details, see Mishra, 1998, 2000).

Natural baculoviruses, although being effective as biopesticides, have some inherent disadvantages, which limit their large-scale applicability and commercialization. Low kill rates and narrow host ranges have been cited as the prominent disadvantages. Natural baculoviruses take about four to seven days to kill their hosts, while some baculoviruses such as *Ld*NPV (*Lymantria dispar* NPV) take up to 18-21 days. The narrow host range is disadvantageous, in the sense that in a pest complex infesting a particular crop, where each different pest needs to be targeted, baculoviruses will not be able to target each one of them. It is precisely here that to overcome these disadvantages, the recombinant DNA technology comes to the fore. The need for taking up molecular biological studies of baculoviruses for their development as expression systems has been of great help in the development of improved baculoviral biopesticides as well.

RECOMBINANT BACULOVIRUS ENGINEERING STRATEGIES

Baculovirus molecular biology is an established research field in its own right. Interest in this area has intensified largely due to the need to use baculoviruses for high level recombinant protein expression, with the expressed protein being antigenically, immunogenically and functionally similar to its native counterpart. Knowledge of the molecular biology of baculoviruses has helped tremendously in the genetic engineering of baculoviruses and thus in the improvement of baculoviral bioinsecticides. The baculovirus expression vector system relies mainly on the replacement of the polyhedrin gene-coding region by foreign gene(s), which then comes under the control of strong polyhedrin promoter to allow for its high level expression. The foreign gene for engineering baculovirus to be used as a bioinsecticide, can be sourced from the coding regions for insect-specific toxins, hormones, enzymes or other gene products which have insecticidal activity. Since the polyhedrin gene has been deleted, polyhedrin protein is not expressed at all resulting in non-occluded progeny virions. Although desirable in terms of biological containment, non-occluded forms of virions are highly unstable under environmental conditions, and therefore, they are not active in the field conditions. Production of genetically engineered baculovirus in an occluded form is, therefore, desirable. Consequently, for the genetic engineering of baculoviruses, other strategies have been considered. In one of these strategies, both the wild type polyhedrin-positive and the genetically engineered polyhedrin-negative viruses are used to coinfect the hosts (Hamblin et al., 1990). The polyhedrin protein produced by the wild type viruses occludes the progeny virions. Thus, genetically modified polyhedrin-negative baculoviruses can be safely delivered to the field in an occluded, stable and infectious form. Individual larvae and cells need to be coinfected with both virus types as the virus is passed from insect to insect. The probability of co-infection determines the persistence of such a virus in the environment. Another very late gene promoter, i.e., *p10* gene promoter, can also be used to produce recombinant baculoviruses (Vlak et al., 1988). Deletion of *p10* gene results in viral polyhedra lacking a polyhedral envelope and the mutant viruses are shown to exhibit significantly reduced LD_{50} (lethal dose of biopesticide that kills 50% of test insects) values as compared to the wild type viruses. As such, polyhedrin-positive viruses with foreign gene inserts in the p10 region would be sufficiently stable for commercial use. Early viral gene promoters can be used to shorten the time of expression of insecticidal transcripts. When early gene promoters are used, polyhedrin expression can be retained resulting in occluded virions (Miller, 1988). A recent study (Regev et al., 2006) showed that the use of a delayed-early 39k promoter rather than the very late p10 promoter significantly enhanced insecticidal activity of recombinant *Ac*NPV carrying AaIT toxin gene. Also, the expression driven by 39K promoter was detected in insect but not in mammalian cells which led to the conclusion that viruses engineered using 39K gene promoter might be able to comply with safety requirements of a genetically modified organism. In another study comparing the insecticidal activity of recombinant baculoviruses harboring the insect-selective *Lq*hIT2 toxin gene under the control of a newly-developed early gene promoter [a polyhedrin upstream (pu)-enhanced minimal CMV (cytomegalovirus) promoter] in one case, and very late *p10* gene promoter in another case, the use of the early gene promoter proved to be economically advantageous by shortening the time-to-paralysis of insect larvae (Tuan et al., 2005). This is in contrast to studies showing relative advantage of the p10 promoter over the early p35 promoter in increasing the infectivity (Gershburg et al., 1998). The use of immediate-early promoters may help in the expression of foreign genes during semi-permissive or non-permissive conditions, so that the host-range specificity may be lost. The insertion of *hr*5 enhancer in ie1 promoter constructs results in high level constitutive expression (Jarvis et al., 1996). Other late gene promoters such as p6.9 have also been utilized in driving the large-scale production of insecticidal proteins earlier in the infection process (Hill-Perkins and Possee, 1990; Lu et al., 1996).

GENETICALLY ENGINEERED BACULOVIRUSES

Increased speed of kill and modification of host specificity of baculoviruses are the areas that have been explored towards the genetic engineering of baculoviruses for better use as biopesticides. Foreign genes encoding insect-specific toxins, hormones, enzymes or other gene products exhibiting insecticidal activity, can be inserted under the baculoviral strong very late gene promoters, such as polyhedrin or p10, in order to overexpress the foreign gene products and/or to shorten the time of kill. The latter condition can also be met using baculoviral late or early gene promoters. Host specificity can also be reduced or expanded for targeting a pest complex.

While the preferred method of production of natural baculoviruses is *in vivo*, genetically engineered baculoviruses are produced *in vitro*. This is so because recombinant baculoviruses have a faster rate of killing and can kill their hosts much before any substantial virus yield can be obtained when using the *in vivo* process. *Ac*NPV and *Ld*NPV have been subjected to *in vitro* production commercially, whereas for other developed cell-virus systems such as *Se*NPV (*Spodoptera exigua* NPV), *Hz*NPV (*Helicoverpa zea* NPV) and *Op*NPV (*Orgyia pseudotsugata* NPV), commercialization aspects still need to be addressed. Novel cell lines that are more efficient in supporting production on a large scale need to be developed. Besides satisfying the EPA regulatory requirements, genetic homogeneity and freedom from human pathogens are also prime requirements of both the viral and cell stocks. Viral genetic stability is dependent on the strain and a minimum of ten amplification passages is required to ensure genetic stability. Production of high levels of budded virus that can retain virulence in storage is also a viral trait. Since the FBS (fetal bovine serum) needed for cell culture is expensive, cell lines such as ExCell401 and Sf900 that can be grown in serum-free media are advantageous for large-scale production. Thus, a complex interplay of virus strain, cell line and medium determines the ability of baculoviruses to generate high budded virus titers for recombinant baculovirus production.

In 1989, a recombinant baculovirus was derived from the introduction of a gene coding for the scorpion *Buthus eupeus* insect toxin-1 (BeIT) into the *Ac*NPV genome under the control of the polyhedrin gene promoter in order to improve insecticidal activity (Carbonell et al., 1988). BeIT is an insect-specific neurotoxin, which induces paralysis and halts feeding in insects. It was found that sufficient toxin was not produced for a detectable biological activity. It was thought to be due to either low toxin concentration or instability of the protein through use of an improper signal sequence. In another study, replacement of *Bombyx mori* nuclear polyhedrosis virus (*Bm*NPV) polyhedrin gene with the diuretic hormone (DH) gene from the tobacco hornworm, *Manduca sexta*, produced a positive result (Maeda, 1989). At four days post-infection, all the larvae infected by the recombinant virus died, whereas the larvae infected by the wild type virus died at five days post-infection. Another hormone, juvenile hormone esterase (JHE), is known to inactivate juvenile hormone (JH) by hydrolysis in the last instar lepidopterous larvae. A reduction in the titre of JH early in the last larval instar initiates metamorphosis and resulted in the feeding cessation. The JHE gene from the tobacco hornworm, *Heliothis virescens*, was inserted into the *Ac*NPV genome under the control of the polyhedrin promoter (Hammock et al., 1990). When these recombinant viruses were fed to the first-instar larvae of *Trichoplusia ni* (*T. ni*), a profound reduction in feeding and growth was seen as compared to the control or wild type virus-infected larvae. However, infections of the later stage larvae showed no effect. It was attributed to a low level of JHE produced in later stages, even though under the control of a very late promoter, which is not sufficient to overcome the level of hormone biosynthesis. It was also speculated that the viral-induced production of ecdysteroid UDP-glucosyltransferase reduced the effects of JHE. Studies using other hormones such as PTTH (prothoracicotropic hormone) (O'Reilly et al., 1995) and eclosion hormone (Eldridge et al., 1992b) proved unsuccessful. A non-occluded derivative of *Ac*NPV expressing insect chitinase gene had more insecticidal activity than the wild type virus on *Spodoptera frugiperda* larvae (Gopalakrishnan et al., 1995).

A relatively strong increase in potency as compared to wild type viruses was, however, not shown by the recombinant baculoviruses carrying DH and JHE. Hence, an insect-specific toxin, AaIT, isolated from the venom of scorpion, *Androctonus australis* was tried (Maeda et al., 1991). This toxin affects only insects and shows no effect on isopods and mammals even at high dosage. It is known to selectively target the insect sodium channel. *AaI*T gene, preceded by a signal sequence from silkworm neuropeptide bombyxin, was inserted in place of polyhedrin gene under the control of polyhedrin gene promoter in *Bm*NPV. Symptoms in insects consistent with sodium channel blocking were seen when *Bombyx mori* larvae were injected with these viruses. It is to be noted that in contrast to the first attempts with *Buthus eupeus* toxin, BeIT, where the signal sequence used was from human interferon-α, the use of the signal sequence from silkworm bombyxin allowed the measurement of a detectable biological activity. Insertion of mite toxin from female mites of the species *Pyemotes tritici* into the *Ac*NPV genome was also found to induce paralysis at a much faster rate compared to that of the wild type virus (Tomalski and Miller, 1991). Further optimization of this system showed that the use of late 6.9K DNA binding protein gene promoter in the place of modified *polh* gene promoter resulted in decreasing the time-to-kill of the insect larvae (Lu et al., 1996). *Ac*NPV carrying tox34 neurotoxin from *P. tritici* under *p10* gene promoter control exhibited its potential in being listed among those modified viruses that show faster rate of killing (Burden et al., 2000).

Natural isolates of *Bacillus thuringiensis (Bt)* have been widely used as bioinsecticides. It is considered useful to insert the delta-endotoxin gene from *Bt* to enhance the insecticidal activity of baculoviruses (Merryweather et al., 1990). The bacterium produces crystalline inclusions containing the protoxin, which are ingested by insects. In the insect midgut, an active toxin is produced from the cleavage of protoxin. The active toxin causes generation of pores in the cell membrane, disruption of osmotic balance, cell lysis and eventual death. Bioassays using second instar *T. ni* larvae which ingested recombinant *Ac*NPV containing the protoxin sequence, showed no detectable enhancement in pesticidal properties. LD_{50} and LT_{50} (lethal time to kill 50% of test insects) for the *Bt* recombinant and wild type viruses did not differ significantly. In another study, *Ac*NPV recombinants expressing the various *Bt* CryIA(b) ICP constructs from *p10* gene promoter were not successful in controlling insect pests (Martens et al., 1995). However, the possibility of effecting a faster death rate of the larvae by the insertion of active toxin sequences in the viral genome cannot be ruled out. Also, *Bt* toxin produced in the larvae when liberated upon death of the larvae could be expected to serve as secondary control. In contrast to the earlier studies involving *Bt* toxin genes, using a different approach in a recent study, *Ac*NPV polyhedrin protein was fused to *Bt* Cry1Ac toxin and tested *in vivo* in *Plutella xylostella* pest (Seo et al., 2005), and this approach resulted in high insecticidal activity. This novel protein-based insecticide could prove to be quite useful when issues such as ecological safety in using living modified organisms (genetically engineered baculoviruses) as bioinsecticides need to be addressed.

The gene for a novel lepidopteran-selective toxin, *Buthus tamulus* insect-selective toxin (ButaIT), fused to the bombyxin signal sequence was engineered into a polyhedrin-positive *Ac*NPV genome under the control of the p10 promoter (Rajendra et al., 2005). Enhanced insecticidal activity on the larvae of *Heliothis virescens* was noted due to a significant reduction in median survival time (ST[50]) and also a greater reduction in feeding damage as compared to the wild type *Ac*NPV. Recombinant baculoviruses harboring *Lq*hIT2 depressant insect toxin from *Leiurus quinquestriatus* hebreus (Black et al., 1997), μ-AgaIV toxin from funnel web spider *Agelenopsis aperta* (Prikhod'ko et al., 1996), T-urf13, a maize mitochondrial gene involved in cytoplasmic male sterility (Korth and Levings, 1993) have also been tested for their improved insecticidal activity towards target insects. It is pertinent to mention that several criteria have been set down for using toxins for an effective enhancement of baculoviral insecticidal activity such as target insect specificity, high activity at low dosage, faster action, little or no cytotoxicity in culture but lethal to host larvae for recombinant virus production, and little or no specialized processing requirements. Among the whole assortment of toxins tested, insect-specific neurotoxins have been the foremost in meeting all the necessary criteria and field studies using recombinant baculoviruses harboring these toxins have shown positive outcomes. These have therefore shown more potential for the development of commercially viable bioinsecticides.

A novel ingenious approach towards improvement of insecticidal activity involved using an antisense gene fragment complementary to the mRNA of a host gene, the protein product of which is presumed to be essential for larval growth and development (Lee et al., 1997). As the two transcripts (sense and antisense) would remain bound to each other, a block would be created in the translation of an essential protein and normal insect physiology would be altered. In this study, a fragment of the human *c-myc* gene was inserted downstream from the polyhedrin promoter of *Ac*NPV and tested in bioassays on *S. frugiperda* larvae. This antisense approach was found to be efficient and 75% of the infected larvae stopped feeding almost immediately as the polyhedrin promoter-driven transcripts began to accumulate. The insecticidal effect was found to be much stronger on smaller larvae, suggesting that the emerging larval population serves as a more effective target.

The above approaches require insertion of a foreign gene into the baculovirus genome and study of the consequences. Deletion of a gene(s) from the viral genome that is not essential for viral replication but reduces virus-mediated killing can also be tested in order to help improve insecticidal activity. Genes which lengthen the life of the infected host by blocking molting, for instance, ecdysteroid UDP-glucosyl transferase can be deleted from the genome (O'Reilly and Miller, 1989). Ecdysteroid UDP-glucosyltransferase (EGT) produced by *Ac*NPV helps maintain the insects in an actively feeding state throughout the infection by blocking molting. Strategies were devised based upon the deletion of the gene encoding EGT, which was supposed to accelerate the virus-induced mortality by allowing the infected larvae to begin molting, resulting in feeding cessation. The results were conclusive, i.e., *S. frugiperda* larvae, infected with *egt*-negative virus, fed less and died more rapidly than those infected with wild type virus (O'Reilly and Miller, 1991). An *egt*-negative baculovirus by itself, or an *egt*-negative baculovirus overexpressing gene such as JHE, can be used for the development of enhanced baculovirus insecticides. Wild type viruses with *egt* gene cannot be used for overexpressing such genes, since they are not expected to have any significant effect due to the inhibition of ecdysis by *egt*. In another experiment, three gene deleted mutants of *Ac*NPV (*pp34*, *egt* and *p10* deletion mutants), wild type *Ac*NPV and wild type *Se*NPV were tested using the droplet feeding assay in *S. exigua* larvae and compared for their LD_{50} and LT_{50} values (Bianchi et al., 2000). Results showed that the wild type *Se*NPV had the lowest LD_{50} and the highest LT_{50} values, while the p10 deletion mutant of *Ac*NPV was found to have a slightly lower speed of action. These results suggest that gene deletion for improving the speed of killing should be assessed critically for each virus-host system before implementation.

In a recent study (Treacy et al., 2000), *Helicoverpa zea* NPV was found to be a better vector than the prototype *Ac*NPV for recombinant baculovirus generation owing to host range differences. *Hz*NPV baculovirus containing foreign AaIT toxin gene was found to be more effective on larval cotton bollworm, *Helicoverpa zea* than *Ac*NPV containing the same foreign toxin gene in laboratory and greenhouse studies. Field trials also demonstrated slightly better control of heliothine species by *Hz*-AaIT than *Bt*. However, the effects of the baculovirus on heliothine species were found to be slightly less than that of pyrethroid and carbamate insecticides in field trials, suggesting the need for better designing of recombinant baculoviruses to serve as more effective methods of pest control as compared to chemical insecticides.

GENETIC ENGINEERING FOR IMPROVING HOST SPECIFICITY

Although the host range of baculoviruses is limited, insects such as mosquito and cells of a non-permissive insect cell line have been found to incorporate baculovirus DNA in their nuclei. In mammalian systems, except hepatocytes (Hofmann et al., 1995; Boyce and Bucher, 1996), there is no evidence of baculovirus DNA being incorporated in their nuclei (Carbonell et al., 1985, 1988; Morris and Miller, 1992), although recent studies have demonstrated that baculoviral genome can be activated in mammalian cells with two baculoviral transactivators (IE1, IE2) playing a central role (Liu et al., 2007). These studies suggest that the control of the host specificity of baculoviruses is at the gene expression level. Among themselves, baculoviruses have been found to differ in the possession of sets of genes with only some genes being common. For example, *Ac*NPV has a gene called *hcf-1*, for host cell-specific factor-1, which other baculoviruses such as *Op*NPV and *Bm*NPV (*Bombyx mori* NPV) do not possess. This gene is found to be responsible for baculoviral host specificity. Reporter gene expression in semi- and non-permissive insect cells is observed mostly via early gene promoters, while results with the late and very late promoters show variable and limited expression levels suggesting that the infection in these cells is blocked subsequent to early gene expression. Some of the genes involved in late viral gene expression and DNA replication are also the factors on which baculoviral host specificity depends. Examples are *p143*, *hcf-1* and *lef-7* involved in the replication process, and other genes such as *p35*, *pp34* and *hrf-1*. P143 is baculoviral DNA helicase and alterations or exchange of *p143* gene to alter the host range may act *via* a mechanism of either directly affecting viral DNA replication or indirectly by disrupting cellular homoeostasis. The role of P143 in host range specificity has been highlighted by the studies done on two baculoviruses *Ac*NPV and *Bm*NPV. *Ac*NPV and *Bm*NPV are highly homologous with 70% general nucleotide sequence similarity and over 90% homology between well-conserved genes. This homology, coupled with their non-overlapping host specificity, is advantageous in the studies of host-range determining mechanisms. A *Bm*NPV-specific cell line, *Bm*N, could be infected by *Ac*NPV following homologous recombination of a 572 bp-region of a *Bm*NPV gene encoding a putative DNA helicase (Maeda et al., 1993). These experiments suggested that *Ac*NPV could be host-range expanded and that the helicase gene could be one of the many genes that control host range. Modifications in host range properties can thus be made by manipulating this helicase gene. In another study, a 79-nucleotide sequence within the *p143* helicase gene capable of extending the *Ac*NPV host range *in vitro* was identified (Croizier et al., 1994). *Ac*NPV and *Bm*NPV are found to differ at six positions corresponding to four amino-acid substitutions in this 79-nucleotide region. Three *Ac*NPV-specific amino acids were replaced by three corresponding *Bm*NPV-specific amino acids of the p143 protein. This resulted in the expansion of *Ac*NPV host range to the *Bombyx mori* larvae. HCF-1 is baculoviral host cell specific factor-1, which functions in a species-specific and tissue-specific manner. HCF-1 is required for efficient virus growth in TN368 cells, but is dispensable for virus replication in *SF*21 cells (Lu and Miller, 1995, 1996). LEF-7 (Late Expression Factor-7) was found to stimulate DNA replication in *Sf*21 cells, but was not required for replication in TN368 cells in replication assays (Lu and Miller, 1995). P35 is an apoptotic suppressor (Clem et al., 1991), while pp34 is a major constituent of the membrane enveloping the polyhedra. Deletion of *pp34* gene causes polyhedra to dissolve in the absence of a membrane and thus increases the infectivity of baculoviruses (Ignoffo et al., 1995). The baculoviral *p35* gene is known to suppress apoptosis of infected cells, thereby promoting viral replication. *Ac*NPV *p35* mutant, known as annihilator, possessed poor replication ability and caused premature cell death in infected *Sf*21 cells; however, there was no effect on replication ability in TN368 cells (Clem and Miller, 1991). Studies have shown that *p35* acts at an upstream step in the reactive oxygen species-mediated cell death pathway (Hasnain et al., 2003). P35 may function in a cell line-specific manner, e. g., in infected *Choristoneura fumiferana* and *Spodoptera littoralis* insect cell lines, P35 was not able to act as an apoptotic suppressor (Chejanovsky and Gershburg, 1995; Palli et al., 1996). In studies conducted to determine infectivity of wild type and *p35* mutant viruses, both the wild type virus and *p35* mutant virus were similar in their infective property in *T. ni* larvae, but in *S. frugiperda* larvae, *p35* mutants were significantly less infectious. These observations were the same using either the budded virus form (Clem and Miller, 1993) or the occluded virus form (Clem et al., 1994). Another gene, the *Ld*NPV hrf-1 (host range factor-1) gene, when introduced into the *Ac*NPV genome, was shown to confer upon *Ac*NPV the ability to replicate in the non-permissive *Ld*652Y cell line (Thiem et al., 1996). Thus, appropriate host range modifications of baculoviruses using host-range determining-gene manipulations have the potential to enhance baculoviral development as bioinsecticides. Several other studies have also shown positive results towards improving the baculoviral host range (Mishra, 2000).

INSECT IMMUNE SYSTEM IN DETERMINING THE HOST RANGE

The role of the insect immune system in host specificity control has also been documented. Studies demonstrated mounting of the immune response in insects against baculoviral infection with high susceptibility in the initial stages, but later infected cells were cleared. In the resistance of *S. littoralis* larvae to *Ac*NPV administration *via* the oral route, both the humoral and cellular immune responses have been implicated (Rivkin et al., 2006). The immune response could be suppressed by the use of chemical inhibitors such as diethyldithiocarbamic acid or biological inhibitors (an endoparasitoid harboring a symbiotic polydnavirus whose gene products are known immunosuppressants). Insertion of polydnavirus gene into the baculovirus genome is a plausible method of increasing host specificity of baculoviruses by suppressing the immune response of the host. Polydnaviruses and their encoded gene products have also been shown to synergize baculovirus infection (Rivkin et al., 2006). In one study, an entomopoxvirus gene encoding a virus enhancing factor was introduced in rice plants (Hukuhara et al., 1999). Armyworm larvae were fed on the transgenic rice and bioassays on these larvae showed increased insect susceptibility to baculoviral infection.

FIELD RELEASE OF RECOMBINANT BACULOVIRUSES: SOME INSIGHTS

Even though recombinant baculoviruses are found to be efficacious under laboratory conditions, their efficacies in field conditions have to be tested before they are to be exploited commercially. Even before field testing, the guidelines for microbial pesticides as laid down by the USEPA, which include information on habitat, host range, non-pathogenicity towards non-target organisms, environmental survival data, etc., are to be followed. In 1986, the first field-release testing of a genetically altered form of *Ac*NPV containing 80 bp non-coding insert downstream of the polyhedrin gene-coding region was implemented (Bishop et al., 1988). These polyhedrin-negative viruses had a lower persistence level in the environment. Field release testing in the USA in 1989 of modified viruses, lacking a functional polyhedrin gene occluded through a co-occlusion process, revealed their persistent nature (Maeda, 1995). By the second and third year, polyhedrin-negative *Ac*NPV accounted for roughly 9 and 6%, respectively, of the population. In 1993 and 1994, field-release trials of a recombinant *Ac*NPV, carrying an insect-specific scorpion-toxin gene (*Aa*HIT) derived from the scorpion *Androctonus australis* Hector, were conducted in England (Cory et al., 1994). The genetically modified virus (*Ac*ST3) and the wild type virus (*Ac*NPV C6) were compared after spraying at low, medium and high doses against third instar *T. ni* larvae on cabbage. Virus treatment significantly reduced insect damage compared to untreated controls. Larvae infected with the recombinant virus died of paralysis at a 10–15% earlier rate than did larvae infected with the wild type virus. A 22% reduction in leaf area was caused by larvae that were untreated, while the virus-treated plots showed a loss of only 12.5%, resulting in reduced crop damage. It was interesting to note that secondary transmission of the modified virus was found to be lower than that of the wild type virus. Although there was no difference in the mortality between the two virus treatments in the initial samplings, in the last samplings the modified virus exhibited significantly less mortality. Wild type NPV-infected insects usually remained on the plant after death where they liquefied, thus releasing a large body of inocula for further infections. Recombinant virus-infected larvae fell onto the soil and did not lyse, thus reducing the virus yield and consequently secondary transmission was also reduced. These results highlighted the importance of speed of action in achieving reduced crop damage.

An *Ac*NPV lacking its ecdysteroid UDP-glucosyl transferase (*egt*) gene had also been field-released in USA in 1994 and the speed of insect killing was found to be faster by about one day, as compared to that of the wild type virus (Black et al., 1997).

BACULOVIRUS AS BIOPESTICIDES: APPLICATION TO HORTICULTURAL CROPS

The majority of baculoviruses have been tested against pests of forest trees as well as against pests of diverse categories of horticultural crops, such as vegetables, fruits, ornamentals and plantation crops, for use as biopesticides. Some of these baculoviruses, as well as the horticultural crops that are amenable to their use, are listed in **Table 7.1**. In India, NPVs of *Helicoverpa armigera* and *Spodoptera litura* are being used against bollworms in cotton with an efficacy percentage of 70-80%. In addition, *S. litura* is also a pest of tobacco, cauliflower, cabbage, castor, cotton, groundnut, potato and lucerne, while *H. armigera* infests vegetables like tomato, okra and dolichos bean. Thus, these crops can also be protected from pest infestation using baculovirus formulations. Root crops like cassava have been treated with a baculovirus effective against cassava hornworm in Columbia with the results being satisfactory and the baculovirus has been developed by CIAT (Centro Internacional de Agricultura Tropical, Cali, Colombia) into a commercial formulation. In

Kenya, larvae of the diamondback moth *Plutella xylostella,* which infect *Brassica* crops, such as kale and cabbage, have been successfully controlled using *Plutella xylostella* granulosis virus applications. Among baculoviruses, celery looper (*Anagrapha falcifera*) NPV is the most versatile in infecting a wide variety of pest insects such as celery looper worm, cabbage looper, tomato hornworm, tobacco hornworm, cotton bollworm and pink bollworm, codling moth and diamondback moth. Another main advantage is that *Af*NPV is highly effective even at low dosage, so it appears as a very promising baculovirus species suitable for widespread use as a biopesticide in IPM programs.

BACULOVIRAL SAFETY FEATURES AND ENVIRONMENTAL PERSISTENCE

The first and foremost concern of a biopesticide application is its level of safety in both the wild type and genetically engineered form. Vertebrates and nontarget invertebrates have proved to be resistant to baculoviral infection. The inability of baculoviral DNA to replicate inside the nucleus of cells from human and other vertebrates, as well as in the acidic gastric environment, are the causes attributed to the safety of baculoviruses. The acidic gastric environment is not suitable for polyhedra dissolution, which needs an alkaline environment like that of the insect's midgut. Also, hepatocytes and some other mammalian cells are an exception in being amenable to *Ac*NPV entry, however the genes are not expressed via *polh* gene promoter, but via the engineered RSV-LTR or cytomegalovirus immediate-early gene promoter (Hofmann et al., 1995; Boyce and Bucher, 1996), or via *Drosophila hsp70* gene promoter (Viswanathan et al., 2003). Also, the DNA is not able to replicate inside, as can be seen from the inability of *hr*1 sequence to function as an *ori* (origin of replication) (Viswanathan et al., 2003), thus rendering these cells infection-resistant.

Beneficial insects such as social wasps, *Polistes metricus*, were infected with a recombinant baculovirus expressing a scorpion toxin or mite toxin under the control of hsp70 promoter (McNitt et al., 1995). Hsp70 promoter is found to be active in a variety of insect cells. Although there was evidence of toxin accumulation; their effect on the growth, fecundity or social behavior of these wasps was negligible. When recombinant baculoviruses carrying a marker gene under the control of baculovirus very late promoters were used to infect the wasps, no expression of reporter gene was found (except in one case), indicating that baculovirus promoters (especially of very late genes) are not active in other nontarget insects. In another study too, the AaIT-recombinant and modified JHE-recombinant viruses had no effect on the mortality of wasps *Microplitis croceipes* (McCutchen et al., 1996).

Perhaps the most compelling evidence for the safety of baculoviruses on humans is found in the fact that humans who eat raw cabbage show no signs of pathogenicity even as about 1.12×10^8 baculovirus polyhedra have been estimated to be present in a 16 in^2 of cabbage.

Natural occluded baculoviruses have a high persistence level in the environment. Both the horizontal and vertical modes of transmission are seen, as well as a latent form of transmission, as documented by a low level of persistent infections in *Mamestra brassicae* (Hughes et al., 1997). Fat body cells from infected *M. brassicae* could also infect other healthy insects. However, in the case of recombinant baculoviruses, there is a need for a balance in persistence level. Low persistence levels are needed for biological containment; yet to kill generations of insect pests, a high persistence level is sought. Secondary transmission of recombinant baculoviruses is reduced as compared to the wild type, since recombinant baculoviruses kill their hosts immediately, leaving little room for more viral progeny to be produced. This is understood by the fact that larvae infected with AaIT-recombinant, *egt*-deleted recombinant and *Lq*hIT2-recombinant produced fewer progeny.

In view of pest resistance to both chemical insecticides and *Bt* endotoxin, with the US National Academy of Sciences reporting a 64-fold increase in pest resistance to pesticides in less than 50 years, baculoviruses need to be developed and used on a large commercial scale. Insect resistance to baculoviruses is relatively less frequent and this advantage coupled with their safety, specificity and efficacy, makes them a particularly alluring option for use as microbial pesticides for pest control in agriculture and horticulture.

ACKNOWLEDGEMENT

I dedicate this book chapter to my mentor and guide, Professor Seyed E. Hasnain for introducing me to baculovirus research and for his unstinting support and encouragement.

REFERENCES

Ayres, M.D., Howard, S.C., Kuzio, J., Lopez-Ferber, M. and Possee, R.D. (1994). The complete DNA sequence of *Autographa californica* nuclear polyhedrosis virus. Virology 202: 586-605.

Behle, R.W., Tamez-Guerra, P. and McGuire, M.R. (2003). Field activity and storage stability of *Anagrapha falcifera* nucleopolyhedrovirus (*Af*MNPV) in spray-dried lignin-based formulations. J. Econ. Entomol. 96: 1066-1075.

Benz, G.A. (1986). Introduction: Historical perspective. In: The Biology of Baculoviruses, Volume. 1 (eds.) R.R. Granados and B.A. Fedirici, CRC Press, Boca Raton, Florida, USA, pp. 1-35.

Bianchi, F.J., Snoeijing, I., van der wert, W., Mans, R.M., Smits, P.H. and Vlak, J.M. (2000). Biological activity of *Se*MNPV, *Ac*MNPV, and three *Ac*MNPV deletion mutants against *Spodoptera exigua* larvae (Lepidoptera: Noctuidae). J. Invertebr. Pathol. 75: 28-35.

Bishop, D.H.L., Entwistle, P.F., Cameron, I.R., Allen, C.J. and Possee, R.D. (1988). Field trial of genetically engineered baculovirus insecticides. In: The Release of Genetically Engineered Organisms (eds.) M. Sussman, C.H. Collins, F.A. Skinner and D.E. Stewart-Tull, Academic Press, New York, USA, pp. 143-179.

Black, C.B., Brennan, L.A., Dierks, P.M. and Gard, I.E. (1997). Commercialization of baculovirus insecticides. In: The Baculoviruses (ed.) L.K. Miller, Plenum Press, New York, USA, pp. 340-386.

Boyce, F.M. and Bucher, N.L.R. (1996). Baculovirus-mediated gene transfer into mammalian cells. Proc. Natl. Acad. Sci., USA 93: 2348-5232.

Burden, J.P., Hails, R.S., Windass, J.D., Suner, M.M. and Cory, J.S. (2000). Infectivity, speed of kill, and productivity of a baculovirus expressing the itch mite toxin txp-1 in second and fourth instar larvae of *Trichoplusia ni*. J. Invertebr. Pathol. 75: 226-236.

Burma, S., Mukherjee, B., Jain, A., Habib, S. and Hasnain, S.E. (1994). An unusual 30-kDa protein binding to the polyhedrin gene promoter of *Autographa californica* nuclear polyhedrosis virus. J. Biol. Chem. 269: 2750-2757.

Carbonell, L.F., Klowden, M.J. and Miller, L.K. (1985). Baculovirus-mediated expression of bacterial genes in dipteran and mammalian cells. J. Virol. 56: 153-160.

Carbonell, L.F., Hodge, M.R., Tomalski, M.D. and Miller, L.K. (1988). Synthesis of a gene coding for an insect-specific neurotoxin and attempts to express it using baculovirus vectors. Gene. 73: 409-418.

Carstens, E.B. and Wu, Y. (2007). No single homologous repeat region is essential for DNA replication of the baculovirus *Autographa californica* multiple nucleopolyhedrovirus. J. Gen. Virol. 88: 114-122.

Chejanovsky, N. and Gershburg, E. (1995). The wild-type *Autographa californica* nuclear polyhedrosis virus induces apoptosis of *Spodoptera littoralis* cells. Virology 209: 519-525.

Clem, R.J., Fechheimer, M. and Miller, L.K. (1991). Prevention of apoptosis by a baculovirus gene during infection of insect cells. Science 254: 1388-1390.

Clem, R.J. and Miller, L.K. (1993). Apoptosis reduces both the in vitro replication and the in vivo infectivity of a baculovirus. J. Virol. 67: 3730-3738.

Clem, R.J., Robson, M. and Miller, L.K. (1994). Influence of infection route on the infectivity of baculovirus mutants lacking the apoptosis-inhibiting gene *p35* and the adjacent gene *p94*. J. Virol. 68: 6759-6762.

Cochran, M.A. and Faulkner, P. (1983). Location of homologous DNA sequences interspersed at five regions in the baculovirus *Ac*MNPV genome. J. Virol. 45: 961-970.

Cory, J.S., Hirst, M.L., Williams, T., Halls, R.S., Goulson, D., Green, B.M., Carty, T.M., Possee, R.D., Cayley, P.J. and Bishop, D.H.L. (1994). Field trial of a genetically improved baculovirus insecticide. Nature 370: 138-140.

Croizier, G., Croizier, L., Argand, O. and Poudevigne, D. (1994). Extension of *Autographa californica* nuclear polyhedrosis virus host range by interspecific replacement of a short DNA sequence in the *p143* helicase gene. Proc. Natl. Acad. Sci., USA 91: 48-52.

Detvisitsakun, C., Cain, E.L. and Passarelli, A.L. (2007). The *Autographa californica* M nucleopolyhedrovirus fibroblast growth factor accelerates host mortality. Virology 365: 70-78.

Eldridge, R., Li, Y. and Miller, L.K. (1992a). Characterization of a baculovirus gene encoding a small conotoxinlike polypeptide. J. Virol. 66: 6563-6571.

Eldridge, R., O'Reilly, D.R. and Miller, L.K. (1992b). Efficacy of a baculovirus insecticide expressing an eclosion hormone gene. Biol. Control 2: 104-110.

Etkins, E., Carp-Weiss, L. and Levi, B.Z. (1994). *Spodoptera frugiperda* Sf-9 cells nuclear factor binds to a specific sequence on the baculovirus polyhedrin promoter. Virus Res. 31: 343-356.

Gershburg, E., Stockholm, D., Froy, O., Rashi, S., Gurevitz, M. and Chejanovsky, N. (1998). Baculovirus-mediated expression of a scorpion depressant toxin improves the insecticidal efficacy achieved with excitatory toxins. FEBS Lett. 422: 132-136.

Ghosh, S., Jain, A., Mukherjee, B., Habib, S. and Hasnain, S.E. (1998). The host factor polyhedrin promoter binding protein (PPBP) is involved in transcription from the baculovirus polyhedrin gene promoter. J. Virol. 72: 7484-7493.

Glocker, B., Hoopes, R.R. Jr., Hodges, L. and Rohrmann, G.F. (1993). In vitro transcription from baculovirus late gene promoters: accurate mRNA initiation by nuclear extracts prepared from infected *Spodoptera frugiperda* cells. J. Virol. 67: 3771-3776.

Gopalakrishnan, K., Muthukrishnan, S. and Kramer, K.J. (1995). Baculovirus-mediated expression of a *Manduca sexta* chitinase gene: Properties of the recombinant protein. Insect Bioche. Mol. Biol. 25: 255-265.

Granados, R.R. and Federici, B.A. (eds.) (1986). The Biology of Baculoviruses, Volumes 1 and 2, CRC Press, Boca Raton, Florida, USA.

Grula, M.A., Buller, P.I. and Weaver, R.F. (1981). α-Amanitin-resistant viral RNA synthesis in nuclei isolated from nuclear polyhedrosis virus-infected *Heliothis zea* larvae and *Spodoptera frugiperda* cells. J. Virol. 38: 916-921.

Habib, S., Pandey, S., Chatterji, U., Burma, S., Jain, A., Ahmad, R. and Hasnain, S.E. (1996). Bifunctionality of the homologous region sequence (hr1): Enhancer and *ori* functions have different sequence requirements. DNA & Cell Biology 15: 737-747.

Habib, S. and Hasnain, S.E. (1996). A 38 kDa host factor interacts with functionally important motifs within the *Ac*MNPV homologous region (*hr1*) DNA sequence. J. Biol. Chem. 271: 28250-28258.

Habib, S. and Hasnain, S.E. (2000). Differential activity of two non-hr origins during replication of the baculovirus *Ac*MNPV genome. J. Virol. 74: 5182-5189.

Hamblin M., van Beek, N.A.M., Hughes, P.R. and Wood, H.A. (1990). Co-occlusion and persistence of a baculovirus mutant lacking the polyhedrin gene. Appl. Environ. Microbiol. 56: 3057-3062.

Hammock, B.D., Bonning, B.C., Possee, R.D., Hanzlik, T.N. and Maeda, S. (1990). Expression and effects of juvenile hormone esterase in a baculovirus vector. Nature 344: 458-461.

Hasnain, S.E, Jain, A., Habib, S., Ghosh, S., Chatterji, U., Ramachandran, A., Das, P., Venkaiah, B., Pandey, S., Liang, B., Ranjan, A., Natarajan, K. and Azim, C.A. (1997). Involvement of host factors in transcription from baculovirus late gene promoters. Gene 190: 113-118.

Hasnain, S.E., Begum, R., Ramaiah, K.V.A., Sahdev, S. Shajil, E.M., Taneja, T.K., Manjari, M., Athar, M., Sah, N.K. and Krishnaveni, M. (2003). Host-pathogen interactions during apoptosis. J. Biosci. 28: 349-358.

Hervas-Stubbs, S., Rueda, P., Lopez, L. and Leclerc, C. (2007). Insect baculoviruses strongly potentiate adaptive immune responses by inducing type I IFN. J. Immunol. 178: 2361-2369. Erratum in: J. Immunol. 2007 178: 6653.

Hill-Perkins, M.S. and Possee, R.D. (1990). A baculovirus expression vector derived from the basic protein promoter of *Autographa californica* nuclear polyhedrosis virus. J. Gen. Virol. 71: 971-976.

Hofmann, C., Sandig, V., Jennings, G., Rudolph, M., Schlag, P. and Strauss, M. (1995). Efficient gene transfer into human hepatocytes by baculovirus vectors. Proc. Natl. Acad. Sci., USA 92: 10099-100103.

Hoopes, R.J. and Rohrmann, G.F. (1991). In vitro transcription of baculovirus immediate early genes: Accurate mRNA initiation by nuclear extracts from both insect and human cells. Proc. Natl. Acad. Sci., USA 88: 4513-4517.

Hughes, D.S., Possee, R.D. and King, L.A. (1997). Evidence for the presence of a low-level, persistent baculovirus infection of *Mamestra brassicae* insects. J. Gen. Virol. 78: 1801-1805.

Huh, N.E. and Weaver, R.F. (1990). Identifying the RNA polymerases that synthesize transcripts of the *Autographa californica* nuclear polyhedrosis virus. J. Gen. Virol. 71: 195-202.

Hukuhara, T., Hayakawa, T. and Wijonarko, A. (1999). Increased baculovirus susceptibility of armyworm larvae feeding on transgenic rice plants expressing an entomopoxvirus gene. Nat. Biotechnol. 17: 1122-1124.

Ignoffo, C.M., Garcia, C., Zuidema, D. and Vlak, J.M. (1995). Relative in vivo activity and simulated sunlight UV stability of inclusion bodies of a wild-type and an engineered polyhedral envelope-negative isolate of nucleopolyhedrovirus of *Autographa californica*. J. Invert. Pathol. 66: 212-213.

Jain, A. and Hasnain, S.E. (1996). A 30-kDa host protein binds to two very-late baculovirus promoter. Eur. J. Biochem. 239: 384-390.

Jarvis, D.L., Reilly, L.M., Hoover, K., Hammock, B. and Guarino, L.A. (1996). Construction and characterization of immediate-early baculovirus insecticides. Biol. Control. 7: 228-235.

Katsuma, S., Horie, S., Daimon, T., Iwanaga, M. and Shimada, T. (2006). *In vivo* and *in vitro* analyses of a *Bombyx mori* nucleopolyhedrovirus mutant lacking functional vfgf. Virology 35: 62-70.

Korth, K.L. and Levings, C.S. III. (1993). Baculovirus expression of the maize mitochondrial protein URF13 confers insecticidal activity in cell cultures and larvae. Proc. Natl. Acad. Sci., USA 90: 3388-3392.

Lee, S-Y., Qu, S., Chen, W., Poloumienko, A., MacAfee, N., Morin, B., Lucarotti, C. and Krause, M. (1997). Insecticidal activity of a recombinant baculovirus containing an antisense c-myc fragment. J. Gen. Virol. 78: 273-281.

Liu, C.Y., Wang, C.H., Wang, J.C. and Chao, Y.C. (2007). Stimulation of baculovirus transcriptome expression in mammalian cells by baculoviral transcriptional activators. J. Gen. Virol. 88: 2176-2184.

Lu, A. and Carstens, E.B. (1992). Transcription analysis of the EcoRI D region of the baculovirus *Autographa californica* nuclear polyhedrosis virus identifies an early 4-kilobase RNA encoding the essential *p143* gene. J. Virol. 66: 655-663.

Lu, A. and Miller, L.K. (1995). Differential requirements for baculovirus late expression factor genes in two cell lines. J. Virol. 69: 6265-6272.

Lu, A. and Miller, L.K. (1996). Species-specific effects of the *hcf-1* gene on baculovirus virulence. J. Virol. 70: 5123-5130.

Lu, A., Seshagiri, S. and Miller, L.K. (1996). Signal sequences and promoter effects on the efficacy of toxin-expressing baculoviruses as biopesticides. Biol. Control. 7: 320-332.

Lu, A., Krell, P.J., Vlak, J.M. and Rohrmann, G.F. (1997). Baculovirus DNA replication. In: The Baculoviruses (ed.) L.K. Miller, Plenum Press, New York, USA, pp. 171-191.

Maeda, S. (1989). Increased insecticidal effect by a recombinant baculovirus carrying a synthetic diuretic hormone gene. Biochem. Biophys. Res. Comm. 165: 1177-1183.

Maeda, S., Volrath, S.L., Hanzlik, T.N., Harper, S.A., Majima, K., Maddox, D.W., Hammock, B. D. and Fowler, E. (1991). Insecticidal effects of an insect-specific neurotoxin expressed by a recombinant baculovirus. Virology 184: 777-780.

Maeda, S., Kamita, S.G. and Kondo, A. (1993). Host range expansion of *Autographa californica* nuclear polyhedrosis virus (NPV) following recombination of a 0.6-kilobase-pair DNA fragment originating from *Bombyx mori* NPV. J. Virol. 67: 6234-6238.

Maeda, S. (1995). Further development of recombinant baculovirus insecticides Curr. Opin. Biotechnol. 6: 313-319.

Martens, J.W., Knoester, M., Weijts, F., Groffen, S.J., Hu, Z., Bosch, D. and Vlak, J.M. *et al.* (1995). Characterization of baculovirus insecticides expressing tailored *Bacillus thuringiensis* Cry1A(b) crystal proteins. J. Invertebr. Pathol. 66: 249-257.

McCutchen, B.F., Herrmann, R., Heinz, K.M., Parella, M.P. and Hammock, B.D. (1996). Effects of recombinant baculoviruses on a non-target endoparasitoid of *Heliothis virescens*. Biol. Control. 6: 45-50.

McNitt, L., Espielie, K.E. and Miller, L.K. (1995). Assessing the safety of toxin producing baculovirus biopesticides to a non-target predator, the social wasp, *Polistes metricus* Say. Biol. Control 5: 267-278.

Merryweather, A.T., Weyer, U., Harris, M.P., Hirst, M., Booth, T. and Possee, R.D. (1990). Construction of genetically engineered baculovirus insecticides containing the *Bacillus thuringiensis* spp. Kurstaki HD73 delta-endotoxin. J. Gen. Virol. 71: 1535-1544.

Miller, L.K. (1988). Baculoviruses for foreign gene expression in insect cells. Biotechnology 10: 457-465.

Mishra, S. (1998). Baculoviruses as biopesticides. Curr. Sci. 75: 1015-1022.

Mishra, S. (2000). Natural and genetically-engineered baculoviruses: Prospects for their use as biological insecticides. In: Microbial Interactions in Agriculture and Forestry, Volume. II. (eds). N.S. Subbarao and Y.R. Dommergues, Science Publishers, Inc. Enfield, NH, USA, pp. 269-294.

Morris, T.D. and Miller, L.K. (1992). Promoter influence on baculovirus-mediated gene expression in permissive and non-permissive insect cell lines. J. Virol. 66: 7397-7405.

Okano, K. Vanarsdall, A.L., Mikhailov, V.S. and Rohrmann, G.F. (2006). Conserved molecular systems of the Baculoviridae. Virology 344: 77-87.

Oliveira, J.V., Wolff, J.L., Garcia-Maruniak, A., Ribeiro, B.M., de Castro, M.E., de Souza, M.L., Moscardi, F., Maruniak, J.E. and Zanotto, P.M. (2006). Genome of the most widely used viral biopesticide: *Anticarsia gemmatalis* multiple nucleopolyhedrovirus. J. Gen. Virol. 87: 3233-3250.

O'Reilly, D.R. and Miller, L.K. (1989). A baculovirus blocks insect molting by producing ecdysteroid UDP-glucosyl transferase. Science 245: 1110-1112.

O'Reilly, D.R. and Miller, L.K. (1991). Improvement of a baculovirus pesticide by deletion of the *egt* gene. Biotechnology 9: 1086-1089.

O'Reilly, D.R., Kelly, T.K., Masler, E.P., Thyagaraja, B.S., Robson, R.M., Shaw, T.C. and Miller, L.K. (1995). Overexpression of *Bombyx mori* prothoracicotropic hormone using baculovirus vectors. Insect Biochem. Mol. Biol. 25: 475-485.

Palli, S.R., Laputo, G.F., Sohi, S.S., Brownright, A.J., Ladd, T.R., Cook, B.J., Primavera, M., Arif, B.M. and Retnakaran, A. (1996). *Cf*MNPV blocks *Ac*MNPV-induced apoptosis in a continuous midgut cell line. Virology 222: 201-213.

Possee, R.D. and Rohrmann, G.F. (1997). Baculovirus genome organization and evolution. In: The Baculoviruses (ed.) L.K. Miller. Plenum Press, New York, USA, pp. 109-138.

Prikhod'ko, G.G., Robson, M., Warmke, J.W., Cohen, C.J., Smith, M.M., Wang, P., Warren, V., Kaczorowski, G., van der Ploeg, L.H.T. and Miller, L.K. (1996). Properties of three baculoviruses expressing genes that encode insect-selective toxins μ-Aga-IVB, AsII and ShI. Biol. Control 7: 236-244.

Rajendra, W., Hackett, K.J., Buckley, E. and Hammock, B.D. (2005). Functional expression of lepidopteran-selective neurotoxin in baculovirus: Potential for effective pest management. Biochim. Biophys. Acta 1760: 158-163.

Ramachandran, A., Jain, A., Arora, P., Bashyam, M.D., Chatterjee, U., Ghosh, S., Parnaik, V.K. and Hasnain, S.E. (2001). Novel Sp family-like transcription factors are present in insect cells and are involved in transcription from the polyhedrin gene initiator promoter. J. Biol. Chem. 276: 23440-23449.

Regev, A., Rivkin, H., Gurevitz, M. and Chejanovsky, N. (2006). New measures of insecticidal efficacy and safety obtained with the 39K promoter of a recombinant baculovirus. FEBS Lett. 580: 6777-6782.

Rivkin, H., Kroemer, J.A., Bronshtein, A., Belausov, E., Webb, B.A. and Chejanovsky, N. (2006). Response of immunocompetent and immunosuppressed *Spodoptera littoralis* larvae to baculovirus infection. J. Gen. Virol. 87: 2217-2225.

Schutz, A., Scheller, N., Breinig, T. and Meyerhans, A. (2006). The *Autographa californica* nuclear polyhedrosis virus *Ac*NPV induces functional maturation of human monocyte-derived dendritic cells. Vaccine 24: 7190-7196.

Seo, J.H., Yeo, J.S. and Cha, H.J. (2005). Baculoviral polyhedrin-*Bacillus thuringiensis* toxin fusion protein: A protein-based bio-insecticide expressed in *Escherichia coli.* Biotechnol. Bioeng. 92: 166-172.

Tamez-Guerra, P., McGuire, M.R., Behle, R.W., Hamm, J.J., Sumner, H.R. and Shasha, B.S. (2000). Sunlight persistence and rainfastness of spray-dried formulations of baculovirus isolated from *Anagrapha falcifera* (Lepidoptera: Noctuidae). J. Econ. Entomol. 93: 210-218.

Thiem, S.M., Du, X., Quentin, M.E. and Berner, M.M. (1996). Identification of baculovirus gene that promotes *Autographa californica* nuclear polyhedrosis virus replication in a nonpermissive insect cell line. J. Virol. 70: 2221-2229.

Tomalski, M.D. and Miller, L.K. (1991). Insect paralysis by baculovirus-mediated expression of a mite neurotoxin gene. Nature 352: 82-85.

Treacy, M.F., Rensner, P.E. and All, J.N. (2000). Comparative insecticidal properties of two nucleopolyhedrovirus vectors encoding a similar toxin gene chimer. J. Econ. Entomol. 93: 1096-1104.

Tuan, S-J., Hou, R.F., Kao, S-S., Lee, C-F. and Chao, Y-C. (2005). Improved plant protective efficacy of a baculovirus using an early promoter to drive insect-specific neurotoxin expression. Bot. Bull. Acad. Sin. 46: 11-20.

Venkaiah, B., Viswanathan, P., Habib, S. and Hasnain, S.E. (2004). An additional copy of the homologous region (*hr1*) sequence in the *Autographa californica* multinucleocapsid polyhedrosis virus genome promotes hyperexpression of foreign genes. Biochemistry 43: 8143-8151.

Viswanathan, P., Venkaiah, B., Kumar, S., Rasheedi, S., Vrati, S., Bashyam, M.D. and Hasnain, S.E. (2003). The homologous region sequence (*hr1*) of *Autographa californica* multinucleocapsid polyhedrosis virus can enhance transcription from non-baculoviral promoters in mammalian cells. J. Biol. Chem. 278: 52564-52571.

Vlak, J.M., Klinkenberg, F.A., Zaal, K.J.M., Usmany, M., Klinge-Roode, E.C., Geervliet, J.B., Roosien, J. and van Lent, J.W. (1988). Functional studies on the *p10* gene of *Autographa californica* nuclear polyhedrosis virus using a recombinant expressing a *p10*-beta-galactosidase fusion gene. J. Gen. Virol. 69: 765-776.

Volkman, L.E., Blissard, G.W., Friesen, P., Keddia, Possee, R. and Theilmann, D.A. (1995). Family Baculoviridae. In: Virus Taxonomy (eds.) F.A. Murphy, C.M. Fauquet, D.H.L. Bishop, S.A. Ghabrial, A.W. Jarvis, G.P. Martelli, M.A. Mayo and M.D. Summers, Springer-Verlag, New York, USA, pp. 104-113.

Weyer, U. and Possee, R.D. (1989). Analysis of the promoter of the *Autographa californica* nuclear polyhedrosis virus *p10* gene. J. Gen. Virol. 70: 203-208.

8

Natural Antimicrobials to Improve Storage and Shelf Life of Fresh Fruits, Vegetables and Cut Flowers

Leonardo Schena, Franco Nigro and Antonio Ippolito*

INTRODUCTION

The natural resistance of fruits, vegetables and flowers to disease generally declines after harvest due to ripening and senescence, leading to decay development and physiological impairment, with severe economical loss. Fungal and bacterial diseases, starting either as latent infections established in the field, or as wound infections during harvesting and subsequent handling, are the major causes of losses during packaging, trading and distribution to the consumers. In cut flowers, blockage of the xylem vessels by microorganisms that accumulate in the vase solution or in the vessels themselves, is another cause of postharvest deterioration (van Doorn et al., 1990). Several methods are currently available to preserve perishable commodities. However, some of these methods are now being questioned. The use of synthetic fungicides to control postharvest diseases is progressively restricted by legislation, social rejection, and development of resistance in disease-causing microorganisms. Furthermore, since the market for postharvest fungicides is relatively small, it has become difficult to sustain the costs of new registration or to support previous ones.

*Corresponding Author

Consumers and vegetable industries are increasingly demanding commodities free of pesticide residues, mycotoxins, harmful microorganisms, and any other contaminant compromising produce quality. Consequent to this request, is the rapid rise in demand for organically produced fruits, vegetables and flowers. All these requirements have increased the demand for natural pesticides and elicited widespread interest in the development of alternatives to synthetic fungicides for controlling postharvest diseases (Suslow, 2000; Tian, 2006, in Volume 1 of this series). Currently, there are no natural antimicrobial compounds utilized in the postharvest industry, however, natural compounds such as pyrrolnitrin and strobilurins produced by bacteria of the genus *Pseudomonas* and by the fungus *Strobilurus tenacellus*, respectively, have served as models for the synthesis of fungicides with low toxicity (Nevill et al., 1988; Ammermann et al., 1992; Godwin et al., 1992).

Microbial control of postharvest diseases of fruits and vegetables by yeasts and bacteria has been discussed elaborately by Shi- Ping Tian in Volume 1 of this series (Tian, 2006). This review summarizes part of the vast amount of research on natural compounds of plant, microbial, and animal origin which seem most promising for future application in the control of postharvest diseases. Other reviews of interest are those reported by Roller (2003), Tripathi and Dubey (2004), Jongen (2005), Holley and Patel (2005), and Vicente et al. (2005). Readers are advised to refer to the original studies for detailed information.

COMPOUNDS OF PLANT ORIGIN

Plants are a source of a number of compounds with antimicrobial activity widely used in human medicine, and in the food, cosmetic, and drug industries. Plant-derived insecticides, including pyrethrins, have been discovered and widely used in agriculture. However, relatively little effort has been devoted towards the development of plant-derived compounds as substitutes for synthetic fungicides.

Plant Extracts

Allium extracts represent one of the first documented applications of a plant extract against plant pathogens (Ark and Thompson, 1959). In this study complete control of brown rot of peaches caused by *Monilinia fructicola* was obtained by treatment with 5, 10 and 20% of a deodorized aqueous extract of commercial powdered garlic. Extracts of *Allium* species inhibited the *in vitro* growth of *Aspergillus parasiticus, A. niger, A. flavus* and *A. fumigatus* (Sharma et al., 1981; Yin and Tsao, 1999), and many other spoilage fungi of grains, legumes, and processed foods. More recently, water and ethanol extracts of garlic cloves applied at 1-5% were effective in controlling *Penicillium digitatum* and *P. italicum* but were less effective compared to the fungicide treatment, which gave a 100% control (Obagwu and Korsten, 2003). The antimicrobial activity of garlic (*Allium sativum* L.) and other *Allium* species [onion (*A. cepa* L.) and leek (*A. porrum* L.)] is due to allicin (2-propenyl-2-propenethiol sulfinate) contained in the tissues of these specie as a precursor (alliin) which is activated by the enzyme alliinase when bulb tissues are disrupted. Allicin readily permeates through phospholipid membranes and probably acts by reacting with critical thiol groups in the cell, affecting several physiological processes including respiration and RNA synthesis (Miron et al., 2000). Other compounds, such as phenolics (Cao et al., 1996; Yin and Tsao, 1999) and antifungal proteins like allivin (Wang and Ng, 2001), are perhaps responsible for the inhibition of fungi. Generally, the activity of *Allium* extracts diminishes during storage and is lost by heating (Sharma et al., 1981; Yin and Tsao, 1999). Crude juices or aqueous extracts are more active than ethyl acetate, ether, chloroform or ethanol extracts (Sharma et al., 1979; Abdou et al., 1992; Obagwu and Korsten, 2003).

Extracts from 345 plants were evaluated *in vitro* by Wilson et al. (1997) for their antifungal activity against *Botrytis cinerea*. A rapid assay was used to determine the antifungal activity in both plant extracts and essential oil. Among the tested extracts, 13 samples applied at 10% dilution of the crude extract showed high levels of antifungal activity, with *Allium* and *Capsicum* extracts being the most active. Petroleum ether extract from *Origanum syriacum* L. (wild marjoram), growing wild in Lebanon, was active against *B. cinerea, Alternaria solani, Penicillium* sp., *Cladosporium* sp., and *Fusarium oxysporum* f. sp. *melonis* (Abou Jawdah et al., 2002). Of the 19 aqueous extracts of leaves and stems from plants indigenous to Mexico, eight were active *in vitro* against sporulation of *Rhizopus stolonifer* and three of them (*Annona cherimola* M., *Bromelia hemisphaerica* L. and *Carica papaya* L.) were also active *in vivo* on 'ciruela' fruits (*Spondias purpurea* L.) in suppressing Rhizopus rot. Interestingly, the leaf extract of *Casimiroa edulis* Llav. et Lex. was not active *in vitro*, but completely inhibited disease development on fruits (Bautista-Baños et al., 2000). The same authors found that powders and aqueous and ethanolic extracts of seeds and monthly harvested leaves of huamuchil [*Pithecellobium dulce* (Roxb.) Benth.] were effective against *B. cinerea, P. digitatum* and *R. stolonifer* infecting strawberry fruits during storage. Highest fungistatic or fungicidal effect, in both *in vitro* and *in vivo* studies, was recorded from extracts of leaves harvested in months characterized by stressful environmental conditions: the cold season (October-February), and the dry, hot season (April-June) (Bautista-Baños et al., 2003a). Extracts from the leaf pulp of *Aloe vera* L., commonly known as *Aloe vera* gel (Saks and Barkai-Golan, 1995), have been tested *in vitro* against *P. digitatum, P. expansum, B. cinerea,* and *Alternaria alternata,* the first and the last being most sensitive. Dipping of *P. digitatum*-inoculated grapefruits in solutions containing 1 mg l^{-1} of gel delayed lesion development and significantly reduced the incidence of infection. Extracts from stems, flowers and leaves of *Euphorbia macroclada* Boiss. were tested *in vitro* against several plant pathogenic fungi including postharvest pathogens, the strongest and weakest inhibitory activity being against *R. stolonifer* and *A. solani,* respectively (Al-Mughrabi, 2003). Extracts from stems had a stronger inhibitory activity than those from flowers or leaves. Aqueous leaf extracts of *Azadirachta indica* Adv. Juss., *Datura fistulosa* L., *Muraya exotica* L., *Lantana camara* L., *Ocimum sanctum* L. and *Catharantes roseus* L. almost completely inhibited the spread of soft rot diseases caused by *Fusarium scirpi* and *Helminthosporium spiciferum* on *Luffa cylindrica* L. (sponge-gourd) when applied after infection. The activity was less evident when leaf extracts were applied before infection (Ahmad and Prasad, 1995). The aqueous extract of *Acacia nilotica* L. showed pronounced antifungal activity against *P. italicum* and enhanced the shelf life of oranges for six days. Among various compounds, kaempferol was the most active in suppressing the growth of the target pathogen (Tripathi et al., 2002). Recently, also extracts from lichens and lichen acids were active against some plant pathogens including *B. cinerea, Evernia prunastri* and *Hypogimnia physodes* with evernic acid exhibiting greatest activity (Halama and Van Haluwin, 2004). Extract of *Pongamia pinnata* (L.) Pierre, actively studied in human medicine for its anti-inflammatory activity and neem extract, were effective in prolonging the shelf life of carnation cut flowers (Chandrashekhar and Gopinath, 2004). Extract of *P. pinnata* (bactosan) also showed preventive activity for controlling fire blight of pears, but only when the inoculum concentration of the pathogen was low (Tsiantos et al., 2003). In a study carried out with extracts from nine plants of Chinese traditional medicine, cinnamon (*Cinnamomum cassia* L.) and aniseed (*Pimpinella anisum* L.) were the most effective, prolonging the vase life of fresh cut roses by three to four days. Both extracts inhibited the growth of microorganisms in the vase solution and promoted water absorption and transport, maintaining the equilibrium of water in leaves and flower petals, and retaining the intactness of cell membranes (Gong et al., 2005).

Propolis

Propolis (bee glue) is a resinous or sometimes wax-like compound collected by *Apis mellifera* bees from the buds and bark of plants. Once collected, this material is enriched with salivary and enzymatic secretions and used for the construction and maintenance of beehives, as a general sealant, draught excluder, antibiotic, and as an embalming substance to cover carcasses of hive invaders. The chemical composition, still insufficiently known, as well as its colour and aroma, varies according to the geographical zones (Bankova, 2005). Among the list of constituents, fatty acids, terpenoids, esters, alcohols, hydrocarbons, aromatic acids, hydroquinones, caffeic acid and its esters, and quercetin are the most representative (Fu et al., 2005; Sahinler and Kaftanoglu, 2005). A study by Bankova et al. (2002) on several propolis samples from Bulgaria, Italy and Switzerland evidenced that most samples contained pinocembrin, pinobanksin and its 3-O-acetate, chrysin, galangin, caffeic acid and ferulic acid esters. The substances identified in propolis until now are used as constituents of food, food additives and are generally regarded as safe (GRAS) substances. The antibacterial, antifungal, antiviral, anti-inflammatory and anticancer properties of extracts of propolis are well-known (Tosi et al., 1996; Ota et al., 2001). These activities seem to be related to the presence of polar compounds, mainly flavonoids, phenolic acids, and their esters (Ghisalberti, 1979). Indeed, the flavonoids galangin, pinocembrin, pinostrobin, and the ferulic and caffeic acids are the most effective antibacterial compounds (Marcucci, 1995). Despite the potential of propolis as a safe antifungal compound, there are a few reports about its applications for controlling plant pathogens. *In vitro* assays of the antifungal activity of propolis against *B. cinerea* were demonstrated (La Torre et al., 1990). In trials aimed at assessing the compatibility of postharvest antagonistic yeasts with additives and agrochemicals, Lima et al. (1998) found that propolis inhibited both antagonists (*Rhodotorula glutinis, Cryptococcus laurentii* and *Aureobasidium pullulans*) and plant pathogens (*B. cinerea* and *P. expansum*), suggesting that it could be used as a safe, natural fungicide. However, field applications of propolis to control postharvest development of grey mould on strawberry were ineffective (Antoniacci et al., 2000). No records are available in literature on the use of propolis to prolong shelf life of cut flowers.

Considering the data available until now on the pharmacological and the antibacterial/antifungal activity of propolis, it seems worthwhile to pursue investigations about its use in controlling postharvest pathogens of fresh fruits and vegetables, as well as other food-spoilage microorganisms on other commodities.

Jasmonates

Jasmonic acid (JA) and its volatile esterified derivative, methyl jasmonate (MJ), are naturally occurring lipid compounds of the plant cell membranes, derived from oxygenase-dependent oxidation of fatty acids by the lipoxygenase pathway. They exert inhibitory and promotory effects, often similar to those of abscissic acid, on many plant physiological processes (Doares et al., 1995; Beno-Moualem et al., 2004). Among these effects is the triggering of various biosynthetic pathways associated with responses to biotic and abiotic stresses, such as wounding and infection by pathogens. The application of JA or MJ to plants induces the expression of genes involved in defensive reactions, such as the genes encoding proteinase inhibitor, and phenylalanine ammonia-lyase (PAL), the key enzyme in the phenylpropanoid pathway (Gundlach et al., 1992; Sharan et al., 1998; Gonzàle-Aguilar et al., 2004). Several jasmonates have been shown to activate genes encoding antifungal proteins, such as thionin (Andresen et al., 1992), osmotin (Xu et al., 1994), and several other genes involved in phytoalexin biosynthesis (Gundlach et al., 1992; Thomma et al., 1998).

Reports on the application of jasmonates to control postharvest decay have recently increased. Preharvest application of MJ in black raspberry cv. Jewel (*Rubus occidentalis* L.) and red raspberry cv. Autumn Bliss (*Rubus idaeus* subsp.) improved fruit quality, increased content of flavonoids and anthocyanin, and enhanced the antioxidant capacities of the fruits (Wang and Zheng, 2005). Postharvest treatment with MJ has been shown to reduce grey mould rot caused by *B. cinerea* in strawberry (Moline et al., 1997). Jasmonates significantly reduced green mould of grapefruit caused by *P. digitatum* after either natural or artificial inoculation, but were ineffective in suppressing *Monilinia fructicola* on sweet cherries (Tsao and Zhou, 2000). MJ vapour at 10^{-4} or 10^{-5} M significantly reduced the fruit surface infected by *Colletotrichum gloeosporioides* and *A. alternata* on papaya (Gonzalez-Aguilar et al., 2003). The same concentration of MJ was effective in activating the fruit defence in guava fruit, increasing lipoxygenase and phenylalanine-ammonia lyase activities, and reducing the chilling injury – a limiting factor for low temperature storage of this tropical fruit (Gonzàle-Aguilar et al., 2004). Raspberries treated with MJ showed a lower percentage of decay and maintained higher levels of sugars, organic acids and oxygen radical absorbance capacity (Wang, 2003). Colour measurements showed that untreated raspberry fruits became darker and less red after storage, but fruits treated with MJ were found to have the highest intensity of red colour (Wang, 2003). Although jasmonates have been reported as having direct antifungal activity against

B. cinerea in vitro, with complete inhibition at 400 μM MJ (Meir et al., 1998), neither JA nor MJ had any direct antifungal effect on *P. digitatum* spore germination or germ tube elongation (Droby et al., 1999). These results suggest that jasmonates act as resistance inducers against green mould decay and the involvement of phytoalexins cannot be excluded, considering that MJ induces the syntheses of scopoletin and scopolin in tobacco cell cultures (Sharan et al., 1998). Systemic protection against *B. cinerea* on various cultivars of cut roses was obtained under laboratory conditions by pulsing flowers with 200 mM MJ (Meir et al., 1998) and, under practical conditions, by pulsing (350 μM MJ) or spraying (500 μM MJ) flowers (Meir et al., 2004). These MJ concentrations did not cause any phytotoxicity on leaves and petals and did not impair flower quality and longevity. Moreover, in yellow, orange and pink cultivars, the MJ treatment improved petal colour by inhibiting colour fading during vase life (Meir et al., 2004). Treatment of cut freesia flowers with MJ vapour suppressed petal specking caused by *B. cinerea* infection. No direct antifungal activity was exerted in an *in vitro* assay, induction of host defense responses was suggested as a possible mechanism of action (Durras et al., 2005).

Glucosinolates

Glucosinolates (GLs) are sulphur-containing plant secondary metabolites occurring mainly in cruciferous crops (Brassicaceae) and in a restricted number of other plant families, among which Capparaceae and Caricaceae are the most important (Fahey et al., 2001). The GL molecule consists of a β-thioglucose moiety, a sulphonated oxime group and a variable side chain, derived from an amino acid. GLs may be enzymatically hydrolyzed by the enzyme myrosinase to yield a variety of biologically-active products, including isothiocyanates, thiocyanates, nitriles, and oxazolidine-2-thiones. The nature of the original GLs present in the plant and the conditions of enzymatic hydrolysis determine the types of compounds produced and their biological activities, depending on both the substrate and reaction conditions, especially pH (Mithen, 2001).

In controlled pH conditions (near neutral) the GL breakdown products are predominantly isothiocyanates (Gil and MacLeod, 1980). These products have a wide range of biological activity, which include both negative and positive nutritional attributes and the effects on the interactions of plants with insects and herbivores. Recent reports about the potential anticarcinogenic activity of GL degradation products have renewed the interest in their possible use as food additives (Verhoeven et al., 1997). Many GL breakdown products are toxic to microorganisms and it has been suggested that these compounds may play a role in plant disease resistance (Mithen et al., 1986; Doughty et al., 1996; Manici et al., 1997).

The activity of six GLs (glucoraphenin, gluconapin, sinigrin, glucotropaeolin, sinalbin, and rapeseed glucosinolates) against the leading postharvest pathogens (*B. cinerea, R. stolonifer, Monilia laxa, Mucor piriformis,* and *P. expansum*) of fruits and vegetables has been extensively tested *in vitro* and *in vivo* trials by Mari et al. (1993, 1996, 2002). They found that the six native GLs were ineffective in inhibiting the conidia germination of the tested pathogens, whereas all the derived isothiocyanates reduced the germination with variable intensity, according to the fungal species and the compound type. The isothiocyanates from glucoraphenin, sinigrin, and sinalbin completely inhibited the germination of the five pathogens tested. None of the tested compounds inhibited the mycelial growth of *M. piriformis* and *R. stolonifer,* whereas isothiocyanates from glucoraphenin proved to be the most effective against *P. expansum, B. cinerea,* and *M. laxa.* The volatile compounds obtained from the enzymatic hydrolysis of sinigrin and gluconapin (2-propenyl and 3-butenyl isothiocyanates, respectively) strongly inhibited conidia germination and/or mycelial growth of *M. laxa, B. cinerea,* and *P. expansum,* thus indicating that antifungal activity is exerted by the volatile fraction of these low molecular weight compounds (Mari et al., 1993). The same GLs and isothiocyanates were tested *in vivo* to evaluate their activity in controlling storage decay of two pear varieties ('Conference' and 'Kaiser') caused by the five fungi mentioned above. Isothiocyanates from glucoraphenin were the most effective against *M. laxa, B. cinerea,* and *M. piriformis* after six days at 20°C, determining a significant reduction of the lesion diameter in artificially inoculated fruits. The concentration of the glucoraphenin derived isothiocyanates strongly affected the antifungal activity. At the highest concentration tested (3.6 mg ml^{-1}), the development of *M. laxa* was completely inhibited even when a high concentration of pathogen spores (10^6 cell ml^{-1}) was used for inoculation. Moreover, the isothiocianate was able to stop *M. laxa* infections already underway, showing a curative effect up to 40 hours after inoculation (Mari et al., 1996).

Allyl-isothiocyanate (AITC), a naturally occurring flavour compound in mustard and horseradish, has a well-documented antimicrobial activity (Ishiki et al., 1992; Delaquis and Mazza, 1995). This volatile compound can be employed successfully in modified atmosphere packaging or as a gaseous treatment before storage (Tripathi and Dubey, 2004). The activity of AITC vapour from pure sinigrin or from *Brassica juncea* against the decay caused by *P. expansum* on pears has recently been tested (Mari et al.,

2002). The best decay control was obtained by exposing fruits for 24 hours at 20°C in an atmosphere enriched with 5 mg l^{-1} of allyl-isothiocyanates. The extent of control was correlated to the inoculum density of the pathogen. Increasing *P. expansum* inoculum concentration at a constant AITC concentration resulted in increasing disease incidence, thus suggesting that inoculum density, occurring on the fruit surface, on packhouse working lines, in floating water, etc., is an important parameter to be determined for proper use of this compound. Moreover, AITC treatments were effective for up to 24 hours after inoculation for cv. 'Conference', and 48 hours for cv. 'Kaiser', and also controlled infection caused by a thiabendazole-resistant strain of *P. expansum* and reduced the incidence of blue mould in both pear cultivars by 90% (Mari et al., 2002). This is particularly relevant since the availability of natural and safe compounds possessing curative effects could address and resolve one of the major limiting factors of alternative methods in controlling postharvest decay, i.e., the inefficacy against active infections. The results of analysis on the skin and pulp of AITC-treated pears indicate the extremely low concentration of residue in the fruit, suggesting the absence of any implications on human health (Mari et al., 2002). However, further evidence is required to validate the effectiveness of this compound at the temperature utilized to store pome and stone fruits and on large-scale treatments.

Perhaps, there are no available records on the use of GLs to control postharvest diseases of cut flowers.

Essential Oils

The antifungal and antibacterial activity of essential oils (EOs) against important plant and human pathogens, as well as food spoilage organisms, has been studied extensively (Roller, 2003). Recently, renewed interest has developed in the application of these substances to control plant pathogens and, in partcular, postharvest diseases (Aligiannis et al., 2001; Arras and Usai, 2001; Thangadurai et al., 2002; Benkeblia, 2004; Palhano et al., 2004; Holley and Patel, 2005). EOs are extremely complex mixtures varying with environmental and genetic factors. Mixtures mainly consist of terpenes and other additional compounds such as aldehydes, fatty acids, phenols, ketones, esters, alcohols, nitrogen, sulphur, etc. (Cowan, 1999; Aligiannis et al., 2001; Arras and Usai, 2001; Thangadurai et al., 2002; Skočibušić et al., 2006). The role played by these substances in the plant has not been fully elucidated. However, it is likely that most of them are involved in chemical defence mechanisms against phytopathogenic microorganisms (Rauha et al., 2000). They also exert their activity on the external environment of producing plants, influencing insects and the microbial composition of the phyllosphere and carposphere. For these reasons, several investigators have considered them as antimicrobial in food and, in particular, against postharvest microbial spoilage of vegetable, fruit and flower commodities. Their mode of action appears to be related to increased cell membrane permeability of microorganisms causing the contents to leak out (Piper et al., 2001). Several EOs have been tested *in vitro* for their activity against a range of bacteria and fungi (Nychas et al., 2003; Benkeblia, 2004; Skočibušić et al., 2006). Among the EOs tested against postharvest pathogens, those from plants of the genus *Thymus* have been particularly active. Thyme EOs have been tested against *P. italicum, P. digitatum, B. cinerea, Alternaria citri, A. alternata, Fusarium oxysporum,* and *R. stolonifer* (Arras and Grella, 1992; Arras et al., 1995; Reddy et al., 1998; Arras and Usai, 2001; Bouchra et al., 2003); EO of *T. capitatus* at concentration of 200-250 μg l^{-1} completely inhibited the growth of four postharvest pathogens (Arras and Usai, 2001), whereas EO of *T. glandulosus* at a concentration of only 100 μg l^{-1} suppressed the growth of *B. cinerea* (Bouchra et al., 2003). On strawberries, *T. vulgaris* EO reduced decay caused by *B. cinerea* and *R. stolonifer* by up to 76% (Reddy et al., 1998). Vapours of thyme EO reduced grey mould development on *Botrytis*-inoculated sweet cherries (Chu et al., 1999) and brown rot in apricots and plums (Liu et al., 2002). EOs from onion and garlic were effective in inhibiting *A. niger* and *Penicillium cyclopium* at concentrations of 50 and 100 ml l^{-1} (Benkeblia, 2004). Thyme, anise and cinnamon EOs have been found to reduce mycotoxins such as aflatoxins, ochratoxins A and fumonisin (Chao et al., 2000). EOs from *O. vulgare* ssp. *vulgare,* beyond the activity against several postharvest fungi, also showed a strong antioxidant activity (Sahin et al., 2004).

Carvacrol has been identified as the substance responsible for the antimicrobial activity in the EO of *T. capitatus* (Arras and Grella, 1992). Thymol, carvacrol and linalool were the active agents in *T. vulgaris* (Reddy et al., 1998). However, other minor components can also contribute to the antimicrobial activity of an EO (Lattaoui and Tantoui-Elaraki, 1994; Cosentino et al., 1999; Karaman et al., 2001). The EO of oregano (*Origanum* spp.), another member of the family of Labiatae containing thymol, carvacrol and other active substances such as caryophyllene and spathulenol, was reported as very active *in vitro* against several fungi (Elgayyar et al., 2001; Lambert et al., 2001; Sahin et al., 2004) and against some citrus disease agents (Arras, 1988). Carvacrol, a-pinene, p-cymene, c-terpinene and thymol methyl ether are the main components of *Satureja subspicata* Vis. EOs collected from Dalmatia (Croatia) were found effective against a number of gram-positive and gram-negative bacteria, and fungi of the genus *Aspergillus* (Skočibušić et al., 2006). Among the complex constituents of citrus EOs, the terpene citral (3,7-dimethyl-2,6-octadiennal) is known to possess strong antifungal properties (Rodov et al., 1995). *In vitro* studies have demonstrated that citral inhibited the spread of *P. expansum, P. italicum* and *P. digitatum*, responsible for severe storage rot of apples and citrus (Caccioni et al., 1995a, 1998; Venturini et al., 2002) and *C. gloeosporioides*, the causal agent of anthracnose of papaya fruit (Palhano et al., 2004). However, because of its phytotoxicity, citral may be difficult to use on fresh fruits and vegetables (Rodov et al., 1995).

Many other EOs are known to possess antimicrobial activity. They have been tested against a wide range of bacteria (Nychas et al., 2003; Sahin et al., 2004), other microbial contaminants of processed food (Beuchat, 2001), and postharvest insect pests (Isman, 2000). However, as reported above, very few of them have been tested against fungi and/or diseases responsible for loss during the postharvest phase of fresh fruits and vegetables. As per information currently available, no EOs have been tested against postharvest diseases of cut flowers. Nevertheless, the abundant data available in the literature can constitute a valid source for new EOs to be tested in the postharvest environment. The strong aroma of EOs limits their application in foods. However, this issue should have limited importance for cut flowers. Their potential use lies in a careful selection and evaluation at low concentration, possibly in synergistic combination with other natural products to improve antimicrobial activity (Lambert et al., 2001).

Plant Phenolic Compounds

Phenolic compounds play important roles in conferring flavour and colour characteristics to plants, flowers, fruits and vegetables, and serve as plant defence mechanisms against the attack by microorganisms, insects, and herbivores. Progress in research on the antimicrobial properties of plant phenolic compounds has been enormous in the last decades and many aspects have been comprehensively reviewed (Nicholson and Hammerschmidt, 1990; Rhodes, 1994; Hammerschmidt, 1999; Dixon et al., 2002; Gibbons, 2005). In this section only the most relevant aspects regarding the relationships between phenolic compounds and postharvest diseases of fresh fruits, vegetables and cut flowers are covered.

Constituents of phenolic compounds are influenced by both internal and external factors. These include genetic variation at the species and cultivar level, maturity at harvest, preharvest agromonic practices and postharvest processing conditions. Several investigations revealed that concentrations of phenolic compounds are generally higher in young fruits and tissues (Macheix et al., 1990; Lattanzio et al., 1994a). Recently, antifungal activity has been found in all types of tissues in strawberry fruit and Thin-Layer Chromatography (TLC) bioassays revealed that all fruit stages yielded antifungal activity due to the occurrence of phenolic compounds (Terry and Joyce, 2004; Terry et al., 2004). Phenolic compounds, which inhibit the growth of fungi, may be present in healthy unchallenged fruits and vegetables (preformed antimicrobial compounds), or may be found only in fruit tissues which have either been infected by pathogens or are weakly stressed (phytoalexins) (Kuć, 1995; Dixon, 2001). The first group includes simple phenols, phenolic acids, flavonols, some isoflavons and dihydrocalchones (phloridzin), whereas the second group comprises phenolics phytoalexins like isoflavonoids, flavans, stilbenes, phenanthrenes, pterocarpans and furocoumarins (Lattanzio et al., 2001). All these compounds originate through different branches of the 'general phenylpropanoid pathway', whose core reaction is the deamination of the phenylalanine by the phenylalanine-ammonia lyase enzyme to produce trans-cinnamic acid. The phenylpropanoid pathway can switch to flavonoid biosynthesis, *via* the condensation of three molecules of malonyl CoA with one molecule of *p*-coumaryl CoA, to yield chalcone. This reaction is catalyzed by chalcone synthase (CHS) (Hahlbrock and Scheel, 1989). Material entering the general phenylpropanoid pathway leads to the formation of a series of hydroxycinnamic acids and hydroxycinnamoyl-CoA esters, varying in their degrees of hydroxylation and O-methylation (Barber et al., 2000). Hydroxycinnamic and cinnamic acid derivatives having antimicrobial activities include caffeic, chlorogenic, *p*-coumaric, ferulic and quinic acid. Depending on the botanical species, hydroxycinnamics may be present at concentrations sufficient to retard microbial invasion and delay rotting of fruits and vegetables. Moulds and yeasts responsible for food spoilage proved sensitive to hydroxycinnamic acid and derivatives (Davidson, 2001). Several studies have also shown that derivatives of benzoic or cinnamic acid inhibit the growth of various filamentous fungi, including *Aspergillus* and *Penicillium*, and food spoilage yeasts, as well as biosynthesis of mycotoxins (Chipley and Uraih, 1980; Tawata et al., 1996; Florianowicz, 1998). A phenolic compound from walnut seed coats, gallic acid, was recently shown to prevent aflatoxin biosynthesis by *A. flavus* (Mahoney and Molyneux, 2004). Moreover, it has recently been demonstrated that cinnamic acid, vanillin, and veratraldehyde inhibit both hyphal growth of *A. flavus* at 5 mM, and spore germination at 10 mM (Kim et al., 2004). Specific *in vitro* trials have been conducted to evaluate the antimicrobial activity of intermediates of the general phenylpropanoid pathway against yeasts and bacteria. Of the three main classes of compounds tested, the hydroxycinnamaldehydes were the most effective, possessing higher antifungal and antibacterial activity than hydroxycinnamic acids and hydroxycinnamyl alcohols (Barber et al., 2000).

Phenolic compounds induced *in vitro* inhibition of *Botryodiplodia theobromae*, the causal agent of Java black rot in sweet potato. Chlorogenic acid provided the highest *in vitro* inhibition followed by pyrogallol, pyrocatechol, phenol and resorcinol (Mohapatra et al., 2000). Among a group of cinnamic acid derivatives tested *in vitro* for their activity against several postharvest pathogens (*B. cinerea, P. digitatum, Sclerotinia sclerotiorum, F. oxysporum* and *Alternaria* spp.), chlorogenic and ferulic acid proved strong inhibitors of *F. oxysporum* and *S. sclerotiorum*, respectively (Lattanzio et al., 1994a). Several strands of evidence indicate that the number of hydroxyl groups in the phenolic ring might affect the level of antifungal activity. Starting from the basic skeleton of the cinnamic acid, the presence of a hydroxyl group in the aromatic ring (i.e., *p*-coumaric acid, *m*-coumaric acid and *o*-coumaric acid) increased the activity against *F. oxysporum* and *Alternaria* spp. An additional hydroxyl group in the benzene ring caused no increase in antifungal activity, making caffeic acid only a moderate inhibitor against *P. digitatum*. Recently, similar results have been reported (Kim et al., 2004) indicating that cinnamic acid (without an -OH group) has the highest antifungal activity, while caffeic acid (with two -OH groups) did not significantly affect the growth of *A. flavus*. Conversely, the presence of a methoxy group increased the activity compared to coumaric acids. Ferulic acids were the best inhibitors amongst the cinnamic derivatives. Benzoic derivatives are reported as the best inhibitors of *B. cinerea, P. digitatum, S. sclerotiorum, F. oxysporum,* and *Alternaria* spp. The presence of an additional hydroxyl group in the ring of p-hydroxybenzoic acid improved the antifungal activity of the monophenol (Lattanzio et al., 1994a).

The antifungal activity of 2,5-dimethoxybenzoic acid (DMBA) in controlling postharvest pathogens has been specifically tested both *in vitro* and *in vivo*. *In vitro* studies demonstrated that DMBA inhibited both spore germination and mycelial growth of *B. cinerea* and *R. stolonifer*. Starting from a 10^{-4} M concentration, spore germination and mycelial growth of both fungi were affected, with *B. cinerea* being more sensitive than *R. stolonifer*. At a concentration of 5×10^{-3} M, DMBA completely inhibited spore germination and mycelial growth of both fungi. *In vivo* studies demonstrated that spraying or dipping into DMBA at 10^{-2} M reduced decay of strawberries stored either at 20 or 3°C. Its practical use on strawberries has also been tested and the best results were obtained when fruits were dipped for one minute in 10^{-2} M DMBA amended with 0.05% (v/v) Tween 20 (Lattanzio et al., 1994b). The antifungal activity seems to be associated with lipophilicity of the compounds permitting penetration of biological membranes. Once inside cells, the fungal cell lipophilic compounds are deacylated to yield an active phenol (Lattanzio et al., 1994a).

Among a group of phenolics from apples cv. Golden delicious, only chlorogenic acid inhibited *Phlyctaena vagabunda* spore germination and mycelial growth *in vitro*, while (+)-catechin, (–)-epicatechin, phloridzin, and quercetin glycosides showed no activity (Lattanzio et al., 2001). Similarly, *in vitro* bioassay of catabolic phloridzin derivatives (phloretin, phloroglucinol, phloretic acid, and p-hydroxybenzoic acid) indicated no inhibitory effects on mycelial growth of *P. vagabunda*.

Flavonoids are synthesized from phenylpropanoid and acetate-derived precursors, and are characterized by a common benzo-γ-pyrone structure (Pietta, 2000). Generally, flavonoids occur as glycosylated derivatives, and play important roles in plant growth and in defence against microorganisms and pests. The antioxidant properties of flavonoids have long been recognized (Schijlen et al., 2004; Cevallos-Casals et al., 2006). They have been reported to: (1) inhibit lipid peroxidation, (2) scavenge free radicals and active oxygen, (3) chelate iron ions, and (4) inactivate lipoxygenase (Pietta, 2000). Usually, flavonoids are divided into several categories, including flavonols, flavones, catechins, proanthocyanidins, anthocyanidins, and isoflavonoids (Skerget et al., 2005). Many studies suggest that flavonoids have biological activities, including antiallergenic, antiviral, antifungal, anti-inflammatory, and vasodilating actions (Bors et al., 1990), and since they show low toxicity in mammals, some are used in human medicine (Cesarone et al., 1992; Hertog et al., 1993; Pietta, 2000). The majority of flavonoids recognized as constitutive antifungal agents in plants are isoflavonoids, flavans, or flavanones. Due to their "*in vitro*" antifungal activity, flavonoid compounds have long been thought to play a role in plant–microorganism interactions as part of the host plant's defensive arsenal (Harborne and Williams, 2000). Stilbene phytoalexins, as flavonoid-type phytoalexins, are formed on the phenylalanine/polymalonate pathway, the last step of this biosynthesis being catalyzed by stilbene synthase (STS). Real time monitoring of STS transcript level indicates that it accumulates selectively in grape skin in response to both biotic and abiotic stresses (Schena et al., 2005). The skeleton of these substances is based on a *trans*-resveratrol structure (3,5,4′-trihydroxy-stilbene) (Jeandet et al., 2002). *Trans*-resveratrol, which possesses an unspecific antifungal character, is among the most studied compound, thus representing a good candidate as a "natural pesticide" against pathogens (Jeandet et al., 1993). In addition, resveratrol is known to possess antioxidant properties that can have positive effects on fruit conservation during storage. It has long been recognized that *trans*-reveratrol enhances the resistance of vine plants to pathogens such as

B. cinerea, Phomopsis viticola (Hoos and Blaich, 1990), *Plasmopara viticola* (Dai et al., 1995), and *R. stolonifer* (Sarig et al., 1997). Several *in vitro* investigations have also been conducted, demonstrating the antifungal activity of this compound (Paul et al., 1998). Recently, direct exogenous application of *trans*-resveratrol to grapes and apples affected their postharvest quality for weeks or months. In addition, it has been demonstrated that the resveratrol application does not alter the fruit organoleptic and biochemical properties (Gonzalez Ureña et al., 2003). Recently the efficacy of 7 phytoalexins (resveratrol, esculetin, quercetin dihydrate, scopoletin, scoparone, umbelliferone, and ferulic acid) has been evaluated against *P. expansum* and its toxin (patulin) in *in vitro* tests and on apples fruits (Sanzani et al., 2006). Quercetin dihydrate and umbelliferone were the most effective phytoalexins completely inhibiting the growth of *P. expansum* and significantly reducing patulin accumulation. Being both phytoalexins more effective in *in vivo* trials, authors speculated a mechanism of action based on the activation of defence responses in the host (Sanzani et al., 2006).

In the recent past, intense research has also been devoted to the role of resveratrol in human health because of its protective effects against cardiovascular diseases and cancer (Doraia and Aggarwal, 2004; Fulda and Debatin, 2004). The valuable therapeutic effect of resveratrol has stimulated investigations on the occurrence of this compound in grapes, other berry fruits (Lyons et al., 2003; Rimando et al., 2004), and various herbs (Cai et al., 2004). Due to its capacity to confer disease resistance to the grapevine, as well as its biological properties, most interest has now centred on STS gene transfer from the grapevine to numerous plants such as rice (Stark-Lorenzen et al., 1997), tomato (Thomzik et al., 1997), apple (Szankowski et al., 2003), and kiwifruit (Kobayashi et al., 2000), with the objective of increasing their tolerance to pathogens and improving the nutritional quality of derived food products. Moreover, increasing levels of flavonoids have recently been reported for tomatoes by overexpressing the petunia gene for chalcone isomerase, leading to an 80-fold increase in the flavonoid content of the tomato peel and a corresponding 20-fold increase in the flavonoid level in tomato paste (Muir et al., 2002). In addition, chalcone synthase and flavonol synthase transgenes were found to act synergistically to up-regulate flavonol biosynthesis significantly in the flesh of tomato fruits (Verhoeyen et al., 2002).

Galangin (3,5,7-trihydroxyflavone) extract from *Helicrysum aureonitens* was active against *P. digitatum* and *P. italicum* (Afolayan and Meyer, 1997), two important storage pathogens of citrus fruit. Four polymethoxylated flavones (3,5,6,7,3',4'-hexamethoxyflavone, 3,5,6,7,8,3',4'-heptamethoxy-flavone, 5,6,7,8,4'-pentamethoxyflavone and 5,6,7,8,3',4'-hexamethoxy-flavone) from cold-pressed orange oil were characterized and evaluated for their antifungal activities against *C. gloeosporioides*, a major pathogen of fruit that causes damage to crops in tropical, subtropical and temperate regions. Methoxylated flavones were effective in inhibiting mycelial growth of the fungus. As flavone concentration increased, mycelial growth decreased. 5,6,7,8,3′,4′-Hexamethoxyflavone completely inhibited the growth of *C. gloeosporioides* at a concentration of 100 μg ml^{-1} (Almada-Ruiz et al., 2003).

The antifungal activity of biflavones from *Taxus baccata* and *Ginkgo biloba,* i.e., amentoflavone, 7-O-methylamentoflavone, bilobetin, ginkgetin, sciadopitysin and 2,3-dihydrosciadopitysin, towards the fungi *A. alternata* and *Cladosporium oxysporum,* has recently been evaluated. Bilobetin exhibited a significant antifungal activity, completely inhibiting the growth of germinating tubes of *C. oxysporum* at a concentration of 100 mM. Activity of ginkgetin and 7-*O*-methylamentoflavone towards *A. alternata* was stronger than that of bilobetin. Moreover, slight structural changes in the cell wall of *A. alternata* exposed to ginkgetin have been reported (Krauze-Baranowska and Wiwart, 2003).

Specific trials about the activity of some synthetic flavonoids against postharvest pathogens have been conducted by Lattanzio et al. (1994a). Generally, the results indicate a low activity at the concentration range tested (10^{-4}- 10^{-5} M). The highest mycelial inhibition was observed with apigenin-7-glucoside and kaempferol-3-rutinoside. All the flavonoids tested, except for kaempferol-3-rutinoside, showed an appreciable activity against *Penicillium* spp. However, these results, far from demonstrating that flavonoids are not important in the resistance against the tested fungi, seem to indicate that the flavonoid/fungus/host combination is of fundamental importance.

Among the constitutive secondary metabolites, those occurring in citrus fruit have been widely investigated. Tangeretin and naringin may act as antifungal compounds in the resistance mechanism against fungal attack. They act as first and second defence barriers, respectively, since polymethoxylated flavones (tangeretin) are mainly localized in the outermost tissue of the fruits, the flavedo, while flavanones (naringin) are located in the albedo (Kanes et al., 1992). Other secondary compounds induced after infection, such as coumarins, also act in the defence mechanism of citrus fruits (Angioni et al., 1998; Arcas et al., 2000). Coumarins are phenolic substances containing fused benzene and alfa-pyrone rings. *In vitro* studies have shown that the lipophilic character is required for rather high antimicrobial activities (Kayser and Kolodziej, 1999). It has been suggested that the avoidance of large side-chains aids in penetration through the cell wall. It is, therefore, likely that the mode of action is attributable to interactions with the cell membrane (Ferrazzoli Deviennea et al., 2005). It is known that some citrus species accumulate coumarins such as xanthyletin, seselin, and scoparone, when infected by pathogenic fungi (Afek and Sztejnberg, 1993; Stange et al., 1993). The nature of the coumarin biosynthesised in this process varies, within a species, according to the pathogen. For example, *Citrus limon* L. accumulated scoparone after inoculation with *P. digitatum* (Kim et al., 1991). However, there was no significant accumulation of any antifungal compounds in the tissues of lemons inoculated with *Geotrichum candidum* (Baudoin and Eckert, 1985). Although coumarin inhibits the germination of spores of *A. niger*, *Penicillium glaucum*, and *Rhizopus nigricans*, other 4-hydroxycoumarins are generally ineffective against fungi. Antibacterial and antifungal effects have been found for umbelliferone and scopoletin (Jurd et al., 1971a, b; Recio et al., 1989; Kwon et al., 1997). *In vitro* tests indicated that pure coumarins have a very modest activity against *A. niger,* whereas *B. cinerea* and *Fusarium* were the most sensitive pathogenic fungi (Ojala et al., 2000).

Limettin (5,7-dimethoxycoumarin), 5-geranoxy-7-methoxycoumarin and isopimpinellin (5,8-dimethoxypsoralen) were shown to be effective antifungal compounds (Rodov et al., 1995). 7-Geranoxycoumarin exhibited inhibitory activity against *P. italicum in vitro,* which was comparable to other naturally occurring compounds like scoparone and scopoletin. Tests *in vivo* indicated that 7-geranoxycoumarin has antifungal activity against *P. italicum* comparable to that of scoparone and scopoletin, only when applied on grapefruit at high concentration (500 mg l^{-1}). However, its activity was much higher against *P. digitatum* (Angioni et al., 1998). Scoparone and scopoletin applications are reported as phytotoxic, causing browning and necrosis to the rind also at very low concentration (50 mg l^{-1}). In contrast, 7-geranoxycoumarin had no adverse effects to the citrus peel when applied at a higher concentration (Rodov et al., 1995; Angioni et al., 1998). For the citrus industry all of these compounds have potential as an alternative means to control postharvest decay, since it has been demonstrated that they can be induced by a number of physical and biological treatments (Ben-Yehoshua et al., 1988; Kim et al., 1991; Wilson et al., 1994).

In conclusion, despite the large number of evidences about the antifungal activity of numerous phenolic compounds, not much data exists regarding their practical application as alternative compounds in controlling postharvest diseases of fresh fruits and vegetables, and in prolonging vase life of cut flowers.

Volatile Compounds

The use of natural volatile aroma compounds as antimicrobial fumigants is an interesting field of investigation still not fully explored. In general, these substances have limited toxicity to mammals and a degree of volatility that allows their application in fumigation of cold storage rooms or in 'active packaging' (Toray Research Center, 1991). Volatile compounds from plants, defined as those compounds with a relatively high vapour pressure, capable of approaching an organism through the liquid and the gas phases, can be either inhibitory or stimulatory to fungal growth and/or spore formation and germination (Fries, 1973). Acetaldehyde, a volatile compound accumulating in fruit during ripening, exhibits a fungicidal effect against various postharvest pathogens. Among nine low molecular weight aliphatic aldehydes produced by sweet cherries, acetaldehyde, together with propanal and butanal, significantly reduced decay of *P. expansum*-inoculated sweet cherries (Mattheis and Roberts, 1993), with acetaldehyde being the most active. Resistance of strawberry fruit to rot in high-CO_2 storage conditions has been ascribed to the production of high levels of acetaldehyde and ethyl acetate by the fruit in response to these conditions. Fumigation with acetaldehyde at 0.1–1% resulted in inhibition of spore germination and reduced mould development on strawberries (Prasad and Stadelbacher, 1974; Pesis and Avissar, 1990), raspberries (Prasad and Stadelbacher, 1973), apples (Stadelbacher and Prasad, 1974), grapes (Avissar and Pesis, 1991) and sweet cherries (Mattheis and Roberts, 1993). However, in some cases acetaldehyde induced phytotoxicity symptoms (Stadelbacher and Prasad, 1974; Stewart et al., 1980; Avissar and Pesis, 1991; Perata and Alpi, 1991; Mattheis and Roberts, 1993) and altered fruit sensory traits (Pesis and Avissar, 1990; Avissar and Pesis, 1991), depending on concentration and exposure duration. Although the mode of action of acetaldehyde has not been fully elucidated, there is evidence that it causes membrane disruption followed by leakage of electrolytes, reducing sugars and amino acids from the cells (Avissar et al., 1990). In spite of the large number of trials carried out against postharvest pathogens and diseases, no commercial application of acetaldehyde is reported in literature and no recent applications have been tested.

Hinokitiol is a natural volatile oil extracted from the root and bark of some Cupressaceae, such as the Hinoki tree (*Hiba arborvitae*), known for their high degree of resistance against wood decay. Hinokitiol showed antimicrobial properties at low dosage and had a wide antimicrobial spectrum against bacteria and fungi. At a dosage of 15-30 $\mu g\ ml^{-1}$ it was effective in reducing by 50% the spore germination of *M. fructicola, Rhizopus oryzae* and *B. cinerea* and prevented decay of harvested peaches (Sholberg and Shimazu, 1991). In *in vivo* trials, hinokitiol at 750 $\mu l\ l^{-1}$ functioned better than prochloraz in suppressing *B. cinerea* and *A. alternata* on eggplants and peppers (Fallik and Grinberg, 1992). Notwithstanding these interesting results and a large number of experiments conducted in the pharmaceutical field, no other tests have been carried out against pre- or postharvest pathogens. Therefore, hinokitiol, as well as a similar substance (Yoshixol), hypothesized as the main compound from neutral wood oil of *Chamaecyparis obtusa* (Kiso-Hinoki, known as the traditional Japanese tree) (Koyama et al., 1997), deserve further investigations. Tests against postharvest diseases of cut flowers should also be included.

Volatiles from 'Isabella' (*Vitis labrusca* L.) grapes revealed a strong inhibitory action on the sporulation and on the sclerotia formation of *B. cinerea,* and significantly limited the incidence of infection on 'Roditis' grapes and 'Haiward' kiwifruits by reducing both the inoculum production and the activity of the pathogen (Kulakiotu et al., 2004a, b). Mycelium of *B. cinerea* grown in presence of volatiles showed endolysis, deformation of the cell wall, and formation of chlamydospores. Studies are in progress to identify the active component(s) of the 'Isabella' volatiles involved in the antifungal activity (Kulakiotu et al., 2004b). *In vitro* trials using 16 volatile compounds from peach and plum demonstrated high efficacy of ethyl benzoate, methyl salicylate and benzaldehyde in inhibiting the growth of *B. cinerea* and *M. fructicola.* Benzaldehyde totally inhibited spore germination of *B. cinerea* at 25 $\mu l\ l^{-1}$ and germination of *M. fructicola* at 125 $\mu l\ l^{-1}$ (Wilson et al., 1987). Tonini and Caccioni (1990) reported similar results for stone fruit. Caccioni et al. (1995b) reported that among eight volatile compounds forming the characteristic aroma of ripe stone fruit, benzaldehyde was, at 5000 $\mu l\ l^{-1}$, one of the most active in reducing decay by *M. laxa* and *R. stolonifer* in inoculated peaches, nectarines, and plums. Archbold et al. (1997) have reported that hexanal, 1-hexanol, (*E*)-2-hexenal, (*Z*)-6-nonenal and (*E*)-3-nonen-2-one, and the aromatic compounds methyl salicylate and methyl benzoate had potential as postharvest fumigants for the control of *B. cinerea* on strawberries, blackberries, and grapes at concentrations as low as 2–12 µl. 250 ml^{-1}. Later, the same authors showed that one of the compounds, (*E*)-2-hexenal, was effective against grey mould on seedless table grapes, but complete mould suppression was not achieved as the level of hexenal declined during the course of the trial (Archbold et al., 1999). Headspace analyses have shown that (*E*)-2-hexenal concentrations below 0.5 $\mu mol\ l^{-1}$ stimulated *B. cinerea* mycelial development *in vitro,* while higher concentrations inhibited growth of the mould (Fallik et al., 1998). (*E*)-2-hexenal is the major volatile compound (accounted for 70-74% of the total peak areas) in the extract of *Poligonum cupsidatum* S. et Z. leaves, a plant commonly used in human medicine for centuries, which is very effective against bacteria, but still not tested against plant pathogenic fungi (Kim et al., 2005). The antifungal activity of hexanal, whose effectiveness seems related to its vapour pressure (Gardini et al., 1997), has been studied on several host-pathogen combinations (Nandi and Fries, 1976; Hamilton-Kemp et al., 1992; Caccioni et al., 1997) and it proved to be active on stone fruit against *R. stolonifer* and *M. laxa*. At 2.5 ml l^{-1}, hexanal produced the same fungistatic effect as 5-10 ml l^{-1} of benzaldehyde, but at higher concentrations it was phytotoxic (Caccioni et al., 1995b). Methyl-salicylate demonstrated some fungistatic activity, but it also gave an unpleasant odour to the fruit (Caccioni et al., 1995b). Recently, the effect of *trans*-2-hexenal vapour has been tested on *P. expansum* control, patulin content and fruit quality in several cultivars of pears (Neri et al., 2006). The best blue mould control (ranging from 50 to 98%, depending on cultivar), and the highest reduction of patulin content were achieved when the fruits were exposed to vapour (12.5 μl l^{-1}) 24 hours after inoculation of the pathogen. Surprisingly treatments applied two hours after inoculation exhibited less efficacy or stimulated *P. expansum* infections. The cause of this stimulation is unknown. Based on findings from other studies, the authors hypothesized that some volatiles produced by fruits in the first hours after *trans*-2-hexenal treatment were able to stimulate pathogen infections (Neri et al., 2006).

Acetic acid, the principal organic compound of vinegar, could also be used in the vapour phase to control postharvest pathogens. Early studies demonstrated its efficacy against conidia of *M. fructicola* on peaches (Roberts and Dunegan, 1932). As little as 1.4 mg l^{-1} of acetic acid vapour prevented decay of peaches inoculated with conidia of *M. fructicola* or *R. stolonifer*. Fumigation with 2.0 or 4.0 mg l^{-1} of acetic acid before wounding prevented decay in apples contaminated with *B. cinerea* or *P. expansum*, respectively (Sholberg and Gaunce, 1995, 1996). Acetic acid was lethal at 0.1 and 0.15% to *B. cinerea* and *P. expansum*, respectively, while 0.7 and 2% acetaldehyde was required to achieve the same effect (Avissar et al., 1990; Stadelbacher and Prasad, 1974). At concentrations of 0.18–0.27% (vol/vol), acetic acid controlled Botrytis and Penicillium decay on two Canadian table grape varieties, to the same extent as SO_2, with no adverse effects on fruit composition (Sholberg et al., 1996). However, acetic acid, like other short-chain organic acids, can be extremely phytotoxic in the vapour form, depending on temperature, concentration and exposure (Sholberg, 1998). After repeated trials with acetic acid vapour on d'Anjou pears, Sholberg et al. (2004) suggested that to effectively reduce stem infection and fruit rot without phytotoxic effect, fumigation needs to be done as soon as possible after harvest, at a rate not over 200 μl l^{-1} and at a temperature of 1°C. A phytotoxic effect on stone fruit, strawberries, and apples has also been avoided using heat-vaporized vinegar, but the volume needed to reduce decay in these fruits was high

(36.6 μl l^{-1} of air) (Sholberg et al., 2000). Since vinegar and its active component, acetic acid, do not penetrate into the fruits, they do not control latent or quiescent infection (Sholberg and Gaunce, 1996). Although informal tasting of the treated fruit has not identified any off-odours, rigorous sensory evaluation of acetic acid-treated fruit has not been undertaken to date (Sholberg et al., 1996, 2000, 2004).

Aqueous ethanol has been extensively used as a dip treatment to control brown rot and Rhizopus rot of peaches (Ogawa and Lyda, 1960; Feliciano et al., 1992) and table grapes (Lichter et al., 2002), or in combination with hot water or other substances to control postharvest decay of table grapes, lemons, and stone fruit (Smilanick et al., 1995; Margosan et al., 1997; Karabulut et al., 2004a, b; Karabulut et al., 2005) with varying degrees of success. To date, the use of vapour of ethanol to control postharvest decay has not been investigated extensively as aqueous solution. Ethanol vapour inhibited decay of oranges by *P. italicum* and *P. digitatum* after five days of continuous exposure (Yuen et al., 1995). Recently, a renewed interest for this treatment lead to optimized applications for preventing *B. cinerea* rot and stem browning in 'Chasselas' table grapes (Chervin et al., 2005). At a dose rate of 2 ml kg^{-1} of grapes, ethanol vapour was as effective as sulphur dioxide pads, with insignificant differences in sensory perception between controls and treated grapes. No other uses of ethanol vapour have been reported on fruits, vegetables and cut flowers, probably because of concern about its inflammability and explosive potential under high pressure.

COMPOUNDS OF MICROBIAL AND ANIMAL ORIGIN

The number of compounds of microbial or animal origin utilized to control postharvest decays is very limited in comparison with compounds of plant origin. Antibiotics secreted by antagonistic bacteria are also among the natural compounds which may suppress postharvest pathogen development. Iturin, an antibiotic produced by several strains of *Bacillus subtilis*, has been effective in controlling brown rot of peaches (Gueldner et al., 1988). Similarly, pyrrolnitrin, purified from a strain of *Pseudomonas cepacia*, has provided effective control of grey mould on raspberry (Goulart et al., 1992), blue mould and grey mould on apples and pears (Janisiewicz et al., 1991), and has delayed rot of strawberries (Takeda et al., 1990). However, the potential for development of antibiotic resistance in humans precludes a more widespread use of these compounds. Six-pentyl-2-pyrone, a secondary metabolite of *Trichoderma* fungi, applied topically at 0.4–0.8 mg $fruit^{-1}$ controlled *B. cinerea* rot on kiwifruit

(Poole and Whitmore, 1997). This compound occurs naturally in ripe peaches and nectarines (Horvat et al., 1990) and is an approved food flavouring additive (Oser et al., 1984). Fusapyrone and deoxyfusapirone, two α-pyrones originally isolated from cultures of *Fusarium semitectum*, but also produced by *Alternaria, Aspergillus, Penicillium*, and *Trichoderma*, inhibited the growth of postharvest pathogens as well as mycotoxigenic and human filamentous fungi *in vitro* (Altomare et al., 2000). Another compound that may have potential as an antimicrobial for fresh fruits and vegetables is Ascopyrone P (APP) produced by the fungi *Anthracobia melanoma, Plicaria anthracina, Plicaria leiocarpa* and *Peziza petersi*. APP was shown to inhibit bacteria but not yeasts (Thomas et al., 2004), therefore, its use can be considered in combination with antagonistic yeasts in biological control of postharvest diseases. Many other antifungal compounds from fungi reviewed by Ng (2004) can be of interest in combatting postharvest diseases.

Xanthan gum is a high molecular weight polysaccharide produced industrially by fermentation of *Xanthomonas campestris* (Kennedy and Bradshaw, 1984). It is a GRAS compound commonly used as a stabilizer and thickener, suspending agent, bodying agent or foam enhancer in foods, and has been recommended as an adjuvant in a commercial wax-based formulation used for control of citrus fruit rot (Cohen and Nussinovitch, 2000). Xanthan gum (XG) reduced the sour rot of grapes when applied prior to harvest at a concentration of 0.05% (w/v) (Ippolito et al., 1998). Against sour rot, a disease even more severe than grey mould in southern Italy, XG was more effective than the antagonist *A. pullulans*, (10^7 cells ml^{-1}), $CaCl_2$ (1% w/v), and the fungicide procymidone at 10 g l^{-1}. Since no synthetic fungicide is currently available to combat sour rot, XG could be an interesting compound to be validated in large-scale trials. The mode of action of XG has not been elucidated, but it is conceivable that the nature of the polysaccharide and its coating properties are involved.

Among the animal-derived compounds, chitosan (poly-N-acetylglucosamine), a biodegradable polymer made commercially by alkaline deacetylation of chitin, has gained particular interest for controlling postharvest diseases. Chitin is an abundant constituent of crustacean shells (e.g., shrimp and crabshell) and fungi (e.g., *A. niger, Mucor rouxii* and *Penicillium notatum*). Chitosan has proved to be useful in many different areas, like flocculating agent in wastewater treatment, additive in food industry, hydrating agent in cosmetics, pharmaceutical agent in biomedicine, and, more recently, in agriculture as edible coating and natural antimicrobial compound capable of inducing plant defence responses (Muzzarelli and Muzzarelli, 2001; Rabea et al., 2003). Due to its film-forming ability, chitosan acts as a barrier against gas diffusion and delays ripening. Tomatoes, bell peppers, cucumbers, strawberries, lettuce and peaches coated with chitosan had reduced weight loss and respiration rates, improved appearance and extended shelf life (El Ghaouth et al., 1992a; Li and Yu, 2001; Devlieghere et al., 2004). The polymer was effective in reducing decay of other fruits including table grapes, mangoes, sweet cherries, papayas, oranges and carrots (El Ghaouth, 1994; Bégin et al., 2001; Romanazzi et al., 2001; Bautista-Baños et al., 2003b; Galed et al., 2004; Molly et al., 2004). The range of chitosan applications generally varied from 0.1 to 2%; the higher the concentration, the greater the control of various diseases. In the above examples, chitosan has been applied as a postharvest treatment, however, the few examples of its application in preharvest treatment of strawberries, sweet cherries and table grapes to control postharvest decay have also been very promising (Romanazzi et al., 1999b, 2000a, b, 2001, 2002; Reddy et al., 2000). Preharvest spray of chitosan to reduce storage decay of table grapes seems to be the best way to apply this compound because postharvest liquid treatments could damage the bloom (Ippolito and Nigro, 2000). The only effect on grape berry appearance was a slight gloss, using the highest concentration of chitosan (1%), while no effects were visible at lower dosage (0.5–0.1%), which were as effective as the higher ones (Romanazzi et al., 2002). Preharvest application of chitosan is also advisable against Botrytis rot on strawberries. Field application of the polymer during flowering can avoid infection via senescent floral parts that later develop into active rots on ripe fruit (Romanazzi et al., 2000a). On strawberries, preharvest application of glycolchitosan, a water-soluble form of chitosan, gave similar results to chitosan (Romanazzi et al., 1999a). Enzymatically hydrolyzed chitosan showed a greater effect than the high molecular weight chitosan on carrots (Molly et al., 2004). Among 15 acids tested to dissolve chitosan, 10 were found to be effective. However, the efficacy of the dissolved polymer against grey mould of table grapes was different. Chitosan acetate was the most active (70% of reduction) followed by chitosan formate, whereas chitosan dissolved in malic and maleic acid was the least effective (Romanazzi et al., 2005). In *in vitro* tests, chitosan-Zn complexes showed up to 8 and 16 times higher antimicrobial activity than those of chitosan and zinc sulphate, respectively (Wang et al., 2004). Compounds based on chitosan have never been reported to induce symptoms of phytotoxicity on treated fruits and vegetables.

The mode of antifungal action of chitosan and its derivatives is still not well understood, but different mechanisms have been proposed. In *B. cinerea*, *R. stolonifer* and *F. oxysporum*, chitosan caused cellular leakage and morphological alterations consisting of excessive branching and cell wall damage (Benhamou, 1992; El Ghaouth et al., 1992b). In tissues of chitosan-treated bell peppers, *B. cinerea* hyphae displayed various levels of cellular disorganization, ranging from wall loosening to cytoplasm disintegration (El Ghaouth et al., 1997). Leakage of proteinaceous and other intracellular constituents has been ascribed to interaction between positively charged chitosan molecules and negatively charged microbial cell membranes (Rabea et al., 2003), presumably mediated by chitosan action on chitin deacetylase (El Ghaouth et al., 1992c).

The eliciting property of chitosan has been demonstrated in several postharvest commodities. Induction of antifungal hydrolases, such as β-1, 3-glucanase, chitinase and chitosanase, has been observed in strawberries, tomatoes and bell peppers (Wilson et al., 1994). In tomato and bell pepper, the activity of these enzymes remained high up to 14 days after treatment. On table grapes, chitosan enhanced the activity of phenylalanine ammonia-lyases (PAL) (Romanazzi et al., 2002). Moreover, it elicited phytoalexin formation in carrot roots, thus restricting *B. cinerea* infection (Reddy et al., 1999). Trullo et al. (2007) demonstrated that chitosan is able to induce resistance in orange tissues although its action is strongly influenced by the level of ripening. The expression of chitinase and β-1,3-glucanase evaluated by real-time PCR in oranges treated with chitosan at colour change was up-regulated by 1790- and 5.4-fold, respectively. As opposed, in full ripe fruits chitinase expression was up-regulated by 259-fold and β-1,3-glucanase expression remained basically unchanged. The induction of lytic enzymes, PAL and phytoalexins in harvested tissues by prestorage treatment with chitosan could supply the tissue with weapons capable of restricting fungal colonization. This could be important in retarding the resumption of quiescent and latent infections, which typically become active when tissue resistance declines. Chitosan treatment also stimulates various structural defence barriers in host tissues, such as thickening of the cell wall, formation of papillae, and deposition of electron opaque materials in the intercellular spaces, presumably, antifungal phenolic-like compounds (Wilson et al., 1994). Chitosan applied as pre- and postharvest treatments on table grapes did not impair the naturally occurring yeasts and yeast-like fungi, among which antagonistic microorganisms are common (Ippolito et al., 1997). This is an important feature because endogenous microflora on fruit surfaces may play an important role in antagonism to pathogens (Wilson and Wisniewski, 1994) and the treatment of fruits and vegetables should avoid any negative effect on the naturally occurring microflora (Nigro et al., 1998, 2000). On the contrary, chitosan treatment reduced the propagules of filamentous fungi naturally occurring on berries (Romanazzi et al., 2002). As naturally occurring filamentous fungi on table grapes include decay-causing species such as *B. cinerea, Penicillium* spp., *Aspergillus* spp. and *Cladosporium* spp., it has been hypothesized that their reduction can result in lowering rot incidence during storage (Ippolito et al., 1998).

Although some commercial formulations are available in the market for controlling foliar and soilborne pathogens, practical applications of chitosan on fruits, vegetables and çut flowers to control postharvest pathogens are not reported to date in the literature.

COMBINED TREATMENTS

Unlike the control of tree, field crop or soilborne diseases, the commercial control of postharvest diseases of fruits and vegetables must be extremely efficient (in the range of 95–98%) (Droby, 2001). None of the natural antimicrobial systems investigated to date offers regular postharvest disease control comparable to that obtained with synthetic fungicides. Attempts to overcome the variable performance and enhance the efficacy of natural compounds have led to the development of combined approaches based on additive and synergistic effects. While many synergistic combinations of antimicrobials have been identified *in vitro*, relatively few investigations of multifactorial systems have focused on organisms or conditions relevant to postharvest storage of fruits and vegetables.

Synergistic effects have been reported for a mixture of *Allium sativum* (0.25%) and *A. cepa* (0.75%) bulb extracts in inhibiting the *in vitro* growth of *Alternaria* spp. (Bokhary, 1985). Similarly, mixtures of *Allium* plant extracts and acetic acid were more active against *A. flavus*, *A. niger* and *A. fumigatus* than single treatments, especially when further combined with high temperature (Yin and Tsao, 1999). In the case of citrus, a remarkable increase in the activity of garlic extracts was observed when extracts were mixed with vegetable (sunflower) cooking oil in controlling *Penicillium* spp. (Obagwu and Korsten, 2003).

Generally, the fungicidal activity of EOs observed *in vitro* was not reproduced *in vivo* because of the volatile nature of the constituents. Therefore, an interesting approach was to combine subatmospheric pressure with essential oils (Arras and Usai, 2001). The fungitoxic activity of *T. capitatus* EOs (75, 150 and 250 μg ml^{-1}) on oranges inoculated with *P. digitatum* was weak at atmospheric pressure (3 to 10% inhibition), while under vacuum (0.5 bar), conidial mortality on the exocarp reached 90–97% (Arras and Usai, 2001). Under subatmospheric conditions (0.2–0.8 atm) two phytoalexins (scoparone and scopoletin) were elicited on orange and mandarin fruit. The biosynthesis of these compounds was also stimulated in fruits treated with thyme oil vapour at a concentration of 50–100 μg ml^{-1}. The simultaneous use of thyme oil and hypobaric pressure on citrus fruits had a synergistic effect eliciting five times as much scoparone (Arras, 1999). This was attributed to increased contact between the EO, the pathogen's conidia and the host tissue. A modified active packaging (MAP) to preserve the quality and safety of table grapes (cv. Crimson Seedless) was developed by Valverde et al. (2005) by adding 0.5 ml of eugenol, thymol, or menthol inside the packages. Addition of these natural antimicrobials improved the beneficial effect of MAP by delaying rachis deterioration, reducing weight loss and colour changes, retarding °Brix/acidity ratio evolution and berry decay, and maintaining firmness. Moreover, the total viable counts of yeast and moulds were significantly reduced in the grapes packaged with the natural antimicrobial compounds (Valverde et al., 2005). In *in vitro* trials, the activity of thymol and carvacrol against mycelial growth of *A. flavus*, a pathogen producing dangerous mycotoxins in food, was enhanced when applied in combination with high a_w (0.99) and low pH (4.5) (López-Malo et al., 2005).

A synergistic effect was observed in sweet cherries under subatmospheric pressure in combination with chitosan. The extent of decay inhibition was, on an average, 20% with subatmospheric pressure alone, 65% with chitosan treatment alone, and 83–89% when both treatments were applied (Romanazzi et al., 2003). Equipment for rapid vacuum cooling of fruits and vegetables is already in use in some packing houses. Therefore, it might be feasible to add some of the natural antifungal compounds reported above to improve the effectiveness of the treatment. Another synergistic effect was obtained on grape berries combining preharvest application of chitosan and postharvest treatments with UV-C for 5 minutes, resulting in reduction of grey mould incidence and increasing *trans*-resveratrol and catechin in two table grape cultivars (Romanazzi et al., 2006). Another promising combination has been reported by Palhano et al. (2004) in which the activity of lemongrass EO against *C. gloeosporioides* spores, significantly increased when applied in combination with a high hydrostatic pressure (150 MPa). The authors suggested that the enhanced effect could be explained by a higher uptake of oil constituents into the spore due to the high pressure, leading to an increase in the number of molecular targets affected. A combination of carvacrol and thymol at low concentration provided as high an inhibition as oregano EOs with a smaller flavour impact, an issue that can limit the application of this EOs in foods (Lambert et al., 2001). 'Bio-Coat', a biocontrol product under commercial development, is a preparation consisting of a water-soluble form of chitosan (glycolchitosan) and the antagonistic yeast *Candida saitoana*. This product was superior to the yeast and glycolchitosan alone in controlling decay of several varieties of citrus fruits (sweet orange and lemons), and apples, and the control level was comparable to that achieved with imazalil or thiabendazole (El Ghaouth et al., 2000). Interesting results were also obtained with preharvest application of Bio-Coat on table grapes (Schena et al., 2004). XG has also been evaluated in combination with *A. pullulans* to control postharvest table grape and strawberry rots. On both commodities, the activity of the antagonist was significantly improved when applied in combination with the polysaccharide at 0.5% (w/v) (Ippolito et al., 1997, 1998). The higher activity of *A. pullulans* combined with XG on table grapes and strawberries, has been related to its greater survival on the fruit surface, probably because of its coating properties. Another natural gum, locust bean gum, extracted from the seed of the carob tree (*Ceratonia siliqua* L.) and applied in combination with antagonistic yeasts, improved their activity with results comparable to the antagonists applied alone, but at 100-fold higher concentration (Lima et al., 2005).

Ethanol applied at concentrations between 8 and 20% and with several strains of *Saccharomyces cerevisiae* (10^8 CFU ml^{-1}), was not effective in reducing grey mould incidence and severity on apples and pears. However, a combination of ethanol and yeasts reduced disease incidence by over 90% (Mari and Carati, 1997). Similar results were obtained against green mould on lemon, by combining a heated solution of ethanol (10%, 45°C) with curing, a physical treatment consisting of keeping fruit at relatively high temperature and humidity (e.g., 32°C, 95–98% RH), and combining ethanol with the yeast *Candida oleophila*. Infections were reduced from 82% (control), 17% (ethanol alone) and 40% (yeast alone) to 3.5% (ethanol-curing) and 3.3% (ethanol-yeast), with no appreciable differences compared with the fungicide imazalil (Lanza et al., 1997). Addition of 0.5 and 1.0% potassium sorbate to 10 and 20% ethanol reduced decay to 10% or less, and was more effective than either potassium sorbate or ethanol alone. After 30 days of storage at 1°C, the combination of 20% ethanol either with 0.5 or 1.0% potassium sorbate was equal in efficacy to commercial SO_2 generator pads in reducing the incidence of grey mould on 'Thompson Seedless' grapes. None of the combinations of ethanol and potassium sorbate injured the berries (Karabulut et al., 2005). The above examples clearly show the advantages in using combination strategies to control postharvest diseases of fresh fruits and vegetables. Many other possible combinations could be explored, such as the use of antagonistic microorganisms, low dosage of fungicides, physical means (UV radiation and modified atmospheres), organic and inorganic salts, nutrients, mixtures of natural substances with different modes of action, etc. The complexity of the mode of action that combined alternatives can produce, should also make the development of pathogen resistance more difficult.

COMMERCIAL DEVELOPMENTS

Chemical fungicides have been used to reduce storage losses for a long time and in many situations, but even with a substantial increase in chemical use, the overall proportion of fruit and vegetable losses and their absolute value appear to have increased over time. Despite this perverse relationship, an increase in fungicides use still appears to be profitable for the chemical industry. Some of these fungicides are persistent enough to be detected after several weeks in fruits, vegetables and the soil. Moreover, as reported in the introduction, inappropriate and excessive fungicide use has increased the development of multi-resistant fungal strains, thus requiring higher dosage to protect products. Progressively, over the 1990s, environmental and health impact issues, both due to direct and indirect impact of chemicals, became increasingly important, resulting in pressure for reduced pesticide use and the abolition of previously useful chemicals. In spite of widespread public concern about the negative effects of synthetic fungicides, especially in Europe and North America, natural fungicides from plants, animals, and microorganisms at present have a limited impact in the marketplace.

The success of large-scale studies on microbial antagonists generated interest by several agrochemical companies, and currently some antagonistic microorganisms to control postharvest diseases of fresh fruits and vegetables are commercially available (http://www.oardc.ohio-state.edu/apsbcc/productlist.htm, 2007). Considering natural insecticides, there is a very active field in developing and applying alternative control method, i.e., *Bacillus thuringiensis*- and pyrethrum-based products, which command 1-2% of the global insecticide market (Isman, 2000). Although the commercial formulation of natural compounds to fight postharvest diseases is still in an initial phase of development, it seems to have potential for market expansion. An additional feature favouring commercial development of several natural compounds with antioxidant properties (essential oil, flavonoids, and other phenolic compounds) is their potential efficacy against the toxic effects of mycotoxins (Atroshi et al., 2002; López-Malo et al., 2005). For certain substances the take-up by industries should be relatively easy, as for those already tested in other systems, especially human medicine and food industry (http://www.vm.cfsam.fda.gov/~dms/eafus.html, 2005). For example, chitosan, owing to its lipid-binding capacity and hypocholesterolemic action (Maezaki et al., 1993), is a safe compound used in human medicine for slimming diets. Similarly, a wide variety of plant extracts, essential oils and plant compounds, are available without prescription through health food stores, herbalists and vitamin retailers, and are commonly used as culinary herbs and spices. XG is a compound commonly used in food, cosmetic, and pharmaceutical industries. It is worth mentioning that in the case of essential oils, since they have a long history of global use by the food and fragrance industries and, recently, in the field of aromatherapy, they are readily available at low-to-moderate cost. Moreover, in a few countries some of these compounds are exempt from the usual data requirements for registration. American companies taking advantage of this situation have been able to bring essential-oil-based pesticides to the market in a far shorter time period than would normally be required for a conventional pesticide. This is the case of Cinnamite™ and Valero™, two miticide/fungicide for glasshouse and horticultural crops, based on cinnamon oil with cinnamaldehyde as the active ingredient (Isman, 2000). A promising development of interest for the plastic films industry could be the incorporation of natural antimicrobials into packaging materials. This will concentrate the antimicrobials in the confined atmosphere around the products and on their surface, which is where most of the noxious organisms grow. Another useful development could be the incorporation of natural substances in edible coatings, as demonstrated by Pranoto et al. (2005), who combined alginate-based edible film with garlic oil at 0.4% (v/v).

Possible barriers to the commercial development of natural compounds as antimicrobials for controlling postharvest diseases are:

(i) Expensive procedures for extracting the active compounds.
(ii) The need for chemical standardization, stability, and quality control.
(iii) Extended studies on toxicological aspects for specific compounds.
(iv) Production at competitive costs with existing pesticides.
(v) The lack of studies on development of resistance.
(vi) Difficulties in registration as pesticides.
(vii) Efficacy sometimes not consistent and acceptable unless in complex integrated approaches.
(viii) Restricted market confined to postharvest environment.
(ix) Limited interest by companies in testing for botanicals, because it is still unclear whether proprietary claims can be made.

The above reported issues should not deter the search for effective antimicrobials of plant, animal and microbial origin. Current 'economic' and 'biological' assessment, upon withdrawal of most of the conventional pesticides registered for postharvest disease control, may change radically, providing new opportunities for the development and commercialization of future pesticides based on natural substances. In this picture, the possibility of using marker-assisted selection or genetic engineering to select or introduce resistant genes from other plant species cannot be discarded.

CONCLUSION AND FUTURE PROSPECTS

What is needed in applying natural antimicrobials should be a change in philosophy of companies and growers, still rooted to the concept that a "stand-alone" treatment has to control a disease completely. In addition, consumers also have to consider the benefits derived from the intake of sound products without residues and with improved quality.

A more sustainable approach against postharvest diseases needs to be based on the use of multifaceted control strategies, including the use of natural compounds, microbial antagonists, physical methods, induced and genetic resistance, low dosage of fungicide, etc. Some evidence reported in this chapter demonstrates the effectiveness of the integrated approach to reach a level of efficacy comparable to that provided by synthetic fungicides.

It has also been demonstrated that some compounds such as isothiocyanates, hold an interesting curative effect against postharvest diseases. The ability to control previously established infections in the postharvest environment is of crucial importance, bearing in mind that under commercial conditions the application of postharvest treatment may be delayed for hours or even days after harvest, allowing the pathogen to penetrate the flesh where control becomes very difficult. Therefore, compounds able to control incipient, latent and quiescent infections should be preferred. As an alternative, for compounds with no curative activity, applications before events that promote infection (harvesting and postharvest handling operations) are suggested (Ippolito and Nigro, 2000). Another interesting feature for some of these compounds is the possibility to act as vapour-phase substances, as in the case of essential oils, vinegar and allyl-isothiocyanates. Apart from the high efficacy, their use against postharvest diseases seems promising considering the capability to penetrate easily inside the mass of staked commodities in a cold room without any further manipulation. This characteristic, exploitable only in the postharvest milieu in a confined environment, seems the most appropriate treatment for those products where postharvest handling reduces their market appeal.

At present, the effectiveness of new emerging alternative biocontrol methods is still evaluated in comparison with that of synthetic pesticides. Natural antimicrobials, as other biocontrol alternatives, may require different application timings, depending on the intrinsic characteristics and the mode of action of the compound. Currently, considering that the majority of natural antimicrobials have no toxic effects on mammals and low phytotoxicity, it should be possible to apply such compounds in a wider range of concentrations and application schedules.

Although scientists have always demonstrated interest in searching for alternative methods to control postharvest diseases, considerable effort has been devoted to this field only since late 1980s. Among natural compounds, several have been tested but only a few are very promising. However, there is an inestimable number of other substances yet to be discovered in nature's store. Many substances effective in an apparently unrelated system could be a source of new compounds to be tested against postharvest diseases. In this regard it is worth mentioning that a rational way of obtaining results at a low cost/benefit ratio should be a collaborative effort among plant pathologists, physicians, chemists and companies, to develop safe commercial products. The public acceptance of these products could be anticipated because, as stated before, most of them are well-known and already used for many other purposes. The possible beneficial effect on human health, for example, in the case of phenolics, which are recognized as antioxidant and anticarcinogenic, should further increase their approval and accelerate the demand.

REFERENCES

Abdou, I.A., Abdou-Zeid, A.A., El-Sherbeeny, M.R. and Abou-El-Gheat, Z.H. (1992). Antimicrobial activity of *Allium sativum, A. cepa, Barbanus sativus, Capsicum frutescens, Eruca sativa* and *Allium kurrat* on bacteria. Qual. Plant Mater. Veget. 22: 29-35.

Abou Jawdah, Y., Soh, A. and Salameh, A. (2002). Antimycotic activity of selected plant flora in Lebanon, against phytopathogenic fungi. J. Agric. Food Chem. 50: 3208-3213.

Afek, U. and Sztejnberg, A. (1993). Temperature and gamma irradiation effects on scoparone, a citrus phytoalexin conferring resistance to *Phytophthora citrophthora*. Phytopathology 83: 753-758.

Afolayan, A.J. and Meyer, J.J.M. (1997). The antimicrobial activity of 3,5,7-trihydroxyflavone isolated from the shoots of *Helicrysum aurenitens*. J. Ethnopharm. 57: 177-181.

Ahmad, S.K. and Prasad, J.S. (1995). Efficacy of foliar extracts against pre- and postharvest diseases of sponge-gourd fruits. Lett. Appl. Microbiol. 21: 375-385.

Aligiannis, N., Kalpoutzakis, E., Chinou, I.B., Mitakou, S.S., Gikas, E. and Tsarbopoulos, A. (2001). Composition and antimicrobial activity of the essential oils of five taxa of *Sideritis* from Greece. J. Agric. Food Chem. 49: 811-815.

Almada-Ruiz, E., Martinez-Tellez, M.A., Hernandez-Alamos, M.M., Vallejo, S., Primo-Yufera, E. and Vargas-Arispuro, I. (2003). Fungicidal potential of methoxylated flavones from citrus for in vitro control of *Colletotrichum gloeosporioides*, causal agent of anthracnose disease in tropical fruits. Pest Manag. Sci. 59: 1245-1249.

Al-Mughrabi, K.I. (2003). Antimicrobial activity of extracts from leaves, stems and flowers of *Euphorbia macroclada* against plant pathogenic fungi. Phytopathol. Mediterr. 42: 245-250.

Altomare, C., Perrone, G., Zonno, M.C., Evidente, A., Pengue, R., Fanti, F. and Ponelli, L. (2000). Biological characterization of fusapyrone and deoxyfusapyrone, two active metabolites of *Fusarium semitectum*. J. Nat. Prod. 63: 1131-1135.

Ammermann, E., Lorenz, G. and Schleberger, K. (1992). A broad-spectrum fungicide with a new mode of action. Proc. Brit. Crop Prot. Conf. Pests Dis. 1: 403-410.

Andresen, I., Becker, W., Schluter, K., Burges, J., Parthier, B. and Apel, K. (1992). The identification of leaf thionin as one of the main jasmonate induced proteins in barley (*Hordeum vulgare*). Plant Mol. Biol. 19: 193-204.

Angioni, A., Cabras, P., D'hallewin, G., Pirisi, F.M., Reneiro, F. and Schirra, M. (1998). Synthesis and inhibitory activity of 7-geranoxycoumarins against *Penicillium* species in citrus fruit. Phytochemistry 47: 1521-1525.

Antoniacci, L., Cobelli, L., De Paoli, E. and Gengotti, S. (2000). Open field control trials against strawberry grey mould. Inf. Fitopatol. 50: 45-51.

Arcas, M.C., Botía, J.M., Ortuño, A.M. and Del Río, J.A. (2000). UV irradiation alters the levels of flavonoids involved in the defence mechanism of *Citrus aurantium* fruits against *Penicillium digitatum*. Eur. J. Plant Pathol. 106: 617-622.

Archbold, D.D., Hamilton Kemp, T.R., Barth, M.M. and Langlois, B.E. (1997). Identifying natural volatile compounds that control grey mould (*Botrytis cinerea*) during postharvest storage of strawberry, blackberry, and grape. J. Agr. Food. Chem. 45: 4032-4037.

Archbold, D.D., Hamilton-Kemp, T.R., Clements, A.M. and Collins, R.W. (1999). Fumigating 'Crimson seedless' table grapes with (*E*)-2-hexenal reduces mould during longterm postharvest storage. Hort. Sci. 34: 705-707.

Ark, P.A. and Thompson, J.P. (1959). Control of certain diseases of plants with antibiotics from garlic (*Allium sativum* L.). Plant Dis. Rep. 43: 276-282.

Arras, G. (1988). Antimicrobial activity of various essential oils against citrus fruit disease agents. Proc. 6th Inter. Citrus Congr. Balaban Publishers, Philadelphia, USA, pp. 787-793.

Arras, G. and Grella, G.E. (1992). Wild Thyme, *Thymus capitatus*, essential oil seasonal changes and antimycotic activity. J. Hort. Sci. 67: 197-202.

Arras, G., Agabbio, M., Piga, A. and D'hallewin, G. (1995). Fungicide effect of volatile compounds of *Thymus capitatus* essential oil. Acta Hortic. 379: 593-600.

Arras, G. (1999). Postharvest response of citrus fruit diseases to natural compounds. In: Advances in Postharvest Diseases, Disorders and Control of Citrus Fruits (ed.) M. Schirra, Pandalai Publisher, Trivandrum, India, pp. 123-131.

Arras, G. and Usai, M. (2001). Fungitoxic activity of twelve essential oils against four postharvest citrus pathogens: chemical analysis of *Thymus capitatus* (L.) Hofmgg oil and its effect in subatmospheric pressure conditions. J. Food Protec. 64: 1025-1029.

Atroshi, F., Rizzo, A., Westermarck, T. and Ali-Vehmas, T. (2002). Antioxidant nutrients and mycotoxins. Toxicology 180: 151-167.

Avissar, I., Droby, S. and Pesis, E. (1990). Characterization of acetaldehyde effects on *Rhizopus stolonifer* and *Botrytis cinerea.* Ann. Appl. Biol. 116: 213-220.

Avissar, I. and Pesis, E. (1991). The control of postharvest decay in table grape using acetaldehyde vapors. Ann. Appl. Biol. 118: 229-237.

Bankova, V., Popova, M., Bogdanov, S. and Sabatini, A.G. (2002). Chemical composition of European propolis: Expected and unexpected results. Z. Naturforsch. 57: 530-533.

Bankova, V. (2005). Recent trends and important developments in propolis research. Evid. Based Complement Alternat. Med. 2: 29-32.

Barber, M.S., Mcconnell, V.S. and Decaux, B.S. (2000). Antimicrobial intermediates of the general phenylpropanoid and lignin specific pathways. Phytochemistry 54: 53-56.

Baudoin, A.B.A.M. and Eckert, J.W. (1985). Development of resistance against *Geotrichum candidum* in lemon peel injuries. Phytopathology 75: 174-179.

Bautista-Baños, S., Hernandez-Lopez, M., Diaz-Perez, J.C. and Cano-Ocha, C.F. (2000). Evaluation of the fungicidal properties of plant extracts to reduce *Rhizopus stolonifer* of 'ciruela' fruit (*Spondias purpurea* L.) during storage. Postharvest Biol. Technol. 20: 99-106.

Bautista-Baños, S., Garcia-Dominguez, E., Barrera-Necha, L.L., Reyes-Chilpa, R. and Wilson, C.L. (2003a). Seasonal evaluation of the postharvest fungicidal activity of powders and extracts of huamuchil (*Pithecellobium dulce*): Action against *Botrytis cinerea, Penicillium digitatum* and *Rhizopus stolonifer* of strawberry fruit. Postharvest Biol. Technol. 29: 81-92.

Bautista-Baños, S., Hernandez-Lopez, M., Bosquez-Molina, E. and Wilson, C.L. (2003b). Effects of chitosan and plant extracts on growth of *Colletotrichum gloeosporioides*, anthracnose levels and quality of papaya fruit. Crop Prot. 22: 1087-1092.

Bégin, A., Dupuis, I., Dufaux, M. and Leroux, G. (2001). Use of chitosan to control growth of *Colletotrichum gloeosporioides in vitro* and on stored mangoes. In: Chitin Enzymology (ed.) R.A.A. Muzzarelli, Atec, Grottammare, Italy, pp. 163-170.

Benhamou, N. (1992). Ultrastructural and cytochemical aspects of chitosan on *Fusarium oxysporum* f. sp. *radici-lycopersici*, agent of tomato crown rot. Phytopathology 82: 1185-1190.

Benkeblia, N. (2004). Antimicrobial activity of essential oil extracts of various onions (*Allium cepa*) and garlic (*Allium sativum*). Lebensm.-Wiss. u.-Technol. 37: 263–268.

Beno-Moualem, D., Gusev, L., Dvir, O., Pesis, E., Meir, S. and Lichter, A. (2004). The effects of ethylene, methyl jasmonate and 1-MCP on abscission of cherry tomatoes from the bunch and expression of endo-1,4-β-glucanases. Plant Sci. 167: 499-507.

Ben-Yehoshua, S., Shapiro, B., Kim, J.J., Sharoni, J., Carmeli, S. and Kashman, Y. (1988). Resistance of citrus fruit to pathogens and its enhancement by curing. Proc. 6th Inter. Citrus Congr. Balaban Publishers, Philadelphia, USA, pp. 1371-1379.

Beuchat, L.R. (2001). Control of food borne pathogen and spoilage microrganisms by naturally occurring antimicrobials. In: Microbial Food Contamination (eds.) C.L. Wilson and S. Droby, CRC Press, Boca Raton, Florida,USA, pp. 149-169.

Bokhary, H.A. (1985). Effects of mixtures of *Allium sativum* L. and *Allium cepa* L. bulb extracts on growth of *Alternaria* spp. and bacteria. J. College Sci. 16: 87-97.

Bors, W., Heller, W., Michel, C. and Saran, M. (1990). Flavonoids as antioxidants: Determination of radical scavenging efficiencies. Methods Enzymol. 186: 343-355.

Bouchra, C., Achouri, M., Hassani, L.M.I. and Hamamouchia, M. (2003). Chemical composition and antifungal activity of essential oils of seven Moroccan Labiatae against *Botrytis cinerea* Pers: Fr. J. Ethnopharm. 89: 165-169.

Caccioni, D.R.L., Deans, S.G. and Ruberto, G. (1995a). Inhibitory effect of citrus fruit essential oil components on *Penicillium italicum* and *P. digitatum*. Petria 5: 177-182.

Caccioni, D.R.L., Tonini, G. and Guizzardi, M. (1995b). Antifungal activity of stone fruit aroma compounds against *Monilinia laxa* and *Rhizopus stolonifer*: in vivo trials. J. Plant Dis. Prot. 102: 518-525.

Caccioni, D.R.L., Gardini, F., Lanciotti, R. and Guerzoni, M.E. (1997). Antifungal activity of natural volatile compounds in relation to their vapor pressure. Sci. Alim. 17: 21-34.

Caccioni, D.R.L., Guizzardi, M., Biondi, D.M., Renda, A. and Ruberto, G. (1998). Relationship between volatile components of citrus fruit essential oils and antimicrobial action on *Penicillium digitatum* and *P. expansum*. Int. J. Food Microbiol. 43: 73-79.

Cai, Y., Luo, Q., Sun, M. and Corke, H. (2004). Antioxidant activity and phenolic compounds of 112 traditional Chinese medicinal plants associated with anticancer. Life Sci. 74: 2157-2184.

Cao, G., Sofic, E. and Prior, R.L. (1996). Antioxidant capacity of tea and common vegetables. J. Agric. Food Chem. 44: 3426-3431.

Cesarone, M.R., Laurora, G., Ricci, A., Belcaco, G. and Pomante, P. (1992). Acute effects of hydroxiethylrutosides on capillary filtration in normal volunteers, patients with various hypotension and in patients with diabetic micro angiopathy. J. Vasc. Dis. 21: 76-80.

Cevallos-Casals, B.A., Byrne, D., Okie, W R and Cisneros-Zevallos, L. (2006). Selecting new peach and plum genotypes rich in phenolic compounds and enhanced functional properties. Food Chem. 96: 273-280.

Chandrashekhar, S.Y. and Gopinath, G. (2004). Influence of chemicals and organic extracts on the postharvest behaviour of carnation cut-flowers. Karnataka J. Agric. Sci. 17: 81-85

Chao, S.C., Young, G.D. and Oberg, C.J. (2000). Screening of inhibitory activity of essential oils on selected bacteria, fungi, and viruses. J. Essent. Oil Res. 12: 639-649.

Chervin, C., Westercamp, P. and Monteils, G. (2005). Ethanol vapours limit *Botrytis* development over the postharvest life of table grapes. Postharvest Biol. Technol. 36: 319-322.

Chipley, J.R. and Uraih, N. (1980). Inhibition of *Aspergillus* growth and aflatoxin release by derivatives of benzoic acid. Appl. Environ. Microbiol. 40: 352-357.

Chu, C.L., Liu, W.T., Zhou, T. and Tsao, R. (1999). Control of postharvest grey mould rot of modified packaged sweet cherry by fumigation with thymol and acetic acid. Can. J. Plant Sci. 79: 685-689.

Cohen, S. and Nussinovitch, A. (2000). The role of xanthan gum in traditional coatings of easy peelers. Food Hydrocolloid 14: 319-326.

Cosentino, S., Tuberoso, C.I.G., Pisano, B., Satta, M., Arzedi, E. and Palmas, F. (1999). In vitro antimicrobial activity and chemical composition of *Sardinian thymus* essential oils. Lett. Appl. Microbiol. 29: 130-135.

Cowan, M.M. (1999). Plant product as antimicrobial agents. Clin. Microbiol. Rev. 12: 564-582.

Dai, G.H., Andary, C., Mondolot-Cosson, L. and Boubals, D. (1995). Histochemical studies on the interaction between three species of grapevine, *Vitis vinifera, V. rupestris* and *V. rotundifolia* and the downy mildew fungus, *Plasmopara viticola*. Physiol. Mol. Plant Pathol. 46: 177-188.

Davidson, P.M. (2001). Chemical preservatives and natural antimicrobial compounds. In: Food Microbiology. Fundamentals and Frontiers, 2nd ed. (eds.) M.P. Doyle, L.R. Beuchat, T.J. Montville, ASM Press, Washington, DC, USA, pp. 593–628.

Delaquis, P.J. and Mazza., G. (1995). Antimicrobial properties of isothiocyanates in food preservation. Food. Technol. 49: 73-84.

Devlieghere, F., Vermeulen, A. and Debevere, J. (2004). Chitosan: Antimicrobial activity, interactions with food components and applicability as a coating on fruit and vegetables. Food Microbiol. 21: 703-714.

Dixon, R.A. (2001). Natural products and plant disease resistance. Nature 411: 843-847.

Dixon, R.A., Achnine, L., Kota, P., Liu, C.J., Reddy, M.S. and Wang, L. (2002). The phenylpropanoid pathway and plant defence - a genomics perspective. Mol. Pl. Pathol. 3: 371-390.

Doares, S.H., Syrovets, T., Weiler, E.W. and Ryan, C.A. (1995). The isoflavonoid phytoalexin pathway: From enzymes to genes to transcription factors. Physiol. Plant 93: 385-392.

Doraia, T. and Aggarwal, B.B. (2004). Role of chemopreventive agents in cancer therapy. Cancer Lett. 215: 129-140.

Doughty, K.J., Blight, M.M., Bock, C.H., Fieldsend, J.K. and Pickett, J.A. (1996). Release of alkenyl isothiocyanates and other volatiles from *Brassica rapa* seedlings during infection by *Alternaria brassicae.* Phytochemistry 43: 371-374.

Droby, S., Porat, R., Cohen, L., Weiss, B., Shapiro, B., Philosoph-Hadas, S. and Meir, S. (1999). Suppressing green mould decay in grapefruit with postharvest jasmonate application. J. Am. Soc. Hort. Sci. 124: 184-188.

Droby, S. (2001). Enhancing biocontrol activity of microbial antagonists of postharvest diseases. In: Enhancing Biocontrol Agents and Handling Risks (eds.) M. Vurro, J. Gressel,

T. Butt, G. Harman, A. Pilgeram, R. St. Leger and D. Nuss, Nato Science Series, Volume 3, IOS Press, Amsterdam, The Netherlands, pp. 77-85.

Durras, A.I., Terry, L.A. and Joyce, D.C. (2005). Methyl jasmonate vapour treatment suppresses specking caused by *Botrytis cinerea* on cut *Freesia hybrida* L. flowers. Postharvest Biol. Technol. 38: 175-182.

El Ghaouth, A., Arul, J. and Asselin, A. (1992a). Potential use of chitosan in postharvest preservation of fruits and vegetables. In: Advances in Chitin and Chitosan (eds.) J.B. Brines, P.A. Sandford and J.P. Zizachis, Elsevier Applied Science, London, UK, pp. 45-52.

El Ghaouth, A., Arul, J., Asselin, A. and Benhamou, N. (1992b). Antifungal activity of chitosan on postharvest pathogen: Induction of morphological and cytochemical alteration in *Rhizopus stolonifer*. Mycol. Res. 96: 769-779.

El Ghaouth, A., Arul, J., Greiner, J. and Asselin, A. (1992c). Effect of chitosan and other polyions on chitin deacetylase in *Rhizopus stolonifer*. Exp. Mycol. 16: 173-177.

El Ghaouth, A. (1994). Manipulation of defence system with elicitors to control postharvest diseases. In: Biological Control of Postharvest Diseases, Theory and Practices (eds.) C.L. Wilson and M.E. Wisniewski, CRC Press, Boca Raton, Florida, USA, pp. 153-167.

El Ghaouth, A., Arul, J., Wilson, C. and Benhamou, N. (1997). Biochemical and cytochemical aspects of the interaction of chitosan and *Botrytis cinerea* in bell pepper fruit. Postharvest Biol. Technol. 12: 183-194.

El Ghaouth, A., Smilanick, J.L., Brown, G.E., Ippolito, A. and Wilson, C.L. (2000). Application of *Candida saitoana* and glycolchitosan for the control of postharvest diseases of apple and citrus fruit under semi-commercial condition. Plant Dis. 84: 243-248.

Elgayyar, M., Draughon, F.A., Golden, D.A. and Mount, J.R. (2001). Antimicrobial activity of essential oils from plants against selected pathogenic and saprophytic microrganisms. J. Food Prot. 64: 1019-1024.

Fahey, J.W., Zalcmann, A.T. and Talalay, P. (2001). The chemical diversity and distribution of glucosinolates and isothiocyanates among plants. Phytochemistry 56: 5-51.

Fallik, E. and Grinberg, S. (1992). Hinokitiol: A natural substance that controls postharvest diseases in eggplant and pepper fruits. Postharvest Biol. Technol. 2: 137-144.

Fallik, E., Archbold, D.D., Hamilton-Kemp, T.R., Clemens, A.M., Collins, R.W. and Barth, M.E. (1998). (*E*)-2-hexenal can stimulate *Botrytis cinerea* in vitro and on strawberry fruit in vivo during storage. J. Am. Soc. Hort. Sci. 123: 875-881.

Feliciano, A., Feliciano, A.J., Vendrusculo, J., Adaskaveg, J.E. and Ogawa, J.M. (1992). Efficacy of ethanol in postharvest benomyl-DCNA treatment for control of brown rot of peach. Plant Dis. 76: 226-229.

Ferrazzoli Deviennea, K., Gonçalves Raddib, M.S., Gomes Coelhoa, R. and Vilegasa, W. (2005). Structure–antimicrobial activity of some natural isocoumarins and their analogues. Phytomedicine 12: 378-381.

Florianowicz, T. (1998). *Penicillium expansum* growth and production of patulin in the presence of benzoic acid and its derivatives. Acta Microbiol. Pol. 47: 45-53.

Fries, N.F. (1973). Effects of volatile organic compounds on the growth and development of fungi. Trans. Br. Mycol. Soc. 60: 1-21.

Fu, S.H., Yang, M.H., Wen, H.M. and Chern, J.C. (2005). Analysis of flavonoids in propolis by capillary electrophoresis. J. Food Drug Anal. 13: 43-50.

Fulda, S. and Debatin, K.M. (2004). Sensitization for anticancer drug-induced apoptosis by the chemopreventive agent resveratrol. Oncogene 23: 6702-6711.

Galed, G., Fernàndez-Valle, M.E., Martìnez, A. and Heras, A. (2004). Application of MRI to monitor the process of ripening and decay in citrus treated with chitosan solutions. Magnetic Resonance Imaging 22: 127-137.

Gardini, F., Lanciotti, R., Caccioni, D.R.L. and Guerzoni, M.E. (1997). Antifungal activity of hexanal as dependent on its vapour pressure. J. Agric. Food Chem. 45: 4297-4302.

Ghisalberti, E.L. (1979). Propolis: a review. Bee World 60: 59-84.

Gibbons, S. (2005). Plants as a source of bacterial resistance modulators and anti-infective agents. Phytochem. Rev. 4: 63-78.

Gil, V. and Macleod, A.J. (1980). The effects of pH on glucosinolate degradation by a thioglucoside preparation. Phytochemistry 19: 2547-2553.

Godwin, J.R., Anthony, V.M., Clough, S.M. and Godfrey, C.R.A. (1992). ICIA5504: a novel broad-spectrum systemic beta-methoxyacrylate fungicide. Brit. Crop Prot. Conf. Pests Dis. 1: 435-442.

Gong, J., Li, S., Zhang, J. and Zhao, F. (2005). Study on prolonged vase-holding life of fresh cut roses by the extracts of Chinese traditional medicine. J. Nanjing Agric. 28: 29-33.

Gonzalez Ureña, A., Orea, J.M., Montero, C. and Jimenéz, J.B. (2003). Improving postharvest resistance in fruits by external application of trans-resveratrol. J. Agric. Food Chem. 51: 82-89.

González-Aguilar, G.A., Buta, J.G., Wang, C.Y. (2003). Methyl jasmonate and modified atmosphere packaging (MAP) reduce decay and maintain post-harvest quality of papaya 'Sunrise'. Postharvest Biol. Technol. 28: 361-370.

González-Aguilar, G.A., Tiznado-Hernandez, M.E., Zavaleta-Gatica, R. and Martýnez-Tellez, M.A. (2004). Methyl jasmonate treatments reduce chilling injury and activate the defence response of guava fruits. Biochem. Biophys. Res. Commun. 313: 694-701.

Goulart Bb, L., Hammer, P.E., Evensen, K.B., Janisiewicz, W. and Takeda, F. (1992). Pyrrolnitrin, captan, benomyl, and high CO_2 enhanced raspberry shelf life. J. Am. Soc. Hort. Sci. 117: 265-270.

Gueldner, R.C., Reilly, C.C., Pusey, P.L., Arrendale, R., Himmelsbach, D.S. and Cutler, H.G. (1988). Isolation and identification of iturin as antifungal peptides in biological control of peach brown rot with *Bacillus subtilis.* J. Agric. Food. Chem. 36: 366-370.

Gundlach, H., Muller, M.J., Kutchan, T.M. and Zenk, M.H. (1992). Jasmonic acid is a signal transducer in elicitor-induced plant cell cultures. Proc. Natl. Acad. Sci., USA 89: 2389-2393.

Hahlbrock, K. and Scheel, D. (1989). Physiology and molecular biology of phenylpropanoid metabolism. Annu. Rev. Plant. Physiol. Plant Mol. Biol. 40: 347-369.

Halama, P. and Van Haluwin C. (2004). Antifungal activity of lichen extracts and lichen acids. BioControl 49: 95-107.

Hamilton-Kemp, T.R., Mckracken, C.T., Lougrin, J.H., Anderson, R.A. and Hildebrand, D.F. (1992). Effects of some natural volatile compounds on the pathogenic fungi *Alternaria alternata* and *Botrytis cinerea.* J. Chem. Ecol. 18: 1083-1091.

Hammerschmidt, R. (1999). Phytoalexins: What have we learned after 60 years? Annu. Rev. Phytopathol. 37: 285-306.

Harborne, J.B. and Williams, C.A. (2000). Advances in flavonoid research since 1992. Phytochemistry 55: 481-504.

Hertog, M.G.L., Hollman, P.C.H., Katan, M.B. and Kiomhout, D. (1993). Intake of potentially anticarcinogenic flavonoids and their determinants in adults in the Netherlands. Nutr. Cancer. 20: 21-29.

Holley, R.A. and Patel, D. (2005). Improvement in shelf-life and safety of perishable foods by plant essential oils and smoke antimicrobials. Food Microbiol. 22: 273-292.

Hoos, G. and Blaich, R.J. (1990). Influence of resveratrol on germination of conidia and mycelial growth of *Botrytis cinerea* and *Phomopsis viticola.* J. Phytopathol. 129: 102-110.

Horvat, R.J., Chapman, G.W., Robertson, J.A., Meredith, F.I., Scorza, R.M., Callahan, A.M. and Morgens, P. (1990). Comparison of the volatile compounds from several commercial peach cultivars. J. Agric. Food Chem. 38: 234-237.

Ippolito, A., Nigro, F., Romanazzi, G. and Campanella, V. (1997). Field application of *Aureobasidium pullulans* against *Botrytis* storage rot of strawberry. In: Non-conventional Methods for the Control of Postharvest Disease and Microbiological Spoilage (eds.) P. Bertolini, P.C. Sijmons, M.E. Guerzoni and F. Serra, COST 914-915, Office for Official Publication of the European Communities, Luxemburg, pp. 127-133.

Ippolito, A., Nigro, F., Lima, G., Romanazzi, G. and Salerno, M. (1998). Xanthan gum as adjuvant in controlling table grape rots with *Aureobasidium pullulans*. J. Plant Pathol. 80: 258 (Abstract).

Ippolito, A. and Nigro, F. (2000). Impact of preharvest application of biological control agents on postharvest diseases of fresh fruits and vegetables. Crop Prot. 19: 715-723.

Ishiki, K., Tokuora, K., Mori, R. and Chiba, S. (1992). Preliminary examination of allyl isothiocyanate vapour for food preservation. Biosci. Biotechnol. Biochem. 56: 1476-1477.

Isman, M.B. (2000). Plant essential oils for pest and disease management. Crop. Prot. 19: 603-608.

Janisiewicz, W., Yourman, L., Roitman, J. and Mahoney, N. (1991). Postharvest control of blue mould and grey mould of apples and pears by dip treatment with pyrrolnitrin, a metabolite of *Pseudomonas cepacia*. Plant. Dis. 75: 490-494.

Jeandet, P., Sbaghi, M. and Bessis, R. (1993). The significance of stilbene-type phytoalexin degradation by culture filtrates of *Botrytis cinerea* in the vine-*Botrytis* interaction. In: Mechanisms of Plant Defence Responses (eds.) B. Fritig and M. Legrand, Kluwer Academic Publishers, Dordrecht, The Netherlands, pp. 84-98.

Jeandet, P., Douillet-Breuil, A.C., Bessis, R., Debord, S., Sbaghi, M. and Adrian, M. (2002). Phytoalexins from the *Vitaceae*: Biosynthesis, phytoalexin gene expression in transgenic plants, antifungal activity, and metabolism. J. Agric. Food Chem. 50: 2731-2741.

Jongen, W. (2005). Improving the safety of fresh fruit and vegetables. Woodhead Publishing, Cambridge, UK.

Jurd, L., Corse, J., King, A.D., Bayne, H. and Mihara, K. (1971a). Antimicrobial properties of 6,7-dihydroxy-, 7,8-dihydroxy-, 6-hydroxy- and 8-hydroxy-coumarins. Phytochemistry 10: 2971-2974.

Jurd, L., King, A.D. and Mihara, K. (1971b). Antimicrobial properties of umbelliferone derivatives. Phytochemistry 10: 2965-2970.

Kanes, K., Tisserat, B., Berhow, M. and Vandercook, C. (1992). Phenolic composition of various tissues of Rutaceae species. Phytochemistry 32: 967-974.

Karabulut, O.A., Arslan, U., Kuruoglu, G. and Ozgenc, T. (2004a). Control of postharvest diseases of sweet cherry with ethanol and hot water. J. Phytopathol. 152: 298-303.

Karabulut, O.A., Gabler, F.M., Mansour, M. and Smilanick, J.L. (2004b). Postharvest ethanol and hot water treatments of table grapes to control gray mold. Postharvest Biol. Technol. 34: 169-177.

Karabulut, O.A., Romanazzi, G., Smilanick, J.L. and Lichter, A. (2005). Postharvest ethanol and potassium sorbate treatments of table grapes to control gray mold. Postharvest Biol. Technol. 37: 129-134.

Karaman, S., Digrak, M., Ravid, U. and Ilcim, A. (2001). Antibacterial and antifungal activity of the essential oils of *Thymus revolutus* Celak from Turkey. J. Ethnopharm. 76: 183-186.

Kayser, O. and Kolodziej, H. (1999). Antibacterial activity of simple coumarins structural requirements for biological activity. Z. Naturforsch. 54: 169-174.

Kennedy, J.F. and Bradshaw, I.J. (1984). Production, properties and applications of xanthan. In: Progress in Industrial Microbiology: Modern Applications of Traditional

Biotechnologies, Volume 19 (ed.) M.F Bushell, Elsevier, Amsterdam, The Netherlands, pp. 319-371.

Kim, J.H., Campbell, B.C., Mahoney, N.E., Chan, K.L. and Molyneux, R.J. (2004). Identification of phenolics for control of *Aspergillus flavus* using *Saccharomyces cerevisiae* in a model target-gene bioassay. J. Agric. Food Chem. 52: 7814-7821.

Kim, J.J., Ben-Yehoshua, S., Shapiro, B., Henis, Y. and Carmeli, S. (1991). Accumulation of scoparone in heat-treated lemon fruit inoculated with *Penicillium digitatum* Sacc. Plant Physiol. 97: 880-885.

Kim, Y.S., Hwang, C.S. and Shin, D.H. (2005). Volatile constituents from the leaves of *Polygonum cupsidatum* and their anti-bacterial activities. Food Microbiol. 22: 139-144.

Kobayashi, S., Ding, C.K., Nakamura, Y., Nakajima, I. and Matsumoto, R. (2000). Kiwifruits (*Actinidia deliciosa*) transformed with a *Vitis* stilbene synthase gene produce piceid (resveratrolglucoside). Plant Cell Rep. 19: 904-910.

Koyama, S., Yamaguchi, Y., Tanaka, S. and Motoyoshiya, J. (1997). A new substance (Yoshixol) with an interesting antibiotic mechanism from wood oil of Japanese traditional tree (Kiso-Hinoki), *Chamaecyparis obtuse*. Gen. Pharmac. 28: 797-804.

Krauze-Baranowska, M. and Wiwart, M. (2003). Antifungal activity of biflavones from *Taxus baccata* and *Ginkgo biloba*. Z. Naturforsch. 58: 65-69.

Kuć, J. (1995). Phytoalexins, stress metabolism and disease resistance in plants. Annu. Rev. Phytopathol. 33: 275-297.

Kulakiotu, E.K., Thanassoulopoulos, C.C. and Sfachiotakis, E.M. (2004a). Biological control of *Botrytis cinerea* by volatiles of 'Isabella' grapes. Phytopathology 94: 924-931.

Kulakiotu, E.K., Thanassoulopoulos, C.C. and Sfachiotakis, E.M. (2004b). Postharvest biological control of *Botrytis cinerea* on kiwifruit by volatiles of 'Isabella' grapes. Phytopathology 94: 1280-1285.

Kwon, Y.S., Kobayashi, A., Kajiyama, S.I., Kawazu, K., Kanzaki, H. and Kim, C.M. (1997). Antimicrobial constituents of *Angelica dahurica* roots. Phytochemistry 44: 887-889.

La Torre, A., Gruccione, M. and Imbroglini, G. (1990). Indagine preliminare sull'azione di preparati a base di propoli nei confronti di *Botrytis cinerea* della fragola. Apicoltura 6: 169-177.

Lambert, R.J.W., Skandamis, P.N., Coote, P.J. and Nychas, G.J.E. (2001). A study of the minimum inhibitory concentration and mode of action of oregano essential oil, thymol and carvacrol. J. Appl. Microbiol. 91: 453-462.

Lanza, G., Di Martino Aleppo, E. and Strano, M.C. (1997). Evaluation of integrated approach to control postharvest green mould of lemons. In: Non-conventional Methods for the Control of Postharvest Disease and Microbiological Spoilage (eds.) P. Bertolini, P.C. Sijmons, M.E. Guerzoni, F. Serra and COST 914-915, Office for Official Publication of the European Communities, Luxemburg, pp. 111-114.

Lattanzio, V., Cardinali, A. and Palmieri, S. (1994a). The role of phenolics in the postharvest physiology of fruits and vegetables: Browning reactions and fungal diseases. Ital. J. Food Sci. 6: 3-22.

Lattanzio, V., De Cicco, V., Di Venere, D., Lima, G. and Salerno, M. (1994b). Antifungal activity of phenolics against different storage fungi. Ital. J. Food Sci. 6: 23-30.

Lattanzio, V., Di Venere, D., Linsalata, V., Bertolini, P., Ippolito, A. and Salerno, M. (2001). Low temperature metabolism of apple phenolics and quiescence of *Phlyctene vagabunda*. J. Agric. Food Chem. 49: 5817-5821.

Lattaoui, N. and Tantoui-Elaraki, A. (1994). Individual and combined antibacterial activity of essential oil components of three thyme essential oil. Rivista Italiana EPPOS 13: 13-19.

Li, H. and Yu, T. (2001). Effect of chitosan coating on incidence of brown rot, quality and physiological attributes of postharvest peach fruit. J. Sci. Food Agric. 81: 269-274.

Lichter, A., Zutkhy, Y., Sonego, L., Dvir, O., Kaplunov, T., Sarig, P. and Ben-Arie, R. (2002). Ethanol controls postharvest decay of table grapes. Postharvest Biol. Technol. 24: 301-308.

Lima, G., De Curtis, F., Castoria, R., Pacifico, S. and De Cicco, V. (1998). Additives and natural products against postharvest pathogens and compatibility with antagonistic yeasts. J. Plant Pathol. 80: 259 (Abstract).

Lima, G., Spina, A.M., Castoria, R., De Curtis, F. and De Cicco, V. (2005). Integration of biocontrol agents and food-grade additives for enhancing protection of stored apples from *Penicillium expansum*. J. Food Prot. 68: 2100-2106.

Liu, W.T., Chu, C.L. and Zhou, T. (2002). Thymol and acetic acid vapors reduce postharvest brown rot of apricots and plums. Hort. Sci. 37: 151-156.

López-Malo, A., Alzamora, S. M. and Palou, E. (2005). *Aspergillus flavus* growth in the presence of chemical preservatives and naturally occurring antimicrobial compounds. Int. J. Food Microbiol. 99: 119-128.

Lyons, M.M., Yu, C., Toma, R.B., Cho, S.Y., Reiboldt, W., Lee, J. and Van Breemen, R.B. (2003). Resveratrol in raw and baked blueberries and bilberries. J. Agric. Food. Chem. 51: 5867-5870.

Macheix, J.J., Fleuriet, A. and Billot, J. (1990). Fruit Phenolics. CRC Press, Boca Raton, Florida, USA.

Maezaki, Y., Tsuji, K., Nakagawa, Y., Kawai, Y., Akimoto, M., Tsugita, T., Takekawa, W., Terada, A., Hara, H. and Mitsuoka, T. (1993). Hypocholesterolemic effect of chitosan in adult males. Biosci. Biotech. Biochem. 57: 1439-1444.

Mahoney, N. and Molyneux, R. (2004). Phytochemical inhibition of aflatoxigenicity in *Aspergillus flavus* by constituents of walnut (*Juglans regia*). J. Agric. Food Chem. 52: 1882-1889.

Manici, L.M., Lazzeri, L. and Palmieri, S. (1997). In vitro fungitoxic activity of some glucosinolates and their enzyme-derived products toward plant pathogenic fungi. J. Agric. Food Chem. 45: 2768-2773.

Marcucci, M.C. (1995). Propolis: chemical composition, biological properties and therapeutic activity. Apidologie 26: 83-99.

Margosan, D.A., Smilanick, J.L., Simmons, G.F. and Henson, D.J. (1997). Combination of hot water and ethanol to control postharvest decay of peach and nectarines. Plant Dis. 81: 1405-1409.

Mari, M., Iori, R., Leoni, O. and Marchi, A. (1993). In vitro activity of glucosinolate-derived isothiocyanates against postharvest fruit pathogens. Ann. Appl. Biol. 123: 155-164.

Mari, M., Iori, R., Leoni, O. and Marchi, A. (1996). Bioassays of glucosinolate-derived isothiocyanates against postharvest pathogens. Plant Pathol. 45: 753-760.

Mari, M. and Carati, A. (1997). Use of *Saccharomyces cerevisiae* with ethanol in the biological control of grey mould on pome fruits. In: Non Conventional Methods for The Control of Postharvest Disease and Microbiological Spoilage (eds.) P. Bertolini, P.C. Sijmons, M.E. Guerzoni and F. Serra, COST 914-915, Office for Official Publication of the European Communities, Luxemburg, pp. 85-91.

Mari, M., Leoni, O., Iori, R. and Cembali, T. (2002). Antifungal vapour-phase activity of allylisothiocyanate against *Penicillium expansum* on pears. Plant Pathol. 51: 231-236.

Mattheis, J.P. and Roberts, R.G. (1993). Fumigation of sweet cherry (*Prunus avium* 'Bing') fruit with low molecular weight aldehydes for postharvest decay control. Plant Dis. 77: 810-814.

Meir, S., Droby, S., Davidson, H., Alsevia, S., Cohen, L., Horev, B. and Philosoph-Hadas, S. (1998). Suppression of *Botrytis* rot in cut rose flowers by postharvest application of methyl jasmonate. Postharvest Biol. Technol. 13: 235-243.

Meir, S., Droby, S., Kochaneck, B., Salim, S. and Philosoph-Hadas, S. (2004). Use of methyl jasmonate for suppression of *Botrytis* rot in various cultivars of cut flowers. Proc. 5th Inter. Postharvest Symp., Verona, Italy, S10-03, pp. 100 (Abstract).

Miron, T., Rabincov, A., Mirelman, D., Wilchek, M. and Weiner, L. (2000). The mode of action of allicin: its ready permeability through phospholipid membranes may contribute to its biological activity. Biochim. Biophys. Acta 1463: 20-30.

Mithen, R. (2001). Glucosinolates - biochemistry, genetics and biological activity. Plant Growth Reg. 34: 91-103.

Mithen, R.F., Lewis, B.G. and Fenwick, G.R. (1986). In vitro activity of glucosinolates and their products against *Leptosphaeria maculans*. Trans. Brit. Mycol. Soc. 87: 433-440.

Mohapatra, N.P., Pati, S.P. and Ray, R.C. (2000). In vitro inhibition of *Botryodiplodia theobromae* (Pat) causing Java black rot in sweet potato by phenolic compounds. Ann. Plant Prot. Sci. 8: 106-109.

Moline, H.E., Buta, J.G., Saftner, R.A. and Maas, J.L. (1997). Comparison of three volatile natural products for the reduction of post harvest diseases in strawberries. Adv. Strawberry Res. 16: 43-48.

Molly, C., Chea, L.H. and Koolaard, J.P. (2004). Induced resistance against *Sclerotinia sclerotiorum* in carrots treated with enzymatically hydrolysed chitosan. Postharvest Biol. Technol. 33: 61-65.

Muir, S.R., Collins, G.J., Robinson, S., Hughes, S., Bovy, A., De Vos, C.H.R., Van Tunen, A.J. and Verhoeyen, M.E. (2002). Overexpression of petunia chalcone isomerase in tomato results in fruits containing increased levels of flavonoids. Nat. Biotechnol. 19: 470-474.

Muzzarelli, R.A.A. and Muzzarelli, C. (2001). Chitosan: Pharmacy and Chemistry, Atec, Grottammare, Italy.

Nandi, B. and Fries, N. (1976). Volatile aldehydes, ketones, esters, and terpenoids as preservatives against storage fungi in wheat. Z. Pflanzenk. Pflanzen. 83: 284-294.

Neri, F., Mari, M., Menniti, A.M., Brigati, B. and Bertolini, P. (2006). Control of *Penicillium expansum* disease in pears and apples by trans-2-hexenal vapors. Postharvest Biol. Technol. (Submitted).

Nevill, D., Nyfeler, R. and Sozzi, D. (1998). CGA 142705 a novel fungicide for seed treatment. Brit. Crop Prot. Conf. Pests Dis. 1: 65-72.

Ng, T.B. (2004). Peptides and proteins from fungi. Peptides 25: 1055-1073.

Nicholson, R.L. and Hammerschmidt, R. (1990). Phenolic compounds and their role in disease resistance. Annu. Rev. Phytopathol. 30: 369-389.

Nigro, F., Ippolito, A. and Lima, G. (1998). Use of UV-C light to reduce *Botrytis* storage rot of table grapes. Postharvest Biol. Technol. 13: 171-181.

Nigro, F., Ippolito, A., Lattanzio, V., Di Venere, D. and Salerno, M. (2000). Effect of ultraviolet-C light on postharvest decay of strawberry. J. Plant Pathol. 82: 29-37.

Nychas, G.J.E., Skandamis, P.N. and Tassou, C.C. (2003). Antimicrobials from herbs and spices. In: Natural Antimicrobials for the Minimal Processing of Foods (ed.) S. Roller, Woodhead Publishing, Cambridge, UK, pp. 176-200.

Ogabwu, J. and Korstern, L. (2003). Control of citrus green and blue molds with garlic extracts. Eur. J. Plant Pathol. 109: 221-225.

Ogawa, J.M. and Lyda, S.D. (1960). Effect of alcohol on spores of *Sclerotinia fructicola* and other peach fruit rotting fungi in California. Phytopathology 50: 790-792.

Ojala, T., Remes, S., Haansuu, P., Vuorela, H., Hiltunen, R., Haahtela, K. and Vuorela, P. (2000). Antimicrobial activity of some coumarins containing herbal plants growing in Finland. J. Ethnopharm. 73: 299-305.

Oser, B.L., Ford, R.A. and Bernard, B.K. (1984). Recent progress in the consideration of flavoring ingredients under the food additive amendments, 13, GRAS substances. Food Technol. 34: 66-89.

Ota, C., Unterkircher, C., Fantinato, V. and Shimizu, M.T. (2001). Antifungal activity of propolis on different species of *Candida*. Mycoses 44: 375-378.

Palhano, F.L., Vilches, T.B., Santos, R.B., Orlando, M.T., Ventura, J.A. and Fernandes, P.M. (2004). Inactivation of *Colletotrichum gloeosporioides* spores by high hydrostatic pressure combined with citral or lemongrass essential oil. Int. J. Food Microbiol. 95: 61- 66.

Paul, B., Chereyathmanjiyil, A., Masih, I., Chapuis, L. and Benoit, A. (1998). Biological control of *Botrytis cinerea* causing grey mould disease of grapevine and elicitation of stilbene phytoalexin (resveratrol) by a soil bacterium. FEMS Microbiol. Lett. 165: 65-70.

Perata, P. and Alpi, A. (1991). Ethanol-induced injuries to carrot cells, the role of acetaldehyde. Plant Physiol. 95: 748-752.

Pesis, E. and Avissar, I. (1990). Effect of postharvest application of acetaldehyde vapour on strawberry decay, taste, and certain volatiles. J. Sci. Food Agric. 52: 377-385.

Pietta, P.G. (2000). Flavonoids as antioxidants. J. Nat. Prod. 63: 1035-1042.

Piper, P., Calderon, C.O., Hatzixanthis, K. and Mollapour, M. (2001). Weak acid adaptation: the stress response that confers resistance to organic acid food preservatives. Microbiology 147: 2635-2642.

Poole, P.R. and Whitmore, K.J. (1997). Effect of topical postharvest application of 6-pentyl-2-pyrone on properties of stored kiwifruit. Postharvest Biol. Technol. 12: 229-237.

Pranoto, Y., Salokhe, V. L. and Rakshit, S. K. (2005). Physical and antibacterial properties of alginate-based edible film incorporated with garlic oil. Food Res. Int. 38: 267–272.

Prasad, Y. and Stadelbacher, G.J. (1973). Control of postharvest decay of fresh raspberries by acetaldehyde vapor. Plant Dis. Rep. 57: 795-797.

Prasad, Y. and Stadelbacher, G.J. (1974). Effect of acetaldehyde vapor on postharvest decay and market quality of fresh strawberries. Phytopathology 64: 948-951.

Rabea, E.I., Badawy, M.E.T., Stevens, C.V., Smagghe, G. and Steurbaut, W. (2003). Chitosan as antimicrobial agent: applications and mode of action. Biomacromolecules 4: 1457-1465.

Rauha, J.P., Remes, S., Heinonen, M., Hopia, A., Kahkonen, M., Kujala, T., Pihlaja, K., Vuorela, H. and Vuorela, P. (2000). Antimicrobial effects of Finnish plant extracts containing flavonoids and other phenolic compounds. Int. J. Food Microbiol. 56: 3-12.

Recio, M.C., Rìos, J.L. and Villar, A. (1989). A review of some antimicrobial compounds isolated from medicinal plants reported in the literature 1978-1988. Phytopthora Res. 3: 117-125.

Reddy, M.V.B., Angers, P., Gosselin, A. and Arul, J. (1998). Characterization and use of essential oil from *Thymus vulgaris* against *Botrytis cinerea* and *Rhizopus stolonifer* in strawberry fruits. Phytochemistry 47: 1515-1520.

Reddy, M.V.B., Corcuff, R., Kasaai, M.R., Castaigne, F. and Arul, J. (1999). Induction of resistance against grey mould rot in carrot roots by chitosan. Phytopathology 89: S6 (Abstract).

Reddy, M.V.B., Belkacemi, K., Corcuff, R., Castaigne, F. and Arul, J. (2000). Effect of preharvest chitosan sprays on post-harvest infection by *Botrytis cinerea* and quality of strawberry fruit. Postharvest Biol. Technol. 20: 39-51.

Rhodes, M.J.C. (1994). Physiological roles for secondary metabolites in plants: some progress, many outstanding problems. Plant Mol. Biol. 24: 1-20.

Rimando, A.M., Kalt, W., Magee, J.B., Dewey, J. and Ballington, J.R. (2004). Resveratrol, pterostilbene, and piceatannol in vaccinium berries. J. Agric. Food Chem. 52: 4713-4719.

Roberts, J.W. and Dunegan, J.C. (1932). Peach brown rot. Technical Bulletin No. 328, US Dept of Agric, Washington DC, USA.

Rodov, V., Ben-Yehoshua, S., Fang, D.Q., Kim, J.J. and Ashkenazi, R. (1995) Preformed antifungal compounds of lemon fruit: Citral and its relation to disease resistance. J. Agric. Food Chem. 43: 1057-1061.

Roller, S. (2003). Natural Antimicrobials for the Minimal Processing of Foods. Woodhead Publishing, Cambridge, UK.

Romanazzi, G., Ippolito, A. and Nigro, F. (1999a). Activity of glicolchitosan on postharvest strawberry rot. J. Plant Pathol. 81: 237 (Abstract).

Romanazzi, G., Schena, L., Nigro, F. and Ippolito, A. (1999b). Preharvest chitosan treatments for the control of postharvest decay of sweet cherries and table grapes. J. Plant Pathol. 81: 237 (Abstract).

Romanazzi, G., Nigro, F. and Ippolito, A. (2000a). Effectiveness of pre- and postharvest chitosan treatments on storage decay of strawberry. Frutticoltura 62: 71-75.

Romanazzi, G., Nigro, F., Ligorio, A. and Ippolito, A. (2000b). Hypobaric and chitosan integrated treatments to control postharvest rots of sweet cherries. Proc. V European Foundation Plant Pathology Congress, Taormina, Italy, pp. 558-560.

Romanazzi, G., Nigro, F. and Ippolito, A. (2001). Chitosan in the control of postharvest decay of some Mediterranean fruits. In: Chitin Enzymology (ed.) R.A.A. Muzzarelli, Atec, Grottammare, Italy, pp. 141-146.

Romanazzi, G., Nigro, F., Ippolito, A., Di Venere, D. and Salerno, M. (2002). Effects of pre and postharvest chitosan treatments to control storage grey mould of table grapes. J. Food Sci. 67: 1862-1867.

Romanazzi, G., Nigro, F. and Ippolito, A. (2003). Short hypobaric treatments potentiate the effect of chitosan in reducing storage decay of sweet cherries. Postharvest Biol. Technol. 29: 73-80.

Romanazzi, G., Mlikota Gabler, F. and Smilanick, J.L. (2006). Preharvest chitosan and postharvest UV irradiation treatments suppress gray mold of table grapes. Plant Dis. 90: 445-450.

Sahin, F., Güllüce, M., Daferera, D., Sökmen, A. , Sökmen, M., Polissiou, M., Agar, G. and Ozer, H. (2004). Biological activities of the essential oils and methanol extract of *Origanum vulgare* ssp. *vulgare* in the Eastern Anatolia region of Turkey. Food Control 15: 549–557.

Sahinler, N. and Kaftanoglu, O. (2005). Natural product propolis: Chemical composition. Nat. Prod. Res. 19: 183-188.

Saks, Y. and Barkai-Golan, R. (1995). *Aloa vera* gel activity against plant pathogenic fungi. Postharvest Biol. Technol. 6: 159-165.

Sanzani, S.M., De Girolamo, A., Schena, L., Solfrizzo, M. and Ippolito, A. (2006). Activity of selected phytoalexins in postharvest control of *Penicillium expansum* and patulin production in apples. J. Plant Pathol. 88: S57 (Abstract).

Sarig, P., Zutkhi, Y., Monjauze, A., Lisker, N. and Ben-Arie, R. (1997). Phytoalexin elicitation in grape berries and their susceptibility to *Rhizopus stolonifer*. Physiol. Mol. Plant Pathol. 50: 337-347.

Schena, L., Nigro, F., Soleti Ligorio, V., Yaseen, T., El Ghaouth, A. and Ippolito, A. (2004). Biocontrol activity of bio-coat and biocure against postharvest rots of table grapes and sweet cherries. Acta Hort. 682: 2115-2119.

Schena, L., Soleti Ligorio, V., Ligorio, A., Pentimone, I., Nigro, F., Ippolito, A. (2005). Expression of phenylalanine ammonia-lyase and stilbene synthase genes in table grapes treated with a biocontrol agent. VII° convegno FISV, Riva del Garda Italy, 22-25 September 2005 (Abstract).

Schijlen, E.G., Ric De Vos, C.H., Van Tunen, A.J. and Bovy, A.G. (2004). Modification of flavonoid biosynthesis in crop plants. Phytochemistry 65: 2631-2648.

Sharan, M., Taguchi, G., Gonda, K., Jouke, T., Shimosaka, M., Hayashida, N. and Okazaki, M. (1998). Effects of methyl jasmonate and elicitor on the activation of phenylalanine ammonia-lyase and the accumulation of scopoletin and scopolin in tobacco cell cultures. Plant Sci. 132: 13-19.

Sharma, A., Tewari, G.M., Shrikhande, A.J., Padwal-Desai, S.R. and Bandyopadhyay, A.G. (1979). Inhibition of aflatoxin-producing fungi by onion extracts. J. Food Sci. 44: 1545-1547.

Sharma, A., Padwal-Desai, S.R., Tewari, G.M. and Bandyopadhyay, G. (1981). Factors affecting antifungal activity of onion extractives against aflatoxins-producing fungi. J. Food Sci. 46: 741-744.

Sholberg, P.L. and Shimazu, B.N. (1991). Use of the natural plant products, hinokitiol, to extend shelf-life of peaches. J. Can. Inst. Food Sci. Technol. 2: 273-276.

Sholberg, P.L. and Gaunce, A.P. (1995). Fumigation of fruit with acetic acid to control postharvest decay. Hort. Sci. 30: 1271-1275.

Sholberg, P.L. and Gaunce, A.P. (1996). Fumigation of stone fruit with acetic acid to control postharvest decay. Crop Prot. 15: 681-686.

Sholberg, P.L., Reynolds, A.G. and Gaunce, A.P. (1996). Fumigation of table grapes with acetic acid to prevent postharvest decay. Plant Dis. 80: 1425-1428.

Sholberg, P.L. (1998). Fumigation of fruit with short-chain organic acids to reduce the potential of postharvest decay. Plant Dis. 82: 689-693.

Sholberg, P.L., Haag, P., Hocking, R. and Bedford, K. (2000). The use of vinegar vapor to reduce postharvest decay of harvested fruit. Hort. Sci. 35: 898-903.

Sholberg, P.L., Shephard, T., Randall, P. and Moyls. L. (2004). Use of measured concentrations of acetic acid vapour to control postharvest decay in d'Anjou pears. Postharvest Biol. Technol. 32: 89-98.

Skerget, M., Kotnik, P., Hadolin, M., Rizner Hras, A., Simonic, M. and Knez, Z. (2005). Phenols, proanthocyanidins, flavones and flavonols in some plant materials and their antioxidant activities. Food Chemistry 89: 191-198.

Skočibušić, M., Bezić, N. and Dunkić, V. (2006). Phytochemical composition and antimicrobial activities of the essential oils from *Satureja subspicata* Vis. growing in Croatia. Food Chem. 96: 20-28.

Smilanick, J.L., Margosan, D.A. and Henson, D.J. (1995). Evaluation of heated solution of sulfur dioxide, ethanol, and hydrogen peroxide to control postharvest green mould of lemons. Plant Dis. 79: 742-747.

Stadelbacher, G.J. and Prasad, Y. (1974). Postharvest decay control of apple by acetaldehyde vapor. J. Am. Soc. Hortic. Sci. 99: 364-368.

Stange, R.R. Jr., Midland, S.L., Eckert, J.W. and Sims, J.J. (1993). An antifungal compound produced by grapefruit and Valencia orange after wounding of the peel. J. Nat. Prod. 56: 1627-1629.

Stark-Lorenzen, P., Nelke, B., Hänssler, G. and Mühlbach Thomzik, J.E. (1997). Transfer of a grapevine stilbene synthase gene to rice (*Oryzae sativa* L.). Plant Cell Rep. 16: 668-673.

Stewart, J.K., Aharoni, Y., Harsten, P.L. and Young, D.K. (1980). Symptoms of acetaldehyde injury on head lettuce. Hort. Sci. 15: 148-149.

Suslow, T. (2000). Postharvest Handling for Organic Crops. University of California, Div. Agriculture and Natural Resources, Publication 7254, USA.

Szankowski, I., Briviba, K., Fleschhut, J., Schonherr, J., Jacobsen, H.J. and Kiesecker, H. (2003). Transformation of apple (*Malus domestica* Borkh.) with the stilbene synthase gene from grapevine (*Vitis vinifera* L.) and a PGIP gene from kiwi (*Actinidia deliciosa*). Plant Cell. Rep. 22: 141-149.

Takeda, F., Janisiewicz, W., Roitman, J., Mahoney, N. and Abeles, F.B. (1990). Pyrrolnitrin delays postharvest fruit rot in strawberry. Hort. Sci. 25: 320-322.

Tawata, S., Taira, S., Kobamoto, N., Zhu, J., Ishihara, M. and Toyama, S. (1996). Synthesis and antifungal activity of cinnamic acid esters. Biosci. Biotechnol. Biochem. 60: 909-910.

Terry, L.A. and Joyce, D.C. (2004). Elicitors of induced disease resistance in postharvest horticultural crops: A brief review. Postharvest Biol. Technol. 32: 1-13.

Terry, L.A., Joyce, D.C., Adikaram, N.K.B. and Khambayd, B.P.S. (2004). Preformed antifungal compounds in strawberry fruit and flower tissues. Postharvest Biol. Technol. 31: 201-212.

Thangadurai, D., Anitha, S., Pullaiah, T., Reddy, P.N. and Ramachandraiah, O.S. (2002). Essential oil constituents *in vitro* antimicrobial activity of *Decalepis hamiltonii* roots against foodborne pathogens. J. Agric. Food Chem. 50: 3147-3149.

Thomas, L.V., Ingram, R.E., Yu, S. and Delves-Broughton, J. (2004). Investigation on the effectiveness of Ascopyrone P as a food preservative. Int. J. Food Microbiol. 93: 319-323.

Thomma, B.P.H.J., Eggermont, K., Penninckx, I.A.M.A., Mauch-Mani, B., Vogelsang, R., Cammue, B.P.A. and Broekaert, W.F. (1998). Separate jasmonate-dependent and salicylate-dependent defense-response pathways in *Arabidopsis* are essential for resistance to distinct microbial pathogens. Proc. Natl. Acad. Sci. USA, 95: 15107-15111.

Thomzik, J.E., Stenzel, K., Stöcker, R., Schreier, P.H., Hain, R. and Stahl, D.J. (1997). Synthesis of a grapevine phytoalexin in transgenic tomatoes (*Lycopersicon esculentum* Mill.) conditions resistance against *Phytophthora infestans.* Physiol. Mol. Plant Pathol. 51: 265-278.

Tian, Shi-Ping (2006). Microbial control of postharvest diseases of fruits and vegetables. In: Microbial Biotechnology in Horticulture, Volume I (eds.) R.C. Ray and O.P. Ward, Science Publishers, Enfield, New Hampshire, USA, pp. 163-201.

Tonini, G. and Caccioni, D.R.L. (1990). The effect of several natural volatiles on *Monilinia laxa.* Proc. Qualita' Dei Prodotti Ortofrutticoli Postraccolta, Fondazione Cesena Agri-cultura, pp. 123-126.

Toray Research Center Inc. (1991). New development in functional packaging materials. In: Information on Frontier Technology and Future Trends (ed.) E. Eumura, Tokyo, Japan, pp. 258-269.

Tosi, B., Donini, A., Romagnoli, C. and Bruni, A. (1996). Antimicrobial activity of some commercial extracts of propolis prepared with different solvents. Phytopthora. Res. 10: 335-336.

Tripathi, P., Dubey, N.K. and Pandey, V.M. (2002). Kaempferol: The antifungal principle of *Acacia nilotica* L. J. Indian. Bot. Soc. 81: 51-54.

Tripathi, P. and Dubey, N.K. (2004). Exploitation of natural products as an alternative strategy to control postharvest fungal rotting of fruit and vegetables. Postharvest Biol. Technol. 32: 235-245.

Trullo, M.C., Schena, L., Dileo, C., Nigro, F., De Giorgi, C. and Ippolito, A. (2007). Transcript levle of chitinase and β-1,3-glucanase genes in orange fruit treated with alternative control means. XIII International Congress on Molecular Plant-Microbe Interactions. Sorrento, Italy, July 21-27 (Abstract).

Tsao, R. and Zhou, T. (2000). Interactions of monoterpenoids, methyl jasmonate, and Ca^{2+} in controlling postharvest brown rot of sweet cherry. Hort. Sci. 35: 1304-1307.

Tsiantos, J., Psallidas, P. and Chatzaki, A. (2003). Efficacy of alternatives to antibiotic chemicals for the control of fire blight of pears. Ann. Appl. Biol. 143: 319-322.

Valverde, J.M., Guillén, F., Martínez-Romero, D., Castillo, S., Serrano, M. and Valero, D. (2005). Improvement of table grapes quality and safety by the combination of modified atmosphere packaging (map) and eugenol, menthol, or thymol. J. Agric. Food Chem. 7458-7464.

van Doorn, W.G., de Witte, Y. and Perik, R.R.J. (1990). Effect of antimicrobial compounds on the number of bacteria in stems of cut rose flowers. J. Appl. Bacteriol. 68: 117-122.

Venturini, M.E., Blanco, D. and Oria, R. (2002). In vitro antifiungal activity of several antimicrobial compounds against *Penicillium expansum*. J. Food Prot. 65: 834-839.

VerhoeVen, D.T.H., Verhagen, H., Goldbohm, R.A., Van Den Brandt, P.A. and Van Poppel, G. (1997). A review of mechanisms underlying anticarcinogenicity by *Brassica* vegetables. Chem. Biol. Int. 103: 79-129.

VerhoeVen, M.E., Bovy, A., Collins, G., Muir, S., Robinson, S., De Vos, C.H.R. and Colliver, S. (2002). Increasing antioxidant levels in tomatoes through modification of the flavonoid biosynthetic pathway. J. Exp. Bot. 53: 2099-2106.

Vicente, A.R., Civello, P.M., Martìnez, G.A., Powel, A.L.T., Labavitch, J.M. and Chaves, A.R. (2005). Control of postharvest spoilage in soft fruit. Steward Postharvest Rev. 1: 1-11.

Wang, C.Y. (2003). Maintaining postharvest quality of raspberries with natural volatile compounds. Int. J. Food Sci. Technol. 38: 869-875.

Wang, S.Y. and Zheng, W. (2005). Preharvest application of methyl jasmonate increases fruit quality and antioxidant capacity in raspberries. Int. J. Food Sci. Technol. 40: 187-195.

Wang, X., Du, Y. and Liu, H. (2004). Preparation, characterization and antimicrobial activity of chitosan-Zn complex. Carbohyd. Polym. 56: 21-26.

Wang, X.G. and Ng, T.B. (2001). Purification of allivin, a novel antifungal protein from bulbs of the round-cloved garlic. Life Sci. 70: 357-365.

Wilson, C.L., Franklin, J.D. and Otto, B. (1987). Fruit volatiles inhibitory to *Monilinia fructicola* and *Botrytis cinerea*. Plant Dis. 71: 316-319.

Wilson, C.L. and Wisniewski, M.E. (1994). Biological Control of Postharvest Diseases – Theory and Practices. CRC Press, Boca Raton, Florida, USA.

Wilson, C.L., El Ghaouth, A., Chalutz, E., Droby, S., Stevens, C., Lu, J.Y., Khan, V. and Arul, J. (1994). Potential of induced resistance to control postharvest diseases of fruits and vegetables. Plant Dis. 78: 837-844.

Wilson, C.L., Solar, J.M., El Ghaouth, A. and Wisniewski, M.E. (1997). Rapid evaluation of plant extracts and essential oils for antifungal activity against *Botrytis cinerea*. Plant Dis. 81: 204-210.

Xu, Y., Chang, P.L., Liu, D., Narasimhan, M.L., Raghothma, K.G., Hasegawa, P.M. and Bressan, R.A. (1994). Plant defense genes are synergistically induced by ethylene and methyl jasmonate. Plant Cell 6: 1077-1085.

Yin, M.C. and Tsao, S.M. (1999). Inhibitory effect of seven *Allium* plants upon three *Aspergillus* species. Int. J. Food Microbiol. 49: 49-56.

Yuen, C.M.C., Paton, J.E., Hanawati, R. and Shen, L.O. (1995). Effect of ethanol, acetaldehyde, and ethyl formate on the growth of *Penicillium italicum* and *P. digitatum* on oranges. J. Hort. Sci. 70: 81-84.

9

Mycotoxins in Fruits and Fruit-derived Products – An Overview

Raffaello Castoria* and
Antonio F. Logrieco

INTRODUCTION

Mycotoxins are low-molecular-weight secondary fungal metabolites, which are toxic to humans and animals. Bennett (1987) defined mycotoxins as "natural products produced by fungi that evoke a toxic response when introduced in low concentration to higher vertebrates and other animals by a natural route". Natural routes may include ingestion, skin contact and inhalation. Most mycotoxins are produced by filamentous fungi, which cause diseases in various crop plants cultivated in the field and in stored commodities, such as cereals, soybeans, nuts and the last but not least, fruits. On the basis of the target organ of their toxicity in humans and animals, mycotoxins can be classified as hepatotoxins, nephrotoxins, neurotoxins, immunotoxins etc. Immunotoxicity and/or immunodepression are the most common and the most serious problems associated with mycotoxins in food production.

Several mycotoxins occur in nature; a fungus may produce various mycotoxins and several different fungi may produce a particular mycotoxin. Generally speaking, the synthesis of mycotoxins by toxigenic fungal strains can occur in the field in preharvest infected plants and may

*Corresponding Author

continue during storage if conditions remain favorable to fungal growth. Fungal attacks that are specific to the postharvest phase and are responsible for mycotoxin accumulation in stored crops are also well-known. It must be underlined that the presence of toxigenic fungi does not necessarily imply mycotoxin contamination.

By definition, secondary metabolites are compounds whose synthesis is not essential for the survival of producing organisms and, in the case of mycotoxins, the function(s) of most of these substances, as well as the possible competitive advantage they may confer in the life cycle of respective fungal producers, have not yet been fully explored. Nevertheless, recent evidence points to some of these toxins as aggressiveness factors in the attack on plant tissue by mycotoxin-producing fungal pathogens. This is the case for some of the mycotoxins produced by *Fusarium* spp. in cereals (Snijders, 1994). Whatever be the biological/ecological role of mycotoxins in the biology of their fungal producers, they represent a serious health hazard worldwide, since they reach the consumer's desk as contaminants of plant-derived and animal-derived products, such as eggs, milk, dairy products and meat, as a consequence of feed contamination (Moss, 1996a).

Mycotoxins display great chemical diversity and approximately 400 of these fungal metabolites are currently considered to be toxic (Moss, 1996a; Sweeney and Dobson, 1998). They encompass a large and heterogeneous group of substances exhibiting acute, sub-acute, and chronic toxicity in animals and humans. Fungal species that are able to synthesize mycotoxins are present in all major taxonomic groups. Some of these compounds are carcinogenic, mutagenic and teratogenic. Further, mycotoxins reduce the quality and acceptability of agricultural products, decrease the nutritional quality of foods and dramatically decrease the viability of plant seeds. Therefore, they are considered a serious global problem from a sanitary, agricultural, economic and social point of view (Anonymous, 1979; Davis and Diener, 1987; Jones, 1992).

Although the first report on human mycotoxicoses dates back to 1100 AD (Gupta and Sharma, 1984), the most explosive acute outbreaks of the mycotoxin issue occurred in the last century and affected both humans and animals. In 1944, many people in the former Soviet Union died due to a disease (subsequently named as Alimentary Toxic Aleukia) caused by ingestion of overwintered cereals. The disease was due to contamination of the ingested cereals with T-2 toxin, a trichothecene synthesized by *Fusarium sporotrichioides* and *F. poae*. In 1960, in Great Britain, the so-called "X disease" caused the death of thousands of turkeys and ducklings, fed with peanuts imported from Brazil. This event led to the discovery of aflatoxins, which are mainly produced by *Aspergillus flavus* and *A. parasiticus*, a group of toxins including aflatoxin B_1, the most carcinogenic natural compound discovered to date. In the 1930s and again in the 1970s, a severe disease called equine leukoencephalomalakia, determining liquefactive necrosis of brain tissue, caused heavy losses in the United States, as a consequence of contamination of corn-based feed with fumonisins, toxins produced by *Fusarium verticillioides* and *F. proliferatum* (ApSimon, 1994).

Since then, mycotoxins have been considered as a major concern mainly for grain crops and derived products. Nevertheless, further research has revealed that mycotoxins are also produced by fungi that attack different fruit crops, contaminating fresh and dried fruits, and derived products such as juices, purees and wine, as shown by literature reports in the last few years (Drusch and Ragab, 2003; Logrieco and Visconti, 2004). Most, if not all, of the fungal species which are mycotoxigenic for fruit crops and which represent an actual health issue – in terms of the frequency of their attacks, the danger and quantity of produced toxins – belong to the genera *Penicillium*, *Aspergillus* and *Alternaria*. The majority of mycotoxigenic *Penicillium* spp. are pathogens of stored fruits, whereas *Aspergillus* and *Alternaria* spp. are both pre- and postharvest pathogens. Therefore, most of the concepts and issues related to mycotoxin contamination of fruits and their respective processed and semiprocessed produce pertain to postharvest plant pathology. Main mycotoxins that are commonly found in fruits are aflatoxins, ochratoxin A, patulin and *Alternaria* toxins.

It must be emphasized that in addition to all other social, economic and health problems caused by mycotoxins, contamination by these compounds undermines one of the major nutritional values and appeal of fruit-derived products (and fruits themselves) to consumers, i.e., their content of antioxidant substances. Most of the mycotoxins, in fact, react with and inactivate antioxidant compounds in attacked fruits and cause oxidative stress in different biological systems. It has been shown that supplementation of diet with natural compounds, such as vitamins, provitamins, carotenoids and phenolics, with antioxidant properties, seems to be very effective in protecting against the toxic effect of mycotoxins (Atroshi et al., 2002).

Another pivotal issue related to mycotoxins is their frequent co-occurrence in the food substrates, including fruits and fruit-derived products. This raises two additional and not secondary points:

(1) The need to take into account the established evidence of the additive and/or synergistic toxic effect exerted by mycotoxins (Speijers and Speijers, 2004).

(2) The need to develop user-friendly analytical methods, which enable personnel dealing with the monitoring of food contamination to easily detect and quantify multiple toxins at the same - reasonably short time.

After presenting chemical and toxicological characteristics and properties of mycotoxins relevant to fruits and fruit-derived products, such as juices and dried fruits, this chapter provides an overview of the present knowledge on the occurrence and origin of mycotoxin contamination of fruits and fruit-derived products, which have been identified as major health issues. It describes the fruit–fungus interactions responsible for the diseases causing mycotoxin contamination, as well as the present strategies and future perspectives for solving the mycotoxin problem in fruits and fruit-derived products.

CHEMICAL NATURE AND TOXICITY OF MYCOTOXINS CONTAMINATING FRUITS AND FRUIT-DERIVED PRODUCTS

Mycotoxins that are found with the highest frequency as natural contaminants of fruits are aflatoxins, ochratoxin A, patulin and *Alternaria* toxins. The following sections explore these metabolites in more detail.

Aflatoxins

Aflatoxins (AFs), like most of the other mycotoxins, are named after the fungal species responsible for their synthesis; in this case *A. flavus*, the filamentous fungus, which along with *A. parasiticus*, are the main species causing contamination of food. Minor AF-producing fungi are *Aspergillus nomius, A. bombycis, A. ochraceoroseus, A. tamari,* and as reported recently, *Emericella astellata* (Goto et al., 1996; Klich et al., 2000; Peterson et al., 2001; Frisvad et al., 2004). However, these fungi neither affect food production nor are encountered in nature as frequently as *A. flavus* and *A. parasiticus.*

The group of AFs include several related compounds, but the most frequently detected toxins are AFB_1, AFB_2, AFG_1 and AFG_2, which are of major concern due to their very high toxicity. AFs are cytotoxic, genotoxic, mutagenic and teratogenic. AFB_1, in particular, is the most carcinogenic natural compound so far identified. The International Agency for Research on Cancer (IARC) has classified AFB_1 as a group 1 carcinogen (i.e., there is enough evidence to conclude that it can cause cancer in humans). The chemical structure of AFB_1 is shown in **Fig. 9.1**.

AFs are difurano-coumarin compounds derived from the polyketide pathway. The intermediate compounds and precursors of their

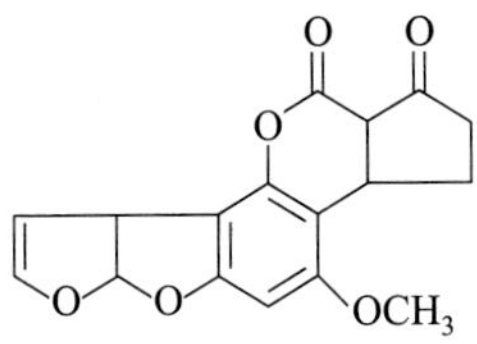

Fig. 9.1. Chemical structure of aflatoxin B_1

biosynthetic pathway, the gene clusters coding for the enzymes involved in their formation as well as some genetic elements acting as regulators of their biosynthesis, have been elucidated. A precursor of AFs, sterigmatocystin, is also a matter of concern both for its toxicity and the relative frequency of its detection in food substrates (Yabe and Nakajima, 2004). AFs are synthesized by a complex of enzymes such as cytochrome P450-type monooxygenases, dehydrogenases, methyltransferases, and polyketide and fatty acid synthases (Bhatnagar et al., 2003; Roze et al., 2004; Yu et al., 2004a, b). The high toxicity of AFs has mainly the liver as the target organ, and liver cancer is actually one of the most frequent consequences of human exposure to AFs in the diet (Bennett and Klich, 2003). Epatocytes, in fact, are the cells in which the activation of AF toxicity occurs, in particular AFB_1, through the action of monooxygenases that cause the formation of a reactive epoxide. This electrophilic compound attacks proteins and nucleic acids, thereby displaying its toxicity (Guengerich, 2003), and binds to thiolic compounds such as glutathione. The latter reaction aims at detoxification, but also causes depletion in cellular antioxidants such as glutathione itself. This event is presumably a cofactor of the oxidative stress/damage on which AFB_1 cytotoxicity and carcinogenicity are based (Towner at al., 2003).

Ochratoxin A

Ochratoxin A (OTA) is a fungal metabolite consisting of a chlorinated isocoumarin derivative linked to L-phenylalanine (*N*-[(5-chloro-3,4-dihydro-8-hydroxy-3-methyl-1-oxo-1*H*-2-benzopyran-7-yl)-carbonyl]-3-phenyl-L-alanine) **(Fig. 9.2)**. Although the biosynthetic pathway for OTA has not yet been completely established, Moss (1996b, 1998) has proposed a hypothetical pathway in which the isocoumarin moiety of OTA is derived from a pentaketide skeleton formed from acetate and malonate via a polyketide-type pathway, and L-phenylalanine is derived from the shikimic acid pathway. Recently, on the basis of the polyketide backbone of OTA, a putative polyketide synthase gene has proved to be involved in the OTA synthesis by *A. ochraceus* (O'Callaghan et al., 2003) and *Penicillium nordicum* (Gaiser et al., 2004).

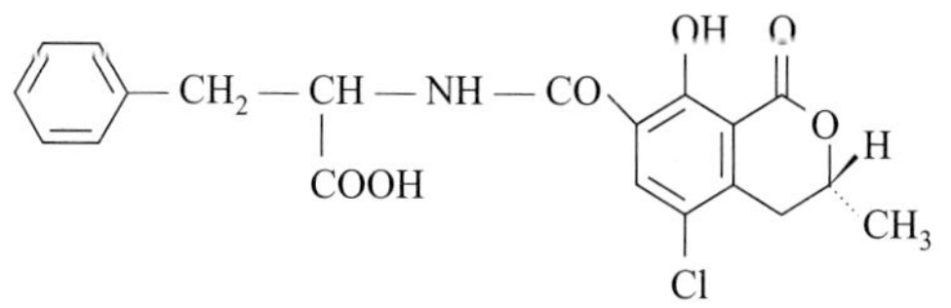

Fig. 9.2. Chemical structure of ochratoxin A

Ochratoxins are synthesized by some Aspergilli and Penicillia, with OTA being the most important in this group of mycotoxins. It is synthesized by *A. ochraceus*, *A. carbonarius* and other Aspergilli of the section Nigri (Accensi et al., 2001). Among Penicillia, *P. verrucosum* is the most important OTA-producing fungus, although it is mainly involved in OTA contamination of wheat. OTA has nephrotoxic activity and is supposedly responsible for urinary tract tumours in patients suffering from Balkan endemic nephropathy. Endemic nephropathy in humans, which affects rural adult populations in several countries in the Balkans, showed striking similarities with the porcine nephropathy initially described in Denmark (Krogh, 1974; Plestina, 1992).

The carcinogenicity of OTA has been established in both male and female rats and male and female mice. In the male rat, OTA has been found to be one of the most potent renal carcinogens (NTP, 1989). The International Agency for Research of Cancer (IARC) has classified OTA as a possible human carcinogen (Group 2B), based on "sufficient evidence" for carcinogenicity in experimental animal studies and "inadequate evidence" in humans (IARC, 1993). Humans and animals can absorb OTA via the gastrointestinal tract after ingestion of contaminated products, and inhalation of airborne toxin can represent a source of additional exposure. Recently, it has also been hypothesized that the mycotoxin is a cause of testicular cancer (Schwartz, 2002). Creppy et al. (1985) showed that OTA causes DNA single strand breaks in mouse spleen cells. These effects were also found in the kidney, liver and spleen of Bal b/c mice having received a single dose of OTA, and in rat kidney and liver, after gavage treatment for 12 weeks at a level equivalent to low dietary concentrations (2 mg kg^{-1}) (Kane et al., 1986). Further studies confirmed the genotoxicity of OTA (Obrecht-Pflumio et al., 1999). OTA has also been shown to have other toxic effects such as teratogenicity (Mayura et al., 1984) immuno-suppression (Creppy et al., 1983; Dwivedi and Burns, 1985) and mutagenicity (Hennig et al., 1991).

Patulin

Patulin (PAT), 4-hydroxy-4H-furo[3,2c]pyran-2(6H)-one, is an unsaturated lactone that was firstly isolated from *Penicillium patulum* (now called *Penicillium griseofulvum*). It is synthesized by several fungal species belonging to different genera such as *Penicillium, Aspergillus, Alternaria* and *Byssochlamis* (Moake et al., 2005), but the postharvest pathogen *Penicillium expansum* is the fungus that is the cause of contamination of fruits, pome fruits in particular, and (pome) fruit-based products. The first step of PAT biosynthesis is the condensation of acetyl CoA (coenzyme A) and 3 malonyl CoA into 6-methyl salicylic acid, catalyzed by 6-methyl salicylic acid synthase, a homotetrameric enzyme, whose respective encoding gene has been cloned and characterized (Gaucher, 1975; Lynen et al., 1978; Beck et al., 1990; Wang et al., 1991). Most of the enzymes of the PAT biosynthetic pathway have been discovered. For one of them, the patulin-specific isoepoxydon dehydrogenase, the respective gene has been sequenced and used to develop a probe for the analysis of PAT production ability by several isolates of the *Penicillium* species (Paterson et al., 2000, 2003).

During the 1940s, this molecule was tentatively used as an antimicrobial active principle, and was tested as both a nose and throat spray for treating the common cold and as an ointment against fungal skin infections (Ciegler et al., 1971; Ciegler, 1977). However, during the following decades it was ascertained that in addition to its beneficial antimicrobial and antiviral activity, PAT is toxic to both plants and animals, and as a consequence, it was re-defined a mycotoxin. The chemical structure of PAT is shown in **Fig. 9.3**.

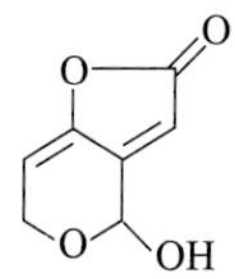

Fig. 9.3. Chemical structure of patulin

The toxicity of this metabolite is relatively low (the 50% lethal dose - LD_{50} - ranges from 10 to 35 mg $kg^{-1}$of body weight), depending on the animal species and the type of administration, especially compared to other mycotoxins such as AFs or OTA. Although PAT has been included by IARC in group 3 (compounds that are unclassifiable as to carcinogenicity in humans), recent findings suggest that it exerts *in vitro* genotoxic effects and oxidative damage to the DNA of mammalian cells, including human peripheral blood lymphocytes and human embryonic kidney cells (Biing-Hui et al., 2003). As in the case of AFs, PAT also appears to induce oxidative stress at the cellular level through depletion of antioxidant glutathione, to which it binds as a consequence of its electrophilic reactivity, and through generation of reactive oxygen species (Barhoumi and Burghardt, 1996; Fliege and Metzler, 2000; Rychlik et al., 2004). As for many other mycotoxins, immunotoxic and immunosuppressive effects were also described (Bourdiol et al., 1990; Wichmann et al., 2002). Although PAT is not as toxic as other mycotoxins, interest in this metabolite is due to the fact that it can be a contaminant of juices and baby foods, mainly from pome fruits, which are frequent components of children's diets.

Alternaria Toxins

Although species of the genus *Alternaria* are known to produce at least 70 secondary metabolites, only seven major toxins belonging to three different structural classes are known as possible food contaminants with a potential toxicological risk (Pero et al., 1973; Griffin and Chu, 1983; Stack and Prival, 1986). These are tenuazonic acid (TA) (3-acetyl-5-sec-butyltetramic acid), a tetramic acid derivative; alternariol (AOH) (3,4′,5-trihydroxy-6′-methyldibenzo[a]-pyrone), alternariol methyl ether (AME) (3,4′-dihydroxy-5-methoxy-6′-methyldibenzo[a]-pirone) and altenuene (ALT) (2′,3′,4′,5′-tetrahydro-3,4′β,5′α-trihydroxy-5-methoxy-2′β-methyldibenzo[a]-pyrone), which are dibenzopyrone derivatives; altertoxin I (ATX-I) (3,6a,7,10-tetrahydroxy 4,9-dioxo-4,5,6,6a,6b,7,8,9-octahydroperylene), altertoxin II (ATX-II) (11R, 12R, 12aS, 12bR)-4,9,12b-trihydroxy-3,10-dioxo-1,2,3,10,11,12,12a,12b-octahydro-11,12-epoxyperylene) and altertoxin III (ATX-III) [Perylo(1,2-b:7,8-b′)bisoxirene-5,10-dione,1a,1b,5a,6a,6b,10a-hexahydro-4,9-dihydroxy-], which are perylene derivatives **(Fig. 9.4)**.

Most information on the biosynthesis of *Alternaria* toxins concerns the polyketides AOH/AME. The currently accepted view is that AOH is synthesised via a classical polyketide mechanism with a single enzyme-bound polyketide chain assembled from one acetyl-CoA and six malonyl-CoA units (Gatenbeck and Hermodsson, 1965). Häggblom and Hiltunen (1992) pointed out that AME dibenzo-pyrone formation is synthesized from AOH by an AOH-methyltransferase. Other pathways are also possible, such as that involving antraquinone rearrangement, which should give rise to a labelling pattern in AOH similar to that shown for the polyketide mechanism (Stinson, 1985).

Alternaria metabolites have been shown to have various biological activities, such as antiviral, antibacterial, antifungal, insecticidal, and are also toxic to several plant and animal systems (Woody and Chu, 1992). Of the *Alternaria* metabolites, ATX-II showed the highest toxicity towards HeLa cells [50% inhibition of cell growth (ID_{50})= 0.5 mg/ml], followed by

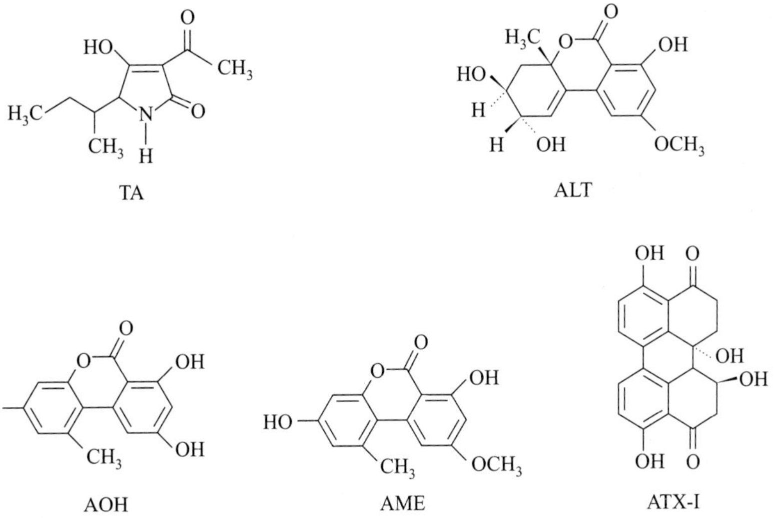

Fig. 9.4. Chemical structures of food-related *Alternaria* toxins. TA = tenuazonic acid; AOH = alternariol; AME = alternariol methyl ether; ALT = altenuene ; ATX-I = altertoxin I

Altertoxin-I (ID_{50} = 20 mg ml^{-1}). A similar relative toxicity was reported for Chinese hamster lung fibroblast (V79): the toxic activity of altertoxins in this assay was almost 100 times higher than that in HeLa cells, and the noncytotoxic levels were found to be less than 0.02, 0.2, and 5 μg ml^{-1} for ATX-II, ATX-III, and ATX-I, respectively (Pero et al., 1973; Boutin et al., 1989). Dibenzo-α-pyrones were reported to be less cytotoxic on HeLa cells than ATX-II, with ID_{50} values of 6, 8-14, and 28 μg ml^{-1} for AOH, AME, and ALT, respectively (Pero et al., 1973). According to Scott and Stoltz (1980) and Woody and Chu (1992), the mutagenic activity of several *Alternaria* metabolites determined, on *Salmonella typhimurium* strains by the Ames test, had the following order: AOH and TA as non-mutagenics, AME as weak mutagenic, and ATX-I, and ATX-II with significant mutagenic activities. In this regard, altertoxins produce mutagenic intensities in the following order: ATX-III > ATX-II > ATX-I, with ATX-III being at least 10 times less mutagenic than aflatoxin B_1 (Stack and Prival, 1986). The mutagenic activity of altertoxins and AME, the most active dibenzo-aa-pyrone derivative, was also confirmed in other systems assessing the potential carcinogenicity. Altertoxins I and II were tested in the Raji cell Epstein-Barr virus early antigen (EBV-EA) induction system, and in murine fibroblast (C3H/10T-1/2) cell transformation system. Both altertoxins significantly enhanced fibroblast transformation and, in addition, increased the activation of EBV-EA expression by eight to ten fold, strongly suggesting that these metabolites play a mutagenic/

carcinogenic role (Osborne et al., 1988). Similar mutagenic/carcinogenic activity was reported in China for culture extracts of strains of *Alternaria alternata* from areas of the Linxian region with high incidence of oesophageal cancer, leading to identification of AME as the main active metabolite (Liu et al., 1988). The mutagenic/carcinogenic activities of AME could explain those exhibited by the culture filtrate of *A. alternata* (An et al., 1989), but other toxicological studies strongly suggest the involvement of other metabolites with more specific mutagenic activity like altertoxins that are as common as alternariols in cultures of the fungus (Stack and Prival, 1986; Osborne et al., 1988; Davis and Stack, 1991). Although ATX-I, ATX-II, TA and AME could have serious effects on human and animal health, no regulations for these Alternaria toxins exist.

THE ATTACK OF MYCOTOXIGENIC FUNGI ON FRUITS

Mycotoxigenic fungi that attack fruits are wound pathogens penetrating their hosts mainly through surface damage, inflicted as a consequence of biotic and/or atmospheric factors and/or improper handling before or during transportation and storage. After penetration, these molds act as necrotrophic fungi, i.e., cause necrosis of the infected plant tissue, and after the death of their host, thrive and multiply on dead plant material. Although the definition "necrotrophic fungi" is generally used for necrotrophic fungal pathogens that attack their host plant(s) in the field, it can also be extended to pre- and postharvest mycotoxigenic pathogens of fruits. Generally, necrotrophic pathogens can live as saprophytes and colonize dead plant material even when they have not been the cause of the plant's death. In other words, necrotrophic (and mycotoxigenic) fungi are facultative parasites that display a relatively low level of specialization. This, in turn, corresponds to a relatively low host specificity; *P. expansum*, for example, is a major pathogen of stored pome fruits, but has a broad host range including cherries, nectarines, and peaches (Marek et al., 2003). However, a certain degree of host range specificity among fruit-infecting pathogens does exist: *P. digitatum*, a postharvest pathogen (not reported to be mycotoxigenic) of citrus fruits, does not attack pome fruits or other deciduous fruits. The level of host specificity displayed by *P. digitatum* can also be referred to as basic compatibility between this pathogen and citrus fruit species - the final outcome of the interaction being the disease and as incompatibility with host fruit species other than citrus - the outcome of the interaction resulting in no infection, i.e., in resistance by non-citrus host(s). Unlike specific fungal pathogen-host plant interactions in which incompatibility (resistance) depends on a single gene pair, as in the case of gene-for-gene interactions based on a host resistance gene (R) and a pathogen avirulence gene (Avr) (Flor, 1971), resistance to postharvest pathogens of fruits (and, generally speaking, to all mycotoxigenic fungi including *Fusarium* spp.) is usually the result of several genes, whose interactions are poorly understood (Prusky, 2003). As a consequence, they cannot be used to develop control programs based on breeding for resistance. Breeding programs for preventing attacks by mycotoxigenic fungi are currently being attempted with encouraging results only against *Fusarium* spp., field pathogens of cereals, by relying on quantitative trait loci (Bai and Shaner 2004; Snijders, 2004).

A pivotal characteristic shared by fruit-infecting fungi is their ability to cope with the acidity of their hosts, the reason for which molds are the main spoilage microorganisms of fruits and fruit products (Splittstoesser, 1987). Fruits have pH values ranging from 5.0 to less than 2.5, and most fungi tolerate these levels of acidity or have their optima for growth within the above mentioned range. In addition, modulation of pH seems to be driven by pathogens during their attack on plant tissue (Prusky and Yakoby, 2003). Acidification of host tissue, in particular, appears as an active process led by postharvest pathogens such as *P. expansum*, *P. digitatum*, *P. italicum* and *Botrytis cinerea* (the agent of gray mold on several plants and fruits) during their attack (Prusky and Yakoby, 2003; Prusky et al., 2004). Modulation of apoplastic pH by the pathogens appears to be aimed at adjusting this parameter to values favoring degradation of plant cell walls by fungal enzymes (Prusky and Yakoby, 2003). For most pathogens of harvested fruits, in fact, infection of the host tissue depends on pectolytic enzymes degrading the polymers of host cell walls, and in particular, the pectic substances of middle lamellae. Middle lamellae are the structures on which the consistency of the fruit relies, because they bind fruit cells together. Pectolysis by fungal enzymes, especially those with an endo-mode of action, leads to tissue maceration due to separation of individual cells, which die from the increase in the permeability of their membranes. Metabolites are thus released and used by the fungus for its further development (Yao et al., 1996; Korsten and Wehner, 2003). The importance of cell-wall-degrading enzymes in the attack on host tissue has also been demonstrated at a molecular level: disruption of the genes *Bcpg*1 of *B. cinerea* and *pec*A of *A. flavus*, both encoding for endo-polygalacturonases, reduced lesion development and cotton ball invasion, respectively (Shieh et al., 1997; Ten Have et al., 1998).

During storage, fruit ripens as its skin layers soften, soluble carbohydrates are formed and defense barriers weaken (Pitt and Hocking, 1997). In this process, endogenous loosening of cell walls also occurs, and this probably contributes to the increased susceptibility of fruit to fungal infection. Ripening (and senescence) of fruits is also characterized by the accumulation of the reactive oxygen species, hydrogen peroxide (H_2O_2) (Halliwell and Gutteridge, 1999). Further, it has been shown that superoxide anion ($O_2^{\bar{}}$) and H_2O_2 are produced at higher levels in fresh wounds of apples stored for longer periods of time than in those at the beginning of the storage period (Castoria et al., 2003). All these conditions could further contribute to the increased susceptibility of stored fruits to necrotrophic fungi. Fungal pathogens such as *B. cinerea* and *P. expansum*, in fact, have been reported to generate/induce reactive oxygen species during their attack on plant and fruit tissue (Prins et al., 2000; Hadas et al., 2004).

MYCOTOXINS IN FRUITS AND FRUIT-DERIVED PRODUCTS

Mycotoxins in Fruits

The natural occurrence of mycotoxins in fruits is elaborated in this section. It should be noted, however, that mycotoxin formation has also been extensively studied following artificial inoculation of fruits, including fruit species that are not usually attacked by certain mycotoxigenic fungi. In some cases, these studies have provided useful information on environmental, nutrional and host factors that either favor or are detrimental to toxin synthesis and penetration of the fungus into the fruit tissue. Furthermore, they collectively suggest that the possibility of fruit contamination with "unexpected" mycotoxins cannot be ruled out. For a comprehensive review of such studies refer to Drusch and Ragab (2003).

Aflatoxins

The growth of aflatoxin (AF)-producing *A. flavus* and *A. parasiticus* is favored by climates characterized by relatively high temperatures. Aspergilli are generally recognized as relatively xerotolerant (can survive and multiply in drier conditions, i.e., in conditions with lower levels of water activity). For these fungi, therefore, humidity, and in particular, water content of attacked substrate (expressed as water activity = a_w), are not as crucial as for other mycetes. This is also true for AF synthesis, which occurs at lower a_w values than for toxins of other mycotoxigenic fungi (**Table 9.1**).

Therefore, natural contamination with AFs is common in regions with warmer climates. Until recently, this contamination was reported as a severe problem for peanuts and tree nuts such as almonds, pecans,

Table 9.1. Minimun a_w values for the growth and mycotoxin formation of some potentially toxic fungi

	Minimum a_w	
Fungus	*Growth*	*Mycotoxin Formation*
Aspergillus flavus	0.78 - 0.82	0.83 - 0.87
A. parasiticus	0.80 - 0.82	0.83
A. ochraceus	0.77	0.88
Penicillium expansum	0.83 - 0.85	0.99

walnuts, Brazil nuts and pistachios (FAO/WHO, 2002a; Mphande et al., 2004). Attacks on pistachios by *A. flavus* and *A. parasiticus* can occur both in the field and during storage, especially if the storage phase is not carried out properly and the drying process is too slow. In the field, nuts affected by early splits (i.e., immature nuts in which shell splitting is accompanied by rupture of the adhering hull) and by the navel orange worm *Amyelios transitella,* are more susceptible to attack by aflatoxigenic fungi (FAO/WHO 2002a).

Contamination with AFs is a major issue for figs too. Like pistachios, figs are susceptible to (high levels of) AF contamination especially during the slow drying process. Further, *A. flavus,* a wound pathogen that readily penetrates wounds (such as those caused by insect feeding), can apparently also penetrate the intact skin of ripe fruits (Buchanan et al., 1975). Natural contamination with AFs has been reported for apples as well: levels of total AFs up to 350 $\mu g\ g^{-1}$ fresh weight were detected in the decaying part of apples from Egyptian markets (Hasan, 2000). Although aflatoxigenic *A. flavus* was able to cause apple rot in artificial inoculation experiments, it wasn't clear whether the natural AF contamination recorded in the survey was due to primary infection by toxigenic aspergilli, or to their "opportunistic" attack on apples already infected by other pathogens. Nevertheless, the danger of AFs should be made evident to people involved in apple (and apple juice) production, whatever the origin of such contamination, especially in warmer areas. In figs examined at different production stages, AF levels up to 141 $\mu g\ kg^{-1}$ (AFB_1 plus AFG_1) have been recorded (Ózay and Alperden, 1991). Examination of data led the authors of this research to conclude that contamination with AFs occurs only if environmental conditions are favorable for contamination. However, the considerations for apples are valid for figs too, especially if one considers that in the same study, co-occurrence of AFs and OTA was reported. Contamination with AFB_1 (11.61 $\mu g\ kg^{-1}$) has also been reported for dates, although at inedible maturation stages only (Shenasi et al., 2002). Further surveys should be carried out to verify the absence of contamination in the dried final product.

Ochratoxin A(OTA)

OTA is considered as one of the main mycotoxins commonly found in food products. Although cereals and cereal-based products normally account for 50-80% of average intake of OTA by consumers (Jorgensen and Jacobsen, 2002), fruits, grape juice, wine and dried fruits are becoming increasingly important as food contaminated by OTA. **Table 9.2** contains a list of fruits proved to be contaminated by this toxin.

Table 9.2. Natural occurrence of Ochratoxin A in fruits

Fruit	*No. of Samples Positive/ No. of Samples Analyzed*	*Maximum OTA Concentration μg kg^{-1}*	*Reference*
Apple	2/4	0.4	Engelhardt et al., 1999
Apricot	0/2	-	Engelhardt et al., 1999
Cherry	6/6	2.7	Engelhardt et al., 1999
Grape	15/52	1.5	Eltem et al., 2003
Nectarine	0/5	-	Engelhardt et al., 1999
Peach	1/9	0.6	Engelhardt et al., 1999
Strawberry	4/10	1.4	Engelhardt et al., 1999
Tomato	6/11	1.4	Engelhardt et al., 1999

As regards fruit contamination with OTA, grape black rot caused by *Aspergillus* is particularly important from a mycotoxicological point of view. Black aspergilli, and primarily strains of *Aspergillus carbonarius* and the *A. niger* aggregate, are considered as the major causes for OTA contamination in grapes and derived products (Abarca et al., 2001; Battilani and Pietri, 2002; Cabanes et al., 2002; Da Rocha Rosa et al., 2002; Magnoli et al., 2003; Battilani et al., 2004). Available evidence indicates that these black aspergilli are mainly saprophytes and responsible for secondary rot (Pitt and Hocking, 1997). Grape black rot is particularly spread in warmer grape-producing areas such as Mediterranean countries (Battilani and Pietri, 2002), Australia (Nair, 1985), as well as in temperate lands such as southern parts of Canada (Jarvis and Traquair, 1984). Infected tissue is generally pale at the first stage of the disease and then becomes black due to the conidial mass. The conidial heads may be seen with the naked eye and the spores are easily liberated when mature, resulting in soot-like deposits on adjacent berries. Cracks and wounds on berries caused by insect or bird feeding may provide entry to berry tissue where fungi find an ideal habitat for their development. A correlation between *Lobesia botrana*-damaged berries and OTA contamination in the field has recently been found (Cozzi et al., 2006). Grape black rot is often associated with OTA occurrence in grape, and consequently, in must and wine. Screening the presence of OTA in freshly harvested grapes, Eltem et al. (2003) showed that 15 out of 52 samples analyzed from various Turkish vineyards, tested positive for OTA, with a concentration of up to 1.5 $\mu g\ kg^{-1}$. In addition, the same authors reported OTA contamination of 20 out of 61 raisin samples (from 0.1 to 25 $\mu g\ kg^{-1}$). A different susceptibility to *A. carbonarius* infection and OTA contamination of 12 different grape varieties has been reported by Battilani et al. (2004).

Patulin (PAT)

Contamination of pome fruits and derived juices with PAT is mainly due to the postharvest disease known as blue mold rot of apple and pear, which is caused by *P. expansum,* a common soilborne fungus, often isolated from the surface of healthy fruits. Under laboratory conditions, PAT can also apparently be produced in grape juice and grains, and has also naturally been found in peaches, strawberries, blueberries, cherries, apricots and grapes (Majerus and Kapp, 2002), but its natural occurrence is essentially a consequence of *P. expansum* attack on apples and pears. However, *P. expansum* is generally considered to be a wound parasite. Mold growth normally occurs where the surface of fruit has been damaged. The rotted areas are soft, watery, and depending on the fruit variety, their color ranges from light to dark brown. The surface of older lesions may be covered by conidial tufts consisting of bluish-green spores, and a typical musty odor is another important characteristic for recognizing infection by *P. expansum.* This fungus has also been reported as a rotting agent of fruits other than apples and pears, and PAT has been detected in various berries, stone fruits, bananas, pineapples, grapes and tomatoes, in some cases at a high concentration (Drusch and Ragab, 2003). Some studies have pointed to the absence of correlation between size of the lesion caused by the fungus, expressed as rotting area or lesion diameter, and PAT concentration (Beretta et al., 2000; Martins et al., 2002). The rotting area or lesion diameter, however, can be considered only as a semi-quantitative index of pathogen growth in decaying fruit tissue. A softer texture of single fruits, in fact, could be coupled to a higher effectiveness of extracellular cell wall depolymerases of the fungus. As a consequence, higher levels of tissue maceration take place, which do not reflect a correspondent fungal growth and toxin synthesis. The level of PAT contamination of apples can range from µg to mg kg^{-1} (Piemontese et al., 2005). This variability probably depends on the biosynthetic ability of *P. expansum* strains and on the attacked apple cultivars. Recently, differences have been reported regarding the levels of PAT contamination in different apple varieties (Martins et al., 2002), but further studies are needed to confirm these results.

Infection of pome fruits commonly follows insect or weather-derived damage during preharvest, rough-gathering and handling at harvest, or strong washing and sorting procedures after harvest. Fungal disease previously occurring in the field, such as apple scab and powdery mildew, can favour infection by *P. expansum* (FAO/WHO, 2002b). Although less frequently, fungal penetration can also occur through the lenticels, the petioles and the pervious calyx tube. In the latter case, the infection is particularly insidious because it is not associated with external symptoms, but can contribute to PAT contamination. Optimal growth of *P. expansum* and the highest levels of toxin synthesis on pome fruits take place at 17-20°C, but during storage infection can occur even at 0°C. At this temperature decay progresses slowly, but it usually develops very rapidly when fruits are brought to higher temperatures. Conidia of *P. expansum* remain viable for long periods of time on pallet-boxes and in the environment of storage facilities. Conidia are easily carried by wind and water. Since fruits often come in contact with water before packing, this represents a major source of inoculum and disease.

Alternaria Toxins

Alternaria toxins (TA) contaminate fruits and other horticultural crops, in particular tomatoes, and consequently, canned tomato-derived sauce and other products. The species that is mainly responsible for such contamination is *A. alternata*. Examples of main fruit diseases and contamination of fruit products by *Alternaria* mycotoxins are reported in **Table 9.3**.

Tomato fruit rot caused by *Alternaria* (blackmold tomatoes) occurs on green and ripe fruits that are affected by some physiological alterations (nutritional deficiency, skin sun-burn etc.). Warm rainy weather or dew formation on the fruit surface are conditions favorable to the disease, the severity of which is greater for infections occurring in the ripe stage than in the green stage. Such disease can frequently cause substantial loss of tomatoes, especially those used for canning. Fruit rot lesions are irregular in shape and slightly sunken. They become light green as sporulation begins. Older lesions are circular and sunken, and dark green to almost black from the abundant sporulation of the fungus. *Alternaria* rot sometimes develops on the fruit beneath the sepals. In this case the fungus can colonize the internal tissues without a clearly visible infection.

Samples of infected tomatoes were found to be contaminated by TA (up to 7.2 $\mu g\ g^{-1}$). All tested strains of *A. alternata* from blackmold tomatoes

Table 9.3. Natural occurrence of main *Alternaria* toxins in fruits

Country	*Fruit*	*Toxins (mg kg⁻¹)*			*Reference*
		AME	*AOH*	*TA*	
Italy	Tomato [(a)]	0.04 - 0.20	1.00	0.02 - 7.00	Visconti et al., 1987
Italy	Olive [(b)]	0.03 - 3.00	0.10 - 2.30	0.10 - 0.20	Visconti et al., 1986
Italy	Melon [(b)]	0.05	—	0.08	Logrieco et al., 1988
Italy	Pepper [(b)]	0.05	0.64	0.05	Logrieco et al., 1988
Italy	Mandarin [(c)]	0.50 - 1.00	1.00 - 5.20	21.00 - 170.00	Logrieco et al., 1988
Italy	Mandarin [(d)]	0.50 - 1.00	1.00 - 5.20	21.00 - 87.00	Logrieco et al., 1990
Italy	Mandarin [(e)]		—	173.00	Logrieco et al., 1990
USA	Tomato	0.30	up to 1.30		Stinson et al., 1981
USA	Tomato [(f)]			Traces	Stack et al., 1985
USA	Apple	<1.00	<1.00		Stinson et al., 1981
———					Wittkowski et al., 1983

(a) Preharvest infected tomato fruits
(b) Mouldy samples at harvest time
(c) Fruit rot of mandarins at harvest time
(d) Black fruit rot of mandarins at harvest time
(e) Gray fruit rot of mandarins at harvest time
(f) Samples of tomatoes from commercial processing lines

produced *in vitro* high amounts of TA (589 to 4,200 $\mu g\ g^{-1}$), while *Alternaria tenuissima* produced low levels of toxin (Visconti et al., 1987). In another survey, selected tomato samples infected by the fungus were contaminated with high concentrations of TA, AME, and AOH (up to 7,200, 260 and 1,200 $\mu g\ g^{-1}$, respectively) (Scott and Kanhere, 1980). Processed tomato pastes were also found to be contaminated with TA (Scott and Kanhere, 1980), as were the tomato samples collected from processing lines (Stack et al., 1985).

Olives can also be easily affected by *Alternaria* species (black olive fruit rot), particularly when the fruits remain on the tree or on the soil for a long time before harvest. Physical damage of the olive surface due to various unfavorable factors (low temperature, insects, etc.) is an important condition for fungal penetration into the fruit pulp and the consequent mycelial growth. During this colonization process the fungus can produce toxic secondary metabolites, which can be found in infected tissues, and consequently, in olive oil (Visconti et al., 1986).

Mandarin fruits with *Alternaria citri* black rot are commonly observed in the field at harvest time. In the first stage of the disease, the fruit does not show any symptoms, but later the surface turns dark starting from the peduncle, and in the advanced stage of the disease the fruit generally falls to the ground. Two kinds of *Alternaria* rots are distinguishable, based on the color of the diseased tissues (gray and black). The causal agents produce gray and black colonies when cultured on potato-sucrose-agar. The darker color is associated with sporulation, whereas the gray color is associated with gray mycelium with conidiophores growing on the aerial hyphae. The two populations attack various citrus fruits, including mandarins, oranges, and lemons to different degrees. They colonize the same percentage of mandarin tissue, but the black population colonizes larger amounts of orange and lemon tissues than the gray. The two kinds of rots differ substantially, both qualitatively and quantitatively, in the production of *Alternaria* toxins. Black rot samples were found to be contaminated by TA, AME and AOH (up to 87, 1.4 and 5.2 $\mu g\ g^{-1}$, respectively), whereas TA was the only detectable toxin in the gray rot samples (173 $\mu g\ g^{-1}$) (Logrieco et al., 1990). Regarding the capacity to produce toxins, the gray strains are reported to produce more ATX-I than black strains, which produce more benzopyrone derivatives (AME and AOH). The occurrence of large amounts of toxins in the naturally infected samples and the toxigenic potential of the strains may represent a potential health hazard, considering that the toxins may also be transferred into processed products. As mentioned above, oranges and lemons can also be attacked by *Alternaria,* and some toxins (TA, AOH and AME) were detected when artificially infected with *Alternaria citri* (Stinson et al. 1981).

Mycotoxins in Dried Fruits

Mycotoxins that have so far been found as contaminants in dried fruits are AFs and OTA. Dried fruits most liable to such contamination are mainly figs, nuts and grapes that are produced in warm, tropical and subtropical areas, where the climate favors both fungal growth and synthesis of these mycotoxins (Drusch and Ragab, 2003). However, recent reports from South America and southern regions of Europe, where ochratoxigenic fungi have been found to attack wine grapes and cause OTA contamination of wine (see below), indicate that contamination with OTA is also affecting warmer areas of temperate regions. Therefore, surveys for this mycotoxin are needed in other areas with analogous climatic characteristics such as California and Australia, both for dried vine fruits and wine. Global warming, caused by release in the atmosphere of CO_2 and other "greenhouse gases" as a result of human activities, has also to be taken into account, since it could favor the diffusion in temperate regions of mycotoxigenic fungi usually detected in tropical and subtropical areas.

Aflatoxins

Dried raisins have been found to contain AFs (Drusch and Ragab, 2003); in a study carried out in Egypt, in particular, two out of 100 samples of dried raisins contained up to 300 μg kg^{-1} of AFB_1 (Youssef et al., 2000). Much of the information on AF contamination in dried fruits relates to figs and pistachios. In 1997, contamination by AFs was so high that it caused the European Union to ban pistachios imported from Iran (FAO/WHO 2002a). A total AF content of 2,200 μg kg^{-1} has been recorded in one sample of pistachios analyzed in Sweden (Thuvander et al., 2001). In the same study on figs under the section on AF contamination of fruits, Ozay and Alperden (1991) detected the presence of all four main AFs, with maximum levels ranging from 63.0 to 78.3 μg kg^{-1}. Homogenization of dried figs for production of fig paste apparently lowered concentration of AFs as well as OTA, which co-occurred in the analyzed samples (Ozay and Alperden, 1991). The high AF levels recorded in dried figs are probably due to the greater susceptibility of these fruits to aspergilli during a slow drying process (refer to the section above on AF contamination of fruits). It is conceivable that a slow drying process creates an environment that is favorable to aspergilli, which can more easily prevail over less xerotolerant potentially competing microflora.

Ochratoxin A

The *Aspergillus* rot of figs is the disease responsible for mycotoxin contamination of dried figs. The rot may be predominantly internal, masses of black powdery conidia being apparent only when the fruit is cut. Alternatively, a very soft sunken water-soaked spot may develop in the peel, later giving rise to black conidia resembling soot. The decay is generally accompanied by an odor of fermentation. Infections can occur whilst the fruits are still on the tree or on the ground.

The formation of OTA in figs is an important mycotoxicological problem for dried figs, especially in the Mediterranean area. The main *Aspergillus* species isolated from dried figs in Turkey were: *A. foetidus* var. *pallidus*, *A. niger*, *A. awamori*, *A. aculeatus*, *A. carbonarius* (Aksoy et al., 2003). Ochratoxin A level and incidence were reported as 5.2-8.3 μg kg^{-1} and in 3% of examined samples (Ozay and Alpenden, 1991). On the other hand, *A. alliaceus* is another important OTA-producing fungus in California, where it may be responsible for the OTA contamination occasionally observed in Californian figs (Bayman et al., 2002).

The occurrence of OTA in raisins has been recently reported in grape Sultanas from Aegean region (Eltem et al., 2003). The most frequently isolated species were *A. foetidus* var. *pallidus*, *A. aculeatus*, *A. tubingensis* and *A. awamori*.

Mycotoxins in Fruit Juices and Other Fruit-derived Products

Contamination of fruit juices and other fruit-derived products with mycotoxins is the obvious consequence of the utilization of contaminated fruits that have been attacked by toxin-producing fungi, especially during storage. **Table 9.4** summarizes the mycotoxins and their maximum concentrations detected in fruit juices. Juice contamination could be

Table 9.4. Mycotoxins in fruit juices (modified from Drusch and Ragab, 2003)

Juice	*Toxin(s)*	*No. of Samples Positive/No. of Samples Analyzed*	*Maximum Concentration µg liter^{-1}*	*Reference*
Apple juice	Aflatoxins B1, G1	5/5	30.00	Abdel-Sater et al., 2001
Guava juice		2/5	12.00	
Grape juice, white	Ochratoxin A	21/27	1.30	Majerus et al., 2000
Grape juice, red		56/64	5.30	
Black currant juice		3/19	0.10	
Tomato juice		3/30	0.03	
Grape must		8/11	0.50	Sage et al., 2002
Grapefruit juice		1/14	1.20	Filali et al., 2001
Apple juice with pulp		-	1,150.00 *	Beretta et al., 2000
Apple and mixed fruit juice	Patulin	101/241	1,130.00	Burda, 1992; Ritieni, 2003
Apple juice concentrate, diluted		215/215	376.00	Gokmen and Acar, 1998
Pear juice		34	20.00*	Ehlers, 1986
Apple juice		27/45	733.00	Yurdun et al., 2001
Apple juice concentrate	Alternariol	17/32	5.42	Delgado and Gómez-Cordovés 1998
Apple juice concentrate	Alternariol methyl ether	17/32	1.71	

*µg kg^{-1}

prevented by the selection of healthy fruits and/or the removal of rotten parts before processing. Unfortunately, this selection is difficult on an industrial scale and the absence of toxins in the final product cannot be ensured. Most of the studies on toxin content in juices focus on PAT, whereas only a few reports exist on the occurrence of AFs, mainly detected in fruits and respective juices produced and processed in warmer climates (Hasan, 2000; Abdel-Sater et al., 2001). However, AFs (and OTA) have recently been found as contaminants of olive oil (Papachristou and Markaki, 2004), although the origin of such contamination has not yet been identified. In the last few years, contamination of wine and grape juice with OTA has also become a matter of concern. For these reasons, this section mainly deals with PAT and OTA. Implications of *Alternaria* diseases of olives and tomatoes for olive oil and tomato-derived products are also discussed.

Ochratoxin A (OTA)

The occurrence of OTA in wine has been widely documented. Wine is receiving increasing attention because after cereals, it is a major source of daily intake, contaminated with OTA, in both developed and developing countries (**Table 9.5**). Various reports evidenced that the highest level of OTA in wines was from wine growing regions of southern Europe and northern Africa, and not in wines from the northern areas (Zimmerli and Dick, 1995; Visconti et al., 1999; Otteneder and Majerus., 2000; Pietri et al., 2001). Surveys on wine revealed higher OTA contamination in red rather than in white wines, and the geographical origin had a significant influence on OTA content (Ottender and Majerus, 2000; Pietri et al., 2001).

Table 9.5. Occurrence of OTA in wine

No of Wine Samples	*Percentage of OTA Contaminated Samples*	*Range ($\mu g\ l^{-1}$ or ppb)*	*Reference*
11	8	0.001-0.500	Sage et al., 2002
30	50	0.010-0.300	Ospital et al., 1998
46	41	0.010-0.200	Ueno, 1998
55	87	0.010-7.600	Visconti et al., 1999
60	56	0.010-0.800	MAFF, 1997, 1999
111	82	0.001-3.800	Pietri et al., 2001
118	70	0.005-0.400	Zimmerli and Dick, 1995
144	42	0.010-7.000	Majerus and Otteneder, 1996
420	48	0.010-3.300	Otteneder and Majerus, 2000
192	82	0.010-0.600	Burdaspal and Legarda, 1999

Visconti et al. (1999) analyzed samples of Italian wines for contamination with OTA. Ochratoxin A was found in 37/38 red wine samples (traces to 7.6 ng ml^{-1}), 7/8 rosè wine samples (traces to 1.1 ng ml^{-1}), in 4/9 white wine sampies (traces to 0.9 ng ml^{-1}), and in 1 sample of dessert wine (0.3 ng ml^{-1}). Maximum tolerable levels for OTA concentration in wine and raisin of 2 and 10 μg kg^{-1}, respectively, have recently been set by the European Union (European Commission, 2005), and a recommendation was made to lower OTA to the lowest technologically feasible levels.

It is worth mentioning that another major fruit-derived product, olive oil, has recently been found to be contaminated with OTA (and AFs) (Papachristou and Markaki, 2004). Although isolates of potentially ochratoxigenic *Aspergillus* spp. have been detected on olives (Gourama and Bullerman, 1988), the fungi responsible for such contamination have not yet been unequivocally identified.

Patulin

Patulin is a "worldwide" mycotoxin, which has predominantly been detected in apple juices, as well as other apple-derived products, even at very high concentrations **(Table 9.4)**. This toxin is stable at acidic pH and, therefore, it can be detected unaltered in these products even after several weeks of juice storage. Pome fruit-derived products have been considered for a long time as the only ones that could be contaminated with this mycotoxin. More recently, however, PAT has been shown to be synthesized in other unfermented juices, such as those derived from blackcurrant, cherry and other small fruits (Larsen et al., 1998). This toxin is detected much less frequently in fermented apple juice (cider), because during fermentative growth of the yeast, *Saccharomyces cerevisiae* lowers PAT levels by producing two major metabolites: E-ascladiol, which is also the immediate precursor of the toxin in its biosynthetic pathway, and the isomer Z-ascladiol (Moss and Long, 2002). E-ascladiol reacts with sulfhydryl-containing compounds, but has a much lower toxicity than the toxin (Suzuki et al., 1971). Some cases of cider contamination have been recorded (Tangni et al., 2003,). These studies, however, do not clarify whether the presence of PAT was due to the specific inability/low efficiency in its degradation by the yeast strains used in cider production, or whether the mycotoxin concentration of apples before processing was so high that it could not be significantly lowered by the fermentation process. After fermentation, cider can support both *P. expansum* growth and PAT formation (McCallum et al., 2002).

An aspect that should not be overlooked is the production by *P. expansum* of additional secondary metabolites other than PAT, such as citrinin, chaetoglobosins, communesins and expansolides, and the presence of most of these metabolites both in apple fruit and in juice from cherries and gooseberries (Andersen et al., 2004). Co-occurrence of citrinin and PAT has been recorded in 19.6% of apple samples examined in Portugal (Martins et al., 2002). Citrinin acts as a nephrotoxin in all animal species tested (Bennett and Klich, 2003), whereas no information is available on the toxicological risks associated to the ingestion of the other metabolites. If toxicological tests for these compounds are positive, a multitoxin approach should be developed to increase the safety of fruit-derived products, such as the one used in a recent mycotoxin survey (Castoria et al., 2005a).

Alternaria Toxins

Mycotoxicological problems related to the potential occurrence of *Alternaria* toxins can take place in various fruit-derived products and particularly in tomato paste, in other canned tomato products (Stinson et al., 1980, 1981) and olive oil (Visconti et al., 1986). As mentioned above in the section on the occurrence of mycotoxins in fruits, processed tomato paste and tomato samples collected from processing lines were also found to be contaminated with TA (Scott and Kanhere, 1980; Stack et al., 1985).

A limited survey on the natural occurrence of the major *Alternaria* mycotoxins was carried out in Southern Italy on olives and related processing products (oil and husks) by Visconti et al. (1986). Some samples were naturally contaminated. Although *Alternaria* species, mostly *A. alternata*, were present on all examined olive samples, mycotoxins were only detected in samples of physically damaged olives. The highest contamination was found in a severely damaged sample that contained AME (2.9 $\mu g\ g^{-1}$), AOH (2.3 $\mu g\ g^{-1}$), ALT (1.4 $\mu g\ g^{-1}$) and ATX-I (0.3 $\mu g\ g^{-1}$). No mycotoxin was detected in olive oil earmarked for human consumption or olive husks collected from oil mills after the first pressing of the fruits. Nevertheless, oil produced in the laboratory by processing the most contaminated olive sample contained low levels of AOH and AME. The estimated mycotoxin percentage transferred into the oil was 4% for AME, 1.8% for AOH and zero for ALT and TA. Since the transfer of mycotoxins from the olive into the oil occurs at quite a low extent, the risk of contamination of olive oil by *Alternaria* mycotoxins does not apparently represent a concern for human health.

THE CONTROL OF MYCOTOXINS IN FRUITS AND FRUIT-DERIVED PRODUCTS

The control strategies of mycotoxin contamination of fruits and fruit-derived products are to be carried out at the different stages of the production line. In this regard, the implementation of the Hazard Analysis Critical Control Points (HACCP) system provides the basis for the establishment of guidelines for good agricultural and manufacturing practices (Park et al., 1999). The key words of mycotoxin control in fruits and their derived products are prevention, monitoring, and possibly, decontamination/detoxification. However, it must be emphasized that prevention is the main strategy for achieving a more satisfactory control of mycotoxigenic fungi and of their respective toxins.

Prevention

Preventative measures should be taken both at the prehavest and postharvest stages, especially for the control of toxigenic *Alternaria* and OTA-producing *Aspergillus* spp., because these fungi can cause infections in the field, as described above for tomatoes and wine grapes, respectively. Besides agronomical strategies, the utilization of biocides (fungicides and insecticides, the latter preventing generation of wounds favoring fungal attacks) still provides a good level of protection, both in the field and during storage. Among recommended practices, the careful handling of fruits is crucial for avoiding the lesions that represent the main penetration sites of mycotoxigenic fungi at harvest and during storage. During the conservation of many types of fruits, in particular, low temperature and modified atmosphere have proved to be valuable tools. Mycotoxigenic fungi, like many necrotrophic fungi, are widespread in the environment because they can live as saprohytes on dead organic matter (of non-plant origin also), and their propagules can remain viable for a long time. Therefore, obvious hygienic measures, such as the cleanliness of storage facilities, are very helpful in preventing storage rots and mycotoxin contamination. Fruits should remain for as short a time as possible in the open air before being stored or dried. As regards dried fruits, the drying process should be as rapid as possible in order to restrict, in particular, xerotolerant aflatoxigenic fungi. A slow drying process could facilitate the prevalence of these fungi over potentially competing, less xerotolerant microflora.

Consumers' increasing concern about chemical pollution in the environment and the presence of chemical (biocides) residues in food has led to the development of novel approaches in the prevention of crop diseases of fungal origin, including the fruit diseases caused by mycotoxigenic fungi. In addition, some fungicides have been withdrawn from the market for legislative actions taken in many countries, or have lost their efficacy (Sholberg et al., 2005). New methods rely on beneficial microorganisms, collectively named biocontrol agents (BCAs), which are able to outcompete fungal pathogens (Chapters 4-6 in Volume 1 of this series). Biocontrol agents can represent an alternative and/or integration to fungicides, especially during storage, where the environmental factors that could affect their survival, persistence and efficacy can be controlled more easily than in the field. Some postharvest BCAs have been commercialized (Wilson and Wisnieswski, 1992; Janisiewicz and Korsten, 2002; Tian, 2006). Many of these biocontrol microorganisms are yeasts, and studies are being carried out on their mechanism of action to enhance their efficacy. Competition for space and nutrients, resistance to oxidative stress to better and timely colonize fruit wounds and production of extracellular enzymes depolymerizing fungal cell walls (glucanases and chitinases) appear to play significant roles in the biological activity of biocontrol yeasts, although the importance of some β-glucanases has recently been questioned (Castoria et al., 1997, 2001, 2003; Droby, 2001; Janisiewicz and Korsten, 2002; Yehuda et al., 2003). Unlike fungicides (especially products like benomyl, that have a single fungal gene product as their target), the complex mechanism of action displayed by BCAs in their biocontrol activity makes the onset of resistance in fungal pathogens unlikely. However, the integration of BCAs with fungicides and/or with soft control methods such as hot water, sodium bicarbonate, calcium chloride, and natural substances used as adjuvants, has also yielded promising results (Janisiewicz and Korsten, 2002; Lima et al., 2003, 2005). Furthermore, the application of postharvest BCAs before harvest has been shown to increase their efficacy during storage (Ippolito and Nigro, 2000).

Monitoring

Monitoring of mycotoxins, especially in fruit-derived products, is necessary both for preventing health hazards arising from the commercialization of contaminated produce, and for complying with the legislative measures that have been taken by many countries across the world. Such measures pertain to AFs, OTA and PAT, but not *Alternaria* toxins as yet. The European Union has recently established the maximum tolerable levels for OTA in wine, grape juice, and raisin (refer above), and for PAT (50 μg kg^{-1} for apple juice and 10 μg kg^{-1} in apple-based baby foods) (European Commission, 2004, 2005), and official methods for sampling and chemical analysis have also been established. For monitoring, the set up of representative samples in the case of juices and other liquid fruit-derived products is not as complicated as for other produce, i.e., stored grain crops, in which an uneven distribution of fungal and mycotoxin contamination (known as spot contamination) takes place. For a comprehensive and exhaustive review of the worldwide legislation on mycotoxins, reference may be made to the work by Van Egmond and Jonker (2004). Monitoring of mycotoxins is based on the availability of reliable, sensitive, specific, and possibly, user-friendly and cost-saving analytical tools and methodologies. Methods based on the use of monoclonal antibodies (Mab) are now widespread. They are used for the cleanup procedures of extracted samples that are necessary prior to the quantification of AFs and OTA by chemical analysis, as well as for rapid and user-friendly methods (e.g., enzyme linked immunosorbent assay - ELISA, etc.) that can be performed to qualitatively detect and even quantify these toxins. For example, the detection of OTA contamination in wine and the establishment by the EU of maximum tolerable levels of OTA in this beverage have led to the development of specific and validated analytical methods, such as that of Visconti et al. (2001) for grapes and wine, which is based on immunoaffinity-based purification of samples prior to HPLC (High Performance Liquid Chromatography) analysis.

Conversely, the apparently lower toxicity of PAT and *Alternaria* toxins has not yet led to the development of analogous Mab-based methodologies for these toxins. An additional problem encountered so far in the development of anti-PAT antibodies is represented by the low molecular mass and the consequent low antigenic activity of this toxin. An official HPLC-based method is currently available for extraction and analysis of PAT in juice (AOAC, 2000), and improvements to it have recently been published (Ritieni et al., 2003; Piemontese et al., 2005). As regards *Alternaria* toxins, they are extracted from naturally contaminated samples by using a variety of methods, depending on the matrices: olives (Visconti et al., 1986), tomatoes (Stack et al., 1985), mandarin (Logrieco et al., 1990). HPLC is generally used either for confirmation of TLC (thin layer chromatography) results and for quantification of these mycotoxins.

The early detection of mycotoxigenic fungi can also help in preventing mycotoxin accumulation in food, although presence of these molds on fruits does not necessarily imply mycotoxin contamination. In this regard, specific molecular tools based on DNA probes and PCR analyses have been developed for the rapid detection/quantification of mycotoxigenic fungi, especially those which attack grain crops. These time-saving techniques do not require the isolation of molds, but only DNA extraction from the possibly infected matrix could be particularly useful for identifying fruits affected by internal (and not visible) infection. Various PCR (polymerase chain reaction) primer pairs have recently been developed for the detection and identification of mycotoxin-producing fungi. These primers are used in pure cultures as well as in contaminated sample matrices in conventional and real-time PCR applications (Niessen et al., 2005; Mulè et al., 2006). Promising methods are also based on molecular markers developed from DNA regions involved in mycotoxin biosynthesis (Bhatnagar et al., 2003; Mayer et al., 2003; Bluhm et al., 2004), as in the case of isoepoxydon dehydrogenase gene of *P. expansum* (Paterson et al., 2003). Finally, a new generation of microsystem technology solutions for monitoring toxigenic fungi and mycotoxins in foodstuffs is under way. They are DNA arrays, electronic noses and electronic tongues for the detection of fungal contaminants (at present mainly developed for feed), and biosensors and chemical sensors based on microfabricated electrode systems, antibodies and novel (molecularly imprinted) synthetic receptors for the specific detection of mycotoxins (Logrieco et al., 2005). At present, however, none of these procedures is widely used because of the high cost of the qualified personnel and equipment needed for analyses.

Decontamination and Detoxification

Procedures of decontamination and detoxification must comply with certain requirements in their practical application: toxins should be destroyed, inactivated or removed with no formation of toxic by-products; the fruit or its derived product should not lose its acceptability and nutritive value; the procedures used should have the lowest possible impact on the production line (Moake et al., 2005).

With regard to natural and dried fruits, the sorting of rotten fruits or removal of their decaying parts can be considered as decontamination procedures. Most of the information on the possibility of reducing mycotoxin levels during juice production through decontamination is referred only to PAT (Drusch and Ragab, 2003; Moake et al. 2005). Sorting of rotten fruits or removal of their decaying parts are commonly suggested for apples, but are not easily carried out on an industrial scale. In addition, removal of rotten parts can lower toxin contamination, but it cannot ensure the absence of PAT in the final produce. This toxin, in fact, diffuses into the healthy apple tissue surrounding the areas affected by the fungus

(Beretta et al., 2000; Rychlik and Schieberle, 2001), and can also be synthesized as a consequence of undetected infections following fungal penetration through the petioles or the calyx tube. The conventional steps of apple juice production are themselves effective in partially reducing PAT contamination. Centrifugation, in particular, can determine a 20% reduction of the initial PAT content, probably for the adsorption of the mycotoxin to microscopic solid particles. Clarification of juice also lowers PAT concentration for similar reasons. Assessment of new materials in this process, also in combination with traditional ones (e.g., charcoal and bentonite), has unfortunately shown that nutritive value and palatability can be negatively affected (Moake et al., 2005). Generally speaking, the use of adsorbent materials (e.g. activated carbon, hydrated sodium and calcium aluminosilicates and special polymers) can be proposed for decontamination of other liquid fruit-derived products also. However, these materials may have undesirable effects on the quality of the treated product, as in the case of OTA removal for wine decontamination (Castellari et al., 2001).

As with apples, the lack of association between visible modifications in fruit and toxin contamination can also occur in dried figs. The utilization of dried figs contaminated by AFs (and OTA) poses a serious health hazard represented by these fungal metabolites. In this regard, a fast and simple toxin-detection system based on blue green yellow fluorescence under UV light (due to the compound kojic acid, synthesized by *A. flavus* along with AFs) has proved to be very helpful in identifying and eliminating contaminated figs, at least the ones showing fluorescence on their surface, thus lowering mycotoxin contamination (Drusch and Ragab, 2003).

The main detoxification procedures assessed so far are chemical and biological, but they are all still at an experimental stage. Post-production chemical procedures have been proposed, as in the case of ozonation for AFs (Proctor et al., 2004). Although preliminary evidence is promising, most of the reaction products and their toxicity have yet to be elucidated (Moake et al., 2005). On the other hand, more information is available on biological detoxification (Karlovsky, 1999; Kakeya et al., 2002; Mishra and Das, 2003). This procedure can be carried out by microorganisms or respective purified enzymes and should lead to the transformation or degradation of mycotoxins to less toxic products. The mycotoxin-producing fungi themselves are able, under particular conditions, to degrade the toxins they produce (Karlovsky, 1999). Detoxification by whole microorganisms could be envisaged for fermentation processes, in which microbes are involved, as in the case of cider and wine production. Further, microbial cells have the potential to bind with mycotoxins, thus permitting their removal (Lahtinen et al., 2004). As mentioned above, *S. cerevisiae* produces two major metabolites from PAT during cider production, E and Z-ascladiol (Moss and Long, 2002). Although E-ascladiol is the immediate precursor of PAT biosynthesis in *P. expansum*, it appears to be less toxic (Suzuki et al., 1971). The biodegradation capacities of strains of *S. cerevisiae* and of malo-lactic bacteria used for fermentation during wine making are also worthy of investigation, to lower the possibility of OTA contamination of wine. Biocontrol agents can also degrade mycotoxins. Recent reports show that BCAs active on apples against *P. expansum* and on wine grapes against *A. carbonarius* are able to degrade PAT and OTA (Castoria et al., 2005b, c). Characterization of the degradation products has been carried out, and in the case of OTA, the respective degradation compound was identified as the less toxic Ochratoxin α (Castoria et al., unpublished results). Microorganisms can represent a source of enzymes/genes to be used in post-production detoxification processes, provided that the respective products of mycotoxin degradation are chemically and toxicologically characterized. Protein engineering aimed at increasing the substrate specificity of mycotoxin–degrading enzymes to prevent undesired reactions, and immobilization of these molecules could be valuable tools for mycotoxin detoxification, although the cost of this procedure(s) needs to be established. The characterization of microbial genes encoding mycotoxin-degrading enzymes could pave the way to their introduction in plants/ fruits, e.g., under the control of pathogen-responsive genetic elements, so that detoxification could be a prevention tool rather than a post-production procedure (Higa et al., 2003).

In the near future, studies on molecular biology of mycotoxin biosynthesis could offer a further opportunity in the prevention of food contamination with these compounds. The cluster organization of genes encoding and regulating the synthesis of important mycotoxins such as trichothecenes, fumonisins and aflatoxins is being unravelled. These discoveries could pave the way to the targeted interruption of the biosynthesis of mycotoxins (Cleveland et al., 2003).

CONCLUDING REMARKS

Mycotoxins in fruits and in their derived products represent a severe health issue. They can seriously affect the income of both growers and the food industry, especially in the light of the increasing awareness of this problem and of the consequent legislative measures that are being taken by many countries. Considerable research is still needed to lower mycotoxin contamination to a more acceptable level, by implementing adequate preventive strategies and monitoring, and by exploiting microbial technology for detoxification at a post-production level. Further studies are also needed to elucidate the fruit-fungal pathogen interaction, in order to enhance possible defense mechanisms deployed by fruits and/ or reduce those mechanisms which interfere with the pathogenic strategies of the (mycotoxigenic) fungi. The possibility that other mycotoxins contaminating fruits and derived products will be discovered and/or that other fungal species may be found responsible for the already known contaminations, cannot be ruled out. The characterization of gene(s) involved in mycotoxin biosynthesis, therefore, could provide the key to screening new fungal species presumably contributing to food contamination with these toxic compounds.

REFERENCES

Abarca, M.L., Accensi, F., Bragulat, M.R. and Cabanẽs, F.J. (2001). Current importance of ochratoxin A-producing *Aspergillus* spp. J. Food Protec. 64: 903-906.

Abdel-Sater, M.A., Zohri, A.A. and Ismail, M.A. (2001). Natural contamination of some Egyptian fruit juices and beverages by mycoflora and mycotoxins. J. Food Sci. Technol. 4: 407-411.

Accensi, F., Abarca, M.L., Cano, J., Figura, L. and Cabanes, F.J. (2001). Distribution of ochratoxin A producing strains in the *Aspergillus niger* aggregate. Antonie Van Leeuwenhoek. J. Microbiol. 79: 365-370.

Aksoy, U., Sabir, E., Eltem, R., Kirac, S., Sarigul, N., Betul, K.B., Ates, M. and Cakir, M. (2003). Researches on potential ochratoxin A contamination in dried figs. I. National Mycotoxin Symposium, Istanbul, Turkey, pp. 41-46.

An, Y.H., Zhao, T.Z., Miao, J., Liu, G.T., Zheng, Y.Z., Xu, Y.M. and Van Etten, R.L. (1989). Isolation, identification and mutagenicity of alternariol monomethyl ether. J. Agric. Food Chem. 37: 1341-1343.

Andersen, B., Smedsgaard, J. and Frisvad, J. C. (2004). *Penicillium expansum*: Consistent production of patulin, chaetoglobosins and other secondary metabolites in culture and their natural occurrence in fruit products. J. Agric. Food Chem. 52: 2421-2428.

Anonymous (1979). Perspectives on mycotoxins. FAO Food and Nutrition Paper No. 13. Food and Agriculture Organization, Rome, Italy, p. 176.

AOAC (Association of Official Analytical Chemists) Official Methods of Analysis (2000). Method 995.10: Patulin in apple juice: Liquid chromatographic method. Nat. Toxins 49: 51-53.

ApSimon J. W. (1994). The biosynthetic diversity of secondary metabolites. In: Mycotoxins in Grain - Compounds other than Aflatoxins (eds.) J.D. Miller and H.L. Trenholm, Eagan Press, St. Paul, Minnesota, USA, pp. 3-18.

Atroshi, F., Rizzo, A., Westermarck, T. and Ali-Vehmas, T. (2002). Antioxidant nutrients and mycotoxins. Toxicol. 180: 151-167.

Bai, G. and Shaner, G. (2004). Management and resistance in wheat and barley to *Fusarium* head blight. Annu. Rev. Phytopathol. 42: 135-161.

Barhoumi, R. and Burghardt, R. C. (1996). Kinetic analysis of the chronology of patulin- and gossypol-induced cytotoxicity in vitro. Fundam. Appl. Toxicol. 30: 290-307.

Battilani, P. and Pietri, A. (2002). Ochratoxin in grape and wine. Eur. J. Plant Pathol. 108: 639-643.

Battilani, P., Logrieco, A., Giorni, P., Cozzi, G., Bertuzzi, T. and Pietri, A. (2004). Ochratoxin A production by *Aspergillus carbonarius* on some grape varieties grown in Italy. J. Sci. Food Agric. 84: 1736-1740.

Bayman, P., Baker, J.L., Doster, M.A., Michailides, T.J. and Mahoney, N.E. (2002). Ochratoxin production by the *Aspergillus ochraceus* group and *Aspergillus alliaceus*. Appl. Environ. Microbiol. 68: 2326-2329.

Beck, J., Ripka, S., Siegner, A., Schiltz, E. and Schweizer, E. (1990). The multifunctional 6-methylsalicylic acid synthase gene of *Penicillium patulum*. Eur. J. Biochem. 192: 487–498.

Bennett, J.W. (1987). Mycotoxins, mycotoxicoses, mycotoxicology and mycopathologia. Mycopathologia 100: 3-5.

Bennett, J. W. and Klich, M. (2003). Mycotoxins. Clin. Microbiol. Rev. 16: 497-516.

Beretta, B., Gaiaschi, A., Galli, C.L. and Restani, P. (2000). Patulin in apple-based foods: Occurrence and safety evaluation. Food Addit. Contam. 17: 399-406.

Bhatnagar, D., Ehrlich, K.C. and Cleveland, T.E. (2003). Molecular genetic analysis and regulation of aflatoxin biosynthesis. Appl. Microbiol. Biotechnol. 61: 83-93.

Biing-Hui, L., Feng-Yih, Y., Ting-Shuan, W., Shuan-Yow, L., Mao-Chang, S., Mei-Chine, W. and Shin-Mei, S. (2003). Evaluation of genotoxic risk and oxidative DNA damage in mammalian cells exposed to mycotoxins, patulin and citrinin. Toxicol. Appl. Pharmacol. 191: 255-263.

Bluhm, B.H., Cousin, M.A. and Woloshuk, C.P. (2004). Multiplex real-time PCR detection of fumonisin-producing and trichothecene-producing groups of *Fusarium* species. J. Food Protec. 67: 536-543.

Bourdiol, D., Escoula, L. and Salvayre, R. (1990). Effect of patulin on microbiocidal activity of mouse peritoneal macrophages. Food Chem. Toxicol. 28: 29-33.

Boutin, B.K., Peeler, J.T. and Twedt, R.M. (1989). Effects of purified altertoxins I, II, and III in the metabolic communication V79 system. Toxicol. Environ. Health. 26: 75-81.

Buchanan, J.R., Sommer, N.E. and Fortlage, R.J. (1975). *Aspergillus flavus* infection and aflatoxin production in fig fruits. Appl. Microbiol. 2: 238-241.

Burda, K. (1992). Incidence of patulin in apple, pear and mixed fruit products marketed in New South Wales. J. Food Protec. 10: 796-798.

Burdaspal, P.A. and Legarda, T.M. (1999). Ochratoxin A in wines and grape products originating from Spain and other European countries. Acta Alimentaria 36: 107-113.

Cabañes, F.J., Accensi, F., Bragulat, M.R., Abarca, M.L., Castella, G., Minguez, S. and Pons, A. (2002) What is the source of ochratoxin A in wine? Int. J. Food Microbiol. 79: 213-215.

Castellari, M., Versari, A., Fabiani, A., Parpinello, G. P. and Galassi, S. (2001). Removal of ochratoxin A in red wines by means of adsorption treatments with commercial fining agents. J. Agric. Food Chem. 49: 3917-3921.

Castoria, R., De Curtis, F., Lima, G. and De Cicco, V. (1997). β-1,3-glucanase activity of two saprophytic yeasts and possible mode of action involved as biocontrol agents against postharvest diseases. Postharvest Biol. Technol. 12: 293-300.

Castoria, R., De Curtis, F., Lima, G., Caputo, L., Pacifico, S. and De Cicco, V. (2001). *Aureobasidium pullulans* (LS-30) an antagonist of postharvest pathogens of fruits: Study on its modes of action. Postharvest Biol. Technol. 22: 7-17.

Castoria, R., Caputo, L., De Curtis, F. and De Cicco, V. (2003). Resistance of postharvest biocontrol yeasts to oxidative stress: A possible new mechanism of action. Phytopathology 93: 564-572.

Castoria, R., Lima, G., Ferracane, R. and Ritieni, A. (2005a). Occurrence of mycotoxins in farro samples from Southern Italy. J. Food Protec. 68: 416-420.

Castoria, R., Morena, V., Caputo, L. Panfili, G., De Curtis F. and De Cicco, V. (2005b). A biocontrol yeast lowers patulin contamination in stored apples and displays potential for the detoxification of the mycotoxin. Phytopathology 95: 1271-1278.

Castoria R., Caputo L., Morena V., De Curtis F., De Cicco V. (2005c). Phenotypic traits of wound competence of postharvest biocontrol yeasts and potential of these microorganisms for prevention/detoxification of mycotoxins. Acta Hortic. 682: 2147-2152.

Ciegler, A., Detroy, R.W. and Lillejoj, E.B. (1971). Patulin, penicillic acid and other carcinogenic lactones. In: Microbial Toxins, Volume. VI. Fungal Toxins (eds.) A. Ciegler, S. Kadis and S.J. Ajl, Academic Press, New York, USA, pp. 409-434.

Ciegler, A. (1977). Patulin. In: Mycotoxins in Human and Animal Health (eds.) J.V. Rodricks, C.W. Hesseltine and M.A. Mehlman, Pathotox Publishers, Park Forest South, Illinois, USA, pp. 609-624.

Cleveland, T.E., Dowd, P.F., Desjardins, A.E., Bhatnagar, D. and Cotty, P.J. (2003). USDA (United States Department of Agriculture) - Agricultural Research Service Research on pre-harvest prevention of mycotoxins and mycotoxigenic fungi in US crops. Pest Manage. Sci. 59: 629-642.

Cozzi, P., Pascale, M., Perrone, G., Visconti, A. and Logrieco, A. (2006). Effect of *Lobesia botrana* damages on black aspergilli rot and ochratoxin A content in grapes. Int. J. Food Microbiol. 111(Suppl. 1): S88-S92.

Creppy, E.E., Stormer, F.C., Roschenthler, R. and Dirheimer, G. (1983). Effects of two metabolites of ochratoxin A (4R)-4hydroxyochratoxin A and ochratoxin α on immune response in mice. Infect. Immun. 39: 1015-1018.

Creppy, E.E., Kane, A., Dirheimer, G., Lafarge-Frayssinet, C., Mousset, S. and Frayssinet, C. (1985). Genotoxicity of ochratoxin A in mice: DNA single-strand break evaluation in spleen, liver and kidney. Toxicol. Lett. 28: 29-35.

Da Rocha Rosa, C.A., Palacios, V., Combina, M., Fraga, M.E., De Oliveira Rekson, A., Magnoli, C.E. and Dalcero, A.M. (2002). Potential ochratoxin A producers from wine grapes in Argentina and Brazil. Food Addit. Contam. 19: 408-414.

Davis, D.F. and Diener, U.L. (1987). Mycotoxins. In: Food and Beverage Mycology (ed.) L.R. Beuchat, Van Nostrand Reinhold, New York, USA, pp. 517-570.

Davis, D.M. and Stack, M.E. (1991). Mutagenicity of stemphyltoxin III, a metabolite of *Alternaria alternata*. Appl. Environ. Microbiol. 57: 180-182.

Delgado, T. and Gómez-Cordovés, C. (1998). Natural occurrence of alternariol and alternariol methyl ether in Spanish apple juice concentrates. J. Chromatogr. A 815: 93-97.

Droby, S. (2001). Enhancing biocontrol activity of microbial antagonists of postharvest diseases. In: Enhancing Biocontrol Agents and Handing Risks (eds.) M. Vurro, J. Gressel, T. Butt, G. Harman, A. Pilgeram, R. St. Leger and D. Nuss, IOS Press, Amsterdam, The Netherlands, pp. 77-85.

Drusch, S. and Ragab, W. (2003). Mycotoxins in fruits, fruit juices, and dried fruits. J. Food Protec. 66: 1514-1527.

Dwivedi, P. and Burns, R.B. (1985). Immunosuppressive effects of ochratoxin A in young turkeys. Avian Pathol. 14: 213-225.

Ehlers, D. (1986). HPLC-Bestimmung von patulin in obstaften-probenaufarbeitung mit einem modifizierten extraktions- und reinigungsverfahren. Lebensmittelchem. Gerichtl. Chem. 40: 1-5.

Eltem, R., Aksoy, U., Altindisli, A, Sarigual, N., Taskin, E., Askun, T., Meyvaci, B., Arasiler, Z., Turgut, H and Kartal, N. (2003). Determination of ochratoxin A in Sultanas of the Aegean region. Proc. I. National Mycotoxin Symposium, Istanbul, Turkey, pp. 54-59.

Engelhardt, G., Ruhland, M. and Wallnöfer, P.R. (1999). Occurrence of ochratoxin A in moldy vegetables and fruits analysed after removal of rotten tissue parts. Adv. Food Sci. 3: 88-92.

European Commission (2004). Regulation (EC) No. 455/2004 amending Regulation (EC) No. 466/2001 as regards patulin. Official Journal of European Union L74: 11.

European Commission (2005). Regulation (EC) No. 123/2005 amending Regulation (EC) No. 466/2001 as regards ochratoxin A. Official Journal of European Union L25: 3.

FAO/WHO (Food and Agriculture Organisation of United Nations/World Health Organisation) (2002a). Discussion paper on aflatoxins in pistachios. Codex Committee on Food Additives and Contaminants, 34th Session, Rotterdam, The Netherlands, p. 5.

FAO/WHO (Food and Agriculture Organisation of United Nations/World Health Organisation) (2002b). Proposed draft code of practice for the prevention of patulin contamination in apple juice and apple juice ingredients in other beverages. Codex Committee on Food Additives and Contaminants, 34th Session, Rotterdam, The Netherlands, p. 8.

Filali, A., Ouammi, L., Betbeder, A.M., Baudrimont, I., Soulay-mani, R., Benayada, A. and Creepy, E.E. (2001). Ochratoxin A in beverages from Morocco: A preliminary survey. Food Addit. Contam. 6: 565-568.

Fliege, R. and Metzler, M. (2000). Electrophilic properties of patulin. N-acetylcysteine and glutathione adducts. Chem. Res. Toxicol. 13: 373-381.

Flor, H.H. (1971). The current status of the gene-for-gene concept. Annu. Rev. Phytopathol. 9: 275-296.

Frisvad, J.C., Samson R.A. and Smedsgaard, J. (2004). *Emericella astellata,* a new producer of aflatoxin B1, B2 and sterigmatocystin. Lett. Appl. Microbiol. 38: 440-445.

Gaiser, R., Mayer, Z., Karolewiez, A. and Farber, P. (2004). Development of real time system for detection of *Penicillium nordicum* and for monitoring ochratoxin A production in foods by targeting the ochratoxin polyketide synthase gene. Syst. Appl. Microbiol. 27: 501-507.

Gatenbeck, S. and Hermodsson, S. (1965). Enzymatic synthesis of aromatic product, alternariols. Acta Chem. Scand. 19: 65-71.

Gaucher, G.M. (1975). m-Hydroxybenzyl alcohol dehydrogenase. Methods Enzymol. 43: 540–548.

Gokmen, V. and Acar, J. (1998). Incidence of patulin in apple juice concentrates produced in Turkey. J. Chromatogr. A 815: 99-102.

Goto, T., Wicklow, D.T. and Ito, Y. (1996). Aflatoxin and cyclopiazonic acid production by a sclerotium-producing *Aspergillus tamarii* strain. Appl. Environ. Microbiol. 62: 4036-4038.

Gourama, H. and Bullerman, L.B. (1988). Mycotoxin production by molds isolated from 'Greek-style' black olives. Int. J. Food Microbiol. 6: 81-90.

Griffin, G.F. and Chu, F.S. (1983). Toxicity of *Alternaria* metabolites alternariol, alternariol methyl ether, altenuene, and tenuazonic acid in the chicken embryo assay. Appl. Environ. Microbiol. 46: 1420-1422.

Guengerich, F.P. (2003). Cytochrome P450 oxidations in the generation of reactive electophiles - epoxidation and related reaction. Arch. Biochem. Biophys. 409: 59-71.

Gupta, P.K. and Sharma, Y.P. (1984). Mycotoxins. In: Modern Toxicology. The Adverse Effects of Xenobiotics, Volume. 2 (eds.) P.K. Gupta and D.K. Salunke, Metropolitan Book, New Dehli, India, pp. 317-340.

Hadas, Y., Goldberg, I., Pines, O. and Prusky, D. (2004). Tissue acidification by *Penicillium* as a mechanism to enhance pathogenicity. Proceedings of 7th European Conference on Fungal Genetics, Workshop IV Signal Transduction, Copenhagen, Denmark, p. 118.

Häggblom, P. and Hiltunen, M. (1992). Regulation of mycotoxin biosynthesis in *Alternaria.* In: *Alternaria* Biology, Plant Disease and Metabolites (eds.) J. Chelkowski and A. Visconti, Elsevier, Amsterdam, The Netherlands, pp. 435-447.

Halliwell, B. and Gutteridge, J.M.C. (1999). Free Radicals in Biology and Medicine, 3rd edn. Oxford University Press, New York, USA, pp. 936.

Hasan, H.A.H. (2000). Patulin and aflatoxin in brown rot lesion of apple fruits and their regulation. World J. Microbiol. Biotechnol. 16: 607-612.

Hennig, A., Fink Gremmels, J. and Leistner, L. (1991). Mutagenicity and effects of ochratoxin A on the frequency of sister chromatid exchange after metabolic activation. In: Mycotoxins, Endemic Nephropathy and Urinary Tract Tumors (eds.) M. Castegnaro, R. Plestina, G. Dirheimer, I. N. Chernozemski and H. Bartsch, IARC Sci. Publ. No. 115, Lyon, USA, pp. 255-260.

Higa, A., Kimura, M., Mimori, K., Ochiai-Fukuda, T., Tokai, T., Takahashi-Ando, N., Nishiuchi, T., Igawa, T., Fujimura, M., Hamamoto, H., Usami, R. and Yamaguchi, I. (2003). Expression in cereal plants of genes that inactivate *Fusarium* mycotoxins. Biosci. Biotechnol. Biochem. 67: 914-918.

IARC (International Agency for Research on Cancer) (1993). Ochratoxin A. IARC Monographs on Evaluation of Carcinogenic Risks to Humans: Some Naturally Occurring substances; Food Items and Constituents, Heterocyclic Aromatic Amines and Mycotoxins, Volume 56, pp. 489-521.

Ippolito, A. and Nigro, F. (2000). Impact of preharvest application of biological control agents on postharvest diseases of fresh fruits and vegetables. Crop Prot. 19: 715-723.

Janisiewicz, W. J. and Korsten, L. (2002). Biological control of postharvest diseases of fruits. Annu. Rev. Phytopathol. 40: 411-441.

Jarvis, W.R. and Traquair, J.A. (1984). Bunch rot of grapes caused by *Aspergillus aculeatus*. Plant Dis. 68: 718-719.

Jones, J. M. (1992). Food Safety. Eagan Press, St. Paul, Minnesota, USA, pp. 453.

Jorgensen, K. and Jacobsen, J.S. (2002). Occurrence of ochratoxin A in Danish wheat and rye, 1992-1999. Food Addit. Contam. 19: 1184-1189.

Kakeya, H., Takahashi-Ando, N., Rimura, M., Onose, R., Yamaguchi, I. and Osada, H. (2002). Biotransformation of the mycotoxin zearalenone to a non-estrogenic compound by a fungal strain of *Clonostachys* sp. Biosci. Biotechnol. Biochem. 66: 2723-2726.

Kane, A., Creppy, E.E., Roth, A., Röschenthaler, R. and Dirheimer, G. (1986). Distribution of the [3H]-label from low doses of radioactive ochratoxin A ingested by rats, and evidence for DNA single-strand breaks caused in liver and kidneys. Arch. Toxic. 58: 219-224.

Karlovsky, P. (1999). Biological detoxification of fungal toxins and its use in plant breeding, feed and food production. Nat. Toxins 7: 1-23.

Klich, M.A., Mullaney, E.J., Daly, C.B. and Cary, J.W. (2000). Molecular and physiological aspects of aflatoxin and sterigmatocystin biosynthesis by *Aspergillus tamarii* and *A. ochraceoroseus*. Appl. Microbiol. Biotechnol. 53: 605–609.

Korsten, L. and Wehner, F.C. (2003). Biotic and abiotic factors involved with spoilage fungi. In: Postharvest Physiology and Pathology of Vegetables, 2nd edn. (eds.) J.A. Bartz and J.K. Brecht, Marcel Dekker, New York, USA, pp. 485-518.

Krogh, P. (1974). Mycotoxin porcine nephropathy: A possible model for Balkan enedemic nephropathy. Proc. Sec. Inter. Sym. on Endemic Nephropathy (eds.) A. Puchlev, Publishing House of Bulgarian Academy of Sciences, Sofia, Bulgaria, pp. 266-270.

Larsen, T.O., Frisvad, J.C., Ravn, G. and Skaaning, T. (1998). Mycotoxin production by *Penicillium expansum* on blackcurrant and cherry juice. Food Addit. Contam. 15: 671-675.

Lahtinen, S.J., Haskard, C.A., Ouwehand, A.C., Salminen, S.J. and Ahokas, J.T. (2004). Binding of aflatoxin B_1 to cell wall components of *Lactobacillus rhamnosus* strain GG. Food Add. Contam. 21: 158-164.

Lima, G., De Curtis, F., Castoria, R. and De Cicco, V. (2003). Integrated control of apple postharvest pathogens and survival of biocontrol yeasts in semi-commercial conditions. Eur. J. Plant Pathol. 109: 341-349.

Lima, G., Spina, A.M., Castoria, R., De Curtis, F. and De Cicco, V. (2005). Integration of biocontrol agents and food-grade additives for enhancing protection of apples from *Penicillium expansum* during storage. J. Food Protec. 68: 2100-2106.

Liu, G.T., Miao, J., Zhen, Y.Z., Xu, Y.M., Chen Y.F. Dong, W.H., Dong, Z.M., Dong, Z.G., Zhang, P. and An, Y.H. (1988). Studies of mutagenicity of *Alternaria alternata* in grain from the areas with high incidence of esophageal cancer. Proc. Jpn. Assoc. Mycotoxicol. Special Issue 1: 131-132.

Logrieco, A., Bottalico, A., Visconti, A. and Vurro, M. (1988). Natural occurrence of *Alternaria*-mycotoxins in some plant products. Microbiol. Aliment. Nutrit. 6: 13-17.

Logrieco, A., Visconti, A. and Bottalico, A. (1990). Mandarin fruit rot caused by *Alternaria alternata* and associated mycotoxins. Plant Dis. 74: 415-417.

Logrieco, A. and Visconti, A. (2004). An Overview on Toxigenic Fungi and Mycotoxins in Europe. Kluwer Academic Publishers, Dordrecht, The Netherlands, pp. 259.

Logrieco, A., Arrigan, D.W.M., Brengel-Pesce, K., Siciliano, P. and Tothill, I. (2005). DNA arrays, electronic noses and tongues, biosensors and receptors for rapid detection of toxigenic fungi and mycotoxins: A review. Food Addit. Contam. 22: 335-344.

Lynen, F. H., Engeser, J., Freidrich, W., Schindlbeck, R., Seyffert, J. and Wieland, F. (1978). Fatty acid synthetase of yeast and 6-Methylsalicylate synthetase of *Penicillium patulum* - Two multi-enzyme complexes. In: Microenvironments and Metabolic Compartmentalization (eds.) P.A. Srere and R.W. Estabrook, Academic Press, New York, USA, pp. 283–303.

MAFF (Ministry of Agriculture, Fisheries and Food – UK) (1997). Food survey of aflatoxins and ochratoxin A in cereals and retail products. Surveillance Information Sheet No. 13 : 1-30.

MAFF (Ministry of Agriculture, Fisheries and Food – (UK) (1999). 1998 Survey of retail products for ochratoxin A. Food-Surveillance-Information-Sheet No. 185: 1-36.

Magnoli, C., Violante, M., Combina, M., Palacio, G. and Dalcero, A. (2003). Mycoflora and ochratoxin-producing strains of *Aspergillus* section *Nigri* in wine grapes in Argentina. Lett. Appl. Microbiol. 37: 179-184.

Majerus, P. and Otteneder, H. (1996). Detection and occurrence of ochratoxin A in wine and grape juice. Deut. Leben.-giene 51: 388-390.

Majerus, P., Bresch, H. and Otteneder, H. (2000). Ochratoxin A in wines, fruit juices and seasonings. Arch. Lebensmittelhyg. 51: 95-98.

Majerus, P. and Kapp, K. (2002). Reports on tasks for scientific cooperation, task 3.2.8. Assessment of dietary intake of patulin by the population of EU Member States (Brussels: SCOOP Report). http://www.europa.eu.int/comm/food/fs/scoop/3.2.8_en.pdf

Marek, P., Annamalai, T. and Venkitanarayanan, K. (2003). Detection of *Penicillium expansum* by polymerase chain reaction. Int. J. Food Microbiol. 89: 139-144.

Martins, M.L., Gimeno, A., Martins, H.M. and Bernardo, E. (2002.) Co-occurrence of patulin and citrinin in Portuguese apples with rotten spots. Food Addit. Contam. 6: 568-574.

Mayer, Z., Bagnara, A., Farber, P. and Geisen, R. (2003). Quantification of the copy number of nor-1, a gene of the aflatoxin biosynthetic pathway by real-time PCR, and its correlation to the cfu of *Aspergillus flavus* in foods. Int. J. Food Microbiol. 82: 143-151.

Mayura, K., Parker, R., Berndt, W.O. and Philips, T.D. (1984). Ochratoxin A induced teratogenesis in rats: Partial protection by phenylalanine. Appl. Environ. Microbiol. 48: 1186-1188.

McCallum, J.L., Tsao, R. and Zhou, T. (2002). Factors affecting patulin production by *Penicillium expansum*. J. Food Protec. 65: 1937-1942.

Mishra, H.N. and Das, C.A. (2003). A review on biological control and metabolism of aflatoxin. Crit. Rev. Food Sci. Nutr. 43: 245-264.

Moake, M.M., Padilla-Zakour, O.I. and Worobo, R.W. (2005). Comprehensive review of patulin control methods in foods. Comp. Rev. Food Sci. Food Safety 1: 8-21.

Moss, M.O. (1996a). Centenary review. Mycotoxins. Mycol. Res. 100: 513-523.

Moss, M.O. (1996b) Mode of formation of ochratoxin A. Food Addit. Contam. 13: 5-9.

Moss, M.O. (1998). Recent studies of mycotoxins. Symp. Ser. Soc. Appl. Microbial. 27: 62S-76S.

Moss, M.O. and Long, M. T. (2002). Fate of patulin in the presence of yeast *Saccharomyces cerevisiae*. Food Addit. Contam. 19: 387-399.

Mphande, F.A., Siame, B.A. and Taylor, J.E. (2004). Fungi, aflatoxins, and cyclopiazonic acid associated with peanut retailing in Botswana. J. Food Protec. 67: 96-102.

Mulè, G., Susca, A., Logrieco, A., Stea, G. and Visconti A. (2006). Development of quantitative real-time PCR assay for the detection of *Aspergillus carbonarius* in grapes. Int. J. Food Microbiol. 111(Suppl. 1): S28-S34.

Nair, N.G. (1985). Fungi associated with bunch rot of grapes in the Hunter Valley. Austr. J. Agric. Res. 36: 435-442.

Niessen, L., Schmidt, H., Mûhlencoert, E., Färber, P., Karolewiez, A. and Gaeisen, R. (2005). Advances in the molecular diagnosis of ochratoxin A-producing fungi. Food Addit. Contam. 22: 324-334.

NTP (National Toxicology Program) (1989). NTP Technical Report on the Toxicology and Carcinogenesis Studies of Ochratoxin A (CAS No 303-47-9) in F344/N Rats (gavage studies), (ed.) by G.A. Boorman (Research Triangle Park, NC:US Department of Health and Human Services, National Institutes of Health), p. 141.

Obrecht-Pflumio, S., Chassat, T., Dirheimer, G. and Marzin, D. (1999). Genotoxicity of ochratoxin A by *Salmonella* mutagenicity test after bioactivation by mouse kidney microsomes. Gen. Tox. Environ. Mut. 446: 95-102.

O'Callaghan, J., Caddick, M.X. and Dobson, A.D.W. (2003). A polyketide synthase gene required for ochratoxin A biosynthesis in *Aspergillus ochraceous*. Microbiology 149: 3485-3491.

Ospital, M., Cazabeil, J.M., Betbeder, A.M., Tricard, C., Creppy, E. and Medina, B. (1998). L'ocratoxine A dans les vins. Revue Francaise d'Oenologie 169: 16-18.

Osborne, L.C., Jones, V.I., Peeler, J.T. and Larkin, E. P. (1988). Transformation of C3H/10T1/2 cells and induction of EBV-early antigen in raji cells by altertoxins I and III. Toxicol. In Vitro 2: 97-102.

Otteneder, H. and Majerus, P. (2000). Occurrence of ochratoxin A in wines: Influence of the type of wine and its geographical origin. Food Addit. Contam. 18: 647-654.

Ózay, G. and Alperden, I. (1991). Aflatoxin and ochratoxin-a con-tamination of dried figs. Mycotoxin Res. 7: 85-91.

Papachristou, A. and Markaki, P. (2004). Determination of ochratoxin A in virgin olive oils of Greek origin by immunoaffinity column clean-up and high-performance liquid chromatography. Food Addit. Contam. 21: 85-92.

Park, D.L., Njapau, H. and Bontrif, E. (1999). Minimizing risks posed by mycotoxins utilizing the HAACP concept. Food Nutrit. Agric. 23: 49-56.

Paterson, R.R.M., Archer, S., Kozakiewicz, Z., Lea, A., Locke, T. and O'Grady, E. (2000). A gene probe for the patulin metabolic pathway with potential use in novel disease control. Biocontrol Sci. Technol. 10: 509-512.

Paterson, R.R.M., Kozakiewicz, Z., Locke, T., Brayford, D. and Jones, S.C.B. (2003). Novel use of the isoepoxydon dehydrogenase gene probe of the patulin metabolic pathway and chromatography to test Penicillia isolated from apple production systems for the potential to contaminate apple juice with patulin. Food Microbiol. 20: 359-364.

Pero, R.W., Posner, H., Blois, M., Harvan, D. and Spalding, J.W. (1973). Toxicity of metabolites produced by the "*Alternaria*". Environ. Health Perspective. 6: 87-94.

Peterson, S.W., Ito, Y., Horn, B.W. and Goto, T. (2001). *Aspergillus bombycis*, a new aflatoxigenic species and genetic variation in its sibling species, *A. nomius*. Mycologia 93: 689–703.

Piemontese, L., Solfrizzo, M. and Visconti, A. (2005). Occurrence of patulin in conventional and organic fruit products in Italy and subsequent exposure assessment. Food Addit. Contam. 22: 437-442.

Pietri, A., Bertuzzi, T., Pallaroni, L. and Piva, G. (2001). Occurrence of ochratoxin A in Italian wines. Food Addit. Contam. 18: 647-654.

Pitt, J.I. and Hocking, A.D. (1997). Fungi and Food Spoilage. Blackie Academic & Professional, London, UK, pp. 593.

Plestina, R. (1992). Some features of Balkan endemic nephropathy. Food and Chem. Toxic. 30: 177-181.

Prins, T., Tudzinsky, P., Von Tiedemann, A., Tudzinsky, B., Ten Have, A., Hansen, M.E., Tenberge, K. and Van Kan J.A.L. (2000). Infection strategies of *Botrytis cinerea* and related necrotrophic pathogens. In: Fungal Pathology (ed.) J.W. Kronstad, Kluwer Academic Publishers, Dordrecht, The Netherlands, pp. 33-63.

Proctor, A.D., Ahmedna, M., Kumar, J.V. and Goktepe, I. (2004). Degradation of aflatoxins in peanut kernels/flour by gaseous ozonation and mild heat treatment. Food Addit. Contam. 21: 786-793.

Prusky, D. (2003). Biotic and abiotic factors involved with spoilage - mechanisms of resistance of fruits and vegetables to postharvest diseases. In: Postharvest Physiology and Pathology of Vegetables, 2nd edn., (eds.) J.A. Bartz and J.K. Brecht, Marcel Dekker, New York, USA, pp. 581-598.

Prusky, D. and Yakoby, N. (2003). Pathogenic fungi: Leading or led by ambient pH? Mol. Pl. Pathol. 4: 509-516.

Prusky, D., McEvoy, J.L., Saftner, R., Conway, W.S. and Jones, R. (2004). The relationship between host acidification and virulence of *Penicillium* spp. on apple and citrus fruit. Phytopathology 94: 44-51.

Ritieni, A. (2003). Patulin in Italian commercial apple products. J. Agric. Food Chem. 51: 6086-6090.

Roze, L.V., Miller, M.J., Rarick, M., Mahanti, N. and Linz, J.E. (2004). A novel cAMP-response element, CRE1, modulates expression of nor-1 in *Aspergillus parasiticus*. J. Biol. Chem. 279: 27428-27439.

Rychlik, M. and Schieberle, P. (2001). Model studies on the diffusion behavior of the mycotoxin patulin in apples, tomatoes, and wheat bread. Eur. Food Res. Technol. 212: 274-278.

Rychlik, M., Kircher, F., Schusdziarra, V. and Lippl, F. (2004). Absorption of the mycotoxin patulin from the rat stomach. Food Chem. Toxicol. 42: 729–735.

Sage, L., Krivobok, S., Delbos, E., Seigle-Murandi, F. and Creppy, E.E. (2002). Fungal flora and ochratoxin a production in grapes and musts from France. J. Agric. Food Chem. 50: 1306-1311.

Schwartz, G.G. (2002). Hypothesis: does ochratoxin A cause testicular cancer? Cancer Causes and Control. 13: 91-100.

Scott, P. and Kanhere, M. (1980). Liquid chromatographic determination of tenuazonic acid in tomato paste. J. Ass. Off. Anal. Chem. 63: 612-621.

Scott, P.M. and Stoltz, D.R. (1980). Mutagens produced by *Alternaria alternata*. Mutat. Res. 78: 33-40.

Shenasi, M., Aidoo, K.E. and Candlish, A.A.G. (2002). Microflora of date fruits and production of aflatoxins at various stages of maturation. Int. J. Food Microbiol. 79: 113-119.

Shieh, M., Brown, R.L., Whitehead, M.P., Carey, J.W., Cotty, P.J., Cleveland, T.E. and Dean, R.A. (1997). Molecular genetic evidence for the involvement of a specific polygalacturonase, P2c, in the invasion and spread of *Aspergillus flavus* in cotton bolls. Appl. Environ. Microbiol. 63: 3548-3552.

Sholberg, P.L., Harlton, C., Haag, P., Lévesque, C.A., O'Gorman, D. and Seifert, K. (2005). Benzimidazole and diphenylamine sensitivity and identity of *Penicillium* spp. that cause postharvest blue mold of apples using α-tubulin gene sequences. Postharvest Biol. Technol. 36: 41-49.

Snijders, C.H.A. (1994). Breeding for resistance to *Fusarium* in wheat and maize. In: Mycotoxins in Grain - Compounds other than Aflatoxins (eds.) J.D. Miller and H.L. Trenholm, Eagan Press, St. Paul, Minnesota, USA, pp. 37-58.

Snijders, C.H.A. (2004). Resistance in wheat to *Fusarium* infection and trichothecene formation. Toxicol. Lett. 153: 37-46.

Speijers, G.J.A. and Speijers, M.H.M. (2004). Combined toxic effects of mycotoxins. Toxicol. Lett. 153: 91-98.

Splittstoesser, D.E. (1987). Fruits and fruit products. In: Food and Beverage Mycology, (ed.) L.R. Beuchat, Van Nostrand Reinhold, New York, USA, pp. 101-128.

Stack, M.E., Mislivec, P.B., Roach, J.A.G. and Pohland, A.E. (1985). Liquid chromatographic determination of tenuazonic acid and alternariol methyl ether in tomatoes and tomato products. J. Assoc. Anal. Chem. 68: 640-642.

Stack, M.E. and Prival, M.J. (1986). Mutagenicity of the *Alternaria* metabolites altertoxins I, II, and III. Appl. Environ. Microbiol. 52: 718-722.

Stinson, E.E., Bills, D.D., Osman, S.F., Siciliano, J., Ceponis, M.J. and Heisler, E.G. (1980). Mycotoxin production by *Alternaria* species grown on apples, tomatoes, and blueberries. J. Agric. Food Chem. 28: 960-963.

Stinson, E.E., Osman, S.F., Heisler, E.G., Siciliano, J. and Bills, D.D. (1981). Mycotoxin production in whole tomatoes, apples, oranges and lemons. J. Agric. Food Chem. 29: 790-792.

Stinson, E.E. (1985). Mycotoxins - their biosynthesis in *Alternaria*. J. Food Protec. 48: 80-91.

Suzuki, T., Takeda, M. and Tanabe, H. (1971). A new mycotoxin produced by *Aspergillus clavatus*. Chem. Pharm. Bull. 19: 1786-1788.

Sweeney, M.J. and Dobson, A.D.W. (1998). Review: mycotoxin production by *Aspergillus, Fusarium* and *Penicillium* species. Int. J. Food Microbiol. 43: 141-158.

Tangni, E.K., Theys, R., Mignolet, E., Madoux, M., Michelet, J.Y. and Larondelle, Y. (2003). Patulin in domestic and imported apple-based drinks in Belgium: Occurrence and exposure assessment. Food Addit. Contam. 20: 482-489.

Ten Have, A., Mulder, W., Visser, J. and van Kan, L.A.J. (1998). The endopolygalacturonase gene *Bcpg1* is required for full virulence of *Botrytis cinerea*. Mol. Plant-Microbe Interact. 11: 1009-1016.

Thuvander, A., Moller, T., Enghardt., H., Jansson, A., Salomonsson, A.-C. and Olsen, M. (2001). Dietary intake of some important mycotoxins by the Swedish population. Food Addit. Contam. 18: 696-706.

Tian, Shi-Ping. (2006). Microbial control of postharvest diseases of fruits and vegetables: Current concepts and future outlook. In: Microbial Biotechnology in Horticulture Vol. 1, (eds.) R.C. Ray and O.P. Ward, Science Publishers, Enfield, NH, USA, pp. 163-201.

Towner, R.A., Qian, S.Y., Kadiiska, M.B. and Mason, R.P. (2003). In vivo identification of Aflatoxin-induced free radicals in rat bile. Free Radic. Biol. Med. 35: 1330-1340.

Ueno, Y. (1998). Residue and risk of ochratoxin A in human plasma and beverages in Japan. Mycotoxins 47: 25-32.

Van Egmond, H.P. and Jonker, M.A. (2004). Worldwide regulations for mycotoxins in food and feed in 2003. Food and Agriculture Organization of the United Nations FAO, Rome, Italy. Food and Nutrition Paper No. 81, pp. 165.

Visconti, A., Logrieco, A. and Bottalico, A. (1986). Natural occurrence of *Alternaria* mycotoxins in olives – their production and possible transfer into the oil. Food Addit. Contam. 3: 323-330.

Visconti, A., Logrieco, A., Vurro, M. and Bottalico, A. (1987). Tenuazonic acid in black-mold tomatoes: Occurrence, production by associated Alternaria species, and phytotoxic properties. Phytopathol. Medit. 26: 125-128.

Visconti, A., Pascale, M. and Centonze, G. (1999). Determination of ochratoxin A in wine by means of immunoaffinity column clean-up and high-performance liquid chromatography. J. Chromatogr. A 864: 89-101.

Visconti, A., Pascale, M. and Centonze, G. (2001). Determination of ochratoxin A in wine and beer by immunoaffinity column cleanup and liquid chromatographic analysis with fluorometric detection: collaborative study. J AOAC Int. 84: 1818-1827.

Wang, I.K., Reeves., C. and Gaucher, G.M. (1991). Isolation and sequencing of a genomic DNA clone containing the 3′ terminus of the 6-methysalicylic acid polyketide synthetase gene of *Penicillium urticae*. Can. J. Microbiol. 37: 86-95.

Wichmann, G., Herbarth, O. and Lehmann, I. (2002). The mycotoxins citrinin, gliotoxin, and patulin affect interferon-γ rather than interleukin-4 production in human blood cells. Environ. Toxicol. 17: 211-218.

Wilson, C.L. and Wisnieswski, M.E. (1992). Future alternatives to synthetic fungicides for the control of postharvest diseases. In: Biological Control of Plant Disease (eds.) E.C. Tjamos, G.C. Papavizas and R.J. Cook, Plenum Press, New York, USA, pp. 133-148.

Wittkowski, M., Baltes, W., Kronert, W. and Weber, R. (1983). Determination of *Alternaria* toxins in fruit and vegetable products. Z. Lebensm. Unters. Forsch. 177: 447-453.

Woody, M.A. and Chu, F.S. (1992). Toxicology of *Alternaria* mycotoxins. In: *Alternaria* – Biology, Plant Disease and Metabolites (eds.) J. Chelkowski and A. Visconti, Elsevier, Amsterdam, The Netherlands, pp. 409-434.

Yao, C., Conway, W.S. and Sarns, C.E. (1996). Purification and characterization of a polygalacturonase produced by *Penicillium expansum* in apple fruit. Phytopathology 86: 1160-1166.

Yabe, K. and Nakajima, H. (2004). Enzyme reactions and genes in aflatoxin biosynthesis. Appl. Microbiol. Biotechnol. 64: 745-755.

Yehuda, H., Droby, S., Bar-Shimon, M. Wisnievski, M. and Goldway, M. (2003). The effect of under- and overexpressed *CoEXG1*-encoded exoglucanase secreted by *Candida oleophila* on the biocontrol of *Penicillium digitatum*. Yeast 20: 771-780.

Youssef, M.S., Abo-Dahab, N.E. and Abou-Seidah, A.A. (2000). Mycobiota and mycotoxin contamination of dried raisins in Egypt. Afr. J. Mycol. Biotechnol. 3: 69-86.

Yu, J., Chang, P.K., Ehrlich, K.C., Cary, J.W., Bhatnagar, D., Cleveland, T.E., Payne, G.A., Linz, J.E., Woloshuk, C.P. and Bennett., J.W. (2004a). Clustered pathway genes in aflatoxin biosynthesis. Appl. Environ. Microbiol. 70: 1253-1262.

Yu, J., Bhatnagar, D. and Cleveland, T.E. (2004b). Completed sequence of aflatoxin pathway gene cluster in *Aspergillus parasiticus*. FEBS Lett. 564: 126-130.

Yurdun, T., Omurtag, G.Z. and Ersoy, O. (2001). Incidence of patulin in apple juices marketed in Turkey. J. Food Protec. 11: 1851-1853.

Zimmerli, B. and Dick, R. (1995). Determination of ochratoxin A at the ppt level in human blood, serum, milk and some foodstuffs by HPLC with enhanced fluorescence detection and immunoaffinity column cleanup: methodology and Swiss data. J. Chromatogr. B. 666: 85-99.

Index